Out of the Mainstream

Out of the Mainstream

Water Rights, Politics and Identity

Edited by
Rutgerd Boelens, David Getches
and Armando Guevara-Gil

earthscan
publishing for a sustainable future

London • Washington, DC

First published in 2010 by Earthscan

Earthscan Ltd, Dunstan House, 14a St Cross Street, London EC1N 8XA, UK

Earthscan LLC, 1616 P Street, NW, Washington, DC 20036, USA

Earthscan publishes in association with the International Institute for Environment and Development

For more information on Earthscan publications, see www.earthscan.co.uk or write to earthinfo@earthscan.co.uk

ISBN: 978-1-84407-676-5

Typeset by MapSet Ltd, Gateshead, UK
Cover design by Susanne Harris

A catalogue record for this book is available from the British Library

Library of Congress Cataloging-in-Publication Data

Out of the mainstream : water rights, politics and identity / edited by Rutgerd Boelens, David Getches and Armando Guevara Gil.
p. cm.
Includes bibliographical references and index.
ISBN 978-1-84407-676-5 (hardback)
1. Water rights—Case studies. 2. Water supply—Case studies. 3. Water resources development—Case studies. 4. Politics and culture—Case studies. 5. Multiculturalism—Case studies. I. Boelens, Rutgerd. II. Getches, David H. III. Guevara Gil, Jorge Armando.
HD1691.O98 2010
333.33'9—dc22

2009032326

At Earthscan we strive to minimize our environmental impacts and carbon footprint through reducing waste, recycling and offsetting our CO_2 emissions, including those created through publication of this book. For more details of our environmental policy, see www.earthscan.co.uk.

Printed and bound in the UK by CPI Antony Rowe.
The paper used is FSC certified.

FSC
Mixed Sources
Product group from well-managed forests and other controlled sources
Cert no. SGS-COC-002953
www.fsc.org
© 1996 Forest Stewardship Council

Contents

PART II – POLITICS OF IDENTITY AND ANDEAN LIVELIHOODS

PART III – TENSIONS AND MERGERS AMONG LOCAL WATER RIGHTS AND NATIONAL POLICIES

List of Figures, Tables and Boxes

Figures

Tables

Boxes

Acknowledgements

The authors of *Out of the Mainstream* examine the multi-scale struggles for cultural justice and socio-economic redistribution of water. The issues arise as communities and user federations seek access to water resources and a fair share in the decision-making power regarding control and management of this vital resource. They are part of the larger and highly dynamic context of unequal globalizing power relations, politics of scale and identity, and the growing encroachment by extractive industries on local livelihoods and the environment. As the authors' narratives and analyses illustrate, water is not only a source of life and culture, but it is also a source of power, conflicting interests and identity battles. Rights to access, organize and control water resources, materially, culturally and politically, are poorly understood by mainstream scientific approaches and hardly addressed by current normative frameworks. These issues become even more challenging when law- and policy-makers and dominant power groups try to handle them in multicultural societies.

Most authors in this volume concentrate on the struggles over the uses, meanings and appropriation of water in the Andean region – particularly Peru, Chile, Ecuador and Bolivia. In this region, throughout history, nation states have attempted to 'civilize' and bring into the mainstream the different cultures and peoples within their borders instead of understanding 'context' and harnessing the strengths and potentials of diversity. Other authors bring in comparative experiences of water rights politics and defence strategies in neighbouring Latin American countries and in the south-west US. They also focus on state reform and multiculturalism across Latin America and the use of international standards in struggles for indigenous water rights.

For their inspiring contributions and their great efforts to make this book's idea into a creative endeavour of committed collective action, we would like to thank our co-authors Hans Achterhuis, Willem Assies, Michiel Baud, Anthony Bebbington, Denise Humphreys Bebbington, Jessica Budds, Jeffrey Bury, Rocío Bustamante, Paul H. Gelles, Jan Hendriks, Gregory A. Hicks, André Hoekema, Tom Perreault, Charles F. Wilkinson, Annelies Zoomers and Margreet Zwarteveen.

Out of the Mainstream shows that, against all odds, people are actively contesting neoliberal globalization and water power plays. In doing so, they construct new hybrid water rights systems, livelihoods, cultures and hydro-political networks, and they dynamically challenge mainstream powers and politics. We express our admiration and our gratitude to the many families, communities, water user organizations, peasant and indigenous federations, and civil society alliances who have collaborated in and motivated the academic and action research carried out for this book. They have shown us the existence of water worlds that go unnoticed in formal water laws and water policy formulation and implementation programmes. They have also brought to light how the latter – not seldom accompanied by the intervention of powerful water interest groups – may severely impact upon everyday water-based livelihoods and territories, as well as the possible alternatives that seek to materialize greater water justice and democracy.

We also thank the many researchers, activists, water users and water professionals of the Water Law and Indigenous Rights (WALIR) alliance and the Concertación action research network for their contributions of creative ideas, experiences, ideals and visions. With enthusiasm we see how the spirit and commitment in these networks is now deepened and broadened in the new, cross-continental alliance Justicia Hídrica. In the same way, we have enormously enjoyed and benefited from the great support we received from our friends and colleagues at Wageningen University, University of Colorado Law School, Pontificia Universidad Católica del Perú, CEDLA, Utrecht University, DGIS, IPROGA, CBC, CAMAREN, CESA, FEPP, Interjuntas and Centro AGUA. Here we need to mention, in particular, the support that we received from Concertación through the fine coordination by Edwin Rap, Patricia Urteaga, Aline Arroyo, Rigel Rocha and Leontien Cremers.

For the permanent contribution to the editing process, we thank Maria Pierce and Gerda de Fauw. And, finally, throughout the years that we have worked on this collective book, for detailed comments, creative support with translations, fine inspiration and warm encouragements, we owe many, special thanks to Samuel Dubois, Douglas Enzor, Matthew Hoppe, Patricio Mena Vásconez and William Wombacher.

Like other journeys, book projects succeed because the energies and talents of many people can be enlisted. *Out of the Mainstream* is no exception. We had the alliance of the most knowledgeable scholars on the subject and excellent assistance from those in our home institutions. But our journey also acquainted us and our authors with people whose experience and inspiration come from being part of the reality of indigenous and *campesino* cultures – people who taught us the importance of water as an organizing force and a giver of life itself.

Rutgerd Boelens, Wageningen, The Netherlands
David Getches, Boulder, Colorado, US
Armando Guevara-Gil, Lima, Peru

January 2010

List of Acronyms and Abbreviations

AIDESEP	Interethnic Association for the Development of the Peruvian Rainforest
ALA	Local Water Authority (Peru)
ALAM	Local Water Authority of Mantaro (Peru)
ANA	National Water Authority (Peru)
ANARESCAPYS	National Association of Irrigators and of Community Drinking Water Systems (Bolivia)
ATDR	Technical Administrator of Irrigation Districts (Peru)
CAMAREN	Training Consortium for the Sustainable Management of Renewable Natural Resources (Ecuador)
CBD	Convention on Biological Diversity
CBC	Centro Bartolomé de las Casas (Peru)
CEDLA	Centre for Latin American Research and Documentation (Netherlands)
Centro AGUA	Andean Centre for the Management and Use of Water (Bolivia)
CERD	United Nations International Convention on the Elimination of All Forms of Racial Discrimination
CESA	Ecuadorian Centre for Agricultural Services
CGIAB	Bolivian Commission for the Integrated Management of Water
CIDOB	Confederation of Indigenous Peoples of the Bolivian Oriente
CIPCA	Centre for Research and Promotion of the Peasantry (Bolivia)
CITES	Convention on International Trade in Endangered Species of Wild Fauna and Flora
CNRA	National Agrarian Reform Council (Bolivia)
CNRH	National Water Resource Council (Ecuador)
COCOPA	Concord and Pacification Committee (Mexico)
CODERECH	Regional Development Corporation for the Central Highlands (previously CORSICEN) (Ecuador)

CONACAMI	National Confederation of Mine Affected Communities in Peru
CONADI	National Corporation of Indigenous Development (Chile)
CONAIE	Confederation of Indigenous Nationalities of Ecuador
CONAMAQ	National Confederation of Ayllus and Markas of Qullasuyu (Bolivia)
CONFENIAE	Confederation of Indigenous Nationalities of the Ecuadorian Amazon
CONIAG	Inter-Institutional Water Council (Bolivia)
CORECAMI	Regional Coordinator of Mine Affected Communities (Peru)
CORSICEN	Regional Development Corporation for the Central Highlands (renamed CODERECH) (Ecuador)
CRIC	Indigenous Regional Council of Cauca Valley (Colombia)
DGA	National Water Directorate (Chile)
DGIS	Directorate General for International Cooperation, The Netherlands
DMI	indigenous municipal district (Bolivia)
DRM	Mantaro Irrigation District (Peru)
ECLAC	United Nations Economic Commission for Latin America and the Caribbean
ECUARUNARI	Confederation of Kichwa Nationalities of Ecuador
ERA	echelons of rights analysis
ESRC	Economic and Social Research Council
ESSAN	Antofagasta Water and Sewage Services Company (Chile)
EU	European Union
FEDECOR	Cochabamba Departmental Irrigators' Federation
FEDURIC	Cotopaxi Irrigators Federation (Ecuador)
FEJUVE-El Alto	Federation of Neighbourhood Organizations–El Alto
FEPP	Fondo Ecuatoriano Populorum Progressio
G10	Group of Ten countries that have agreed to participate in the General Arrangements to Borrow (Belgium, Canada, France, Italy, Japan, The Netherlands, the UK and the US, and the central banks of Germany and Sweden)
GATT	General Agreement on Tariffs and Trade
IACHR	Inter-American Convention on Human Rights
ICCPR	United Nations International Covenant on Civil and Political Rights
ICESCR	United Nations International Covenant on Economic, Social and Cultural Rights
IDB	Inter-American Development Bank
IEP	Institute for Peruvian Studies
IFAD	International Fund for Agricultural Development
ILO	International Labour Organization
IMF	International Monetary Fund

INCORA	Colombian Agrarian Reform Institute
INI	National Indigenista Institute (Mexico)
INRA	National Agrarian Reform Institute (Bolivia)
INRENA	Natural Resources Institute (Peru)
IPROGA	Institute for the Improvement of Water Management (Peru)
ISI	import substitution industrialization
JBNQA	James Bay and Northern Quebec Agreement
JUDRM	Users' Board of the Mantaro Irrigation District
JNUDRP	National Board of Users in Irrigation Districts (Peru)
LPP	People's Participation Law (Bolivia)
MAS	Movimiento al Socialismo (Movement to Socialism) (Bolivia)
NAFTA	North American Free Trade Agreement
NARF	Native American Rights Fund
NGO	non-governmental organization
OAS	Organization of American States
PAN	National Action Party (Mexico)
PCR	Crop and Irrigation Plan
PLACA	Program on Latin America and the Caribbean
PRATEC	Andean Project of Peasant Technologies
PRI	Institutional Revolutionary Party (Mexico)
PROFODUA	Programme to Formalize Water Use Rights (Peru)
PRSP	*Poverty Reduction Strategy Paper*
PSG	Peru Support Group
SAGARPA	Ministry of Agriculture, Livestock, Rural Development, Fisheries and Food (Mexico)
SEDERI	Departmental Irrigation Service (Bolivia)
SEMAPA	Bolivian Public Waterworks Company
SEMARNAT	Ministry of the Environment and Natural Resources (Mexico)
SENARI	National Irrigation Service (Bolivia)
SNRA	National Agrarian Reform Service (Bolivia)
SNV	Netherlands Development Organization
TCO	Original Community Territory (Bolivia)
UMSS	University of San Simón (Bolivia)
UN	United Nations
UNCED	United Nations Conference on Environment and Development
UNESCO	United Nations Educational, Scientific and Cultural Organization
UNRISD	United Nations Research Institute for Social Development
US	United States
WALIR	Water Law and Indigenous Rights Programme
WTO	World Trade Organization

Part I

An Introduction to Water Rights, Power, Identity and Social Struggle

1

Water Struggles and the Politics of Identity

Rutgerd Boelens, David Getches and Armando Guevara-Gil

> … *words coalesce to say:* From water we come – *and to denounce the politicians and technocrats who betray the most ancient memory of the Americas. That memory encourages the life and strivings of pillaged indigenous communities, and teaches us that water, like air, belongs not to those who can pay for it.* Water belongs to those who are thirsty. (Galeano, 2009)

Introduction

Many new water policies in Latin America have been culturally, economically and politically harmful for local water-user collectives, peasant communities and indigenous people. National and international water policies are often well intentioned and are based on models widely accepted in development circles; yet they fail to account for the complexity and diversity of locally prevailing norms for water use and control. The policies are fundamentally objectifying, de-politicizing and de-contextualizing as policy-makers consecrate universalistic principles of water development and planning. As such, they lack the capacity to understand or even imagine the cumbersome real-life human impact of their policies. The last decades have shown how, across the globe and, in particular, in Latin America, water officialdom embraces mainstream water science, officially endorsed rational 'water culture' and policies built on universalistic recipes without considering existing water societies, identities and practices that have succeeded locally in their own countries.[1]

Despite notions of 'inclusion' and 'integration' ostensibly embraced by current policies or the apparently neutral ideal of 'demand-driven water policy', policy-makers tend to neglect these on-the-ground water user practices and water-control experiences. Whether the approaches are more state based, neoliberal market driven or based on 'participatory' dogmas of integrated water management, they have in common the presumption that expert knowledge and 'modern' water rights systems need to be the core building blocks of policy. The homogenizing effect of these policies results from the apparently benign ideal of equality: equal water users whose norms, desires and practices neatly fit into a formally accredited model of water law and policy. In this way, rather than understanding the existence of 'difference', they construct powerful towers of indifference. These towers, by creating distanced views, tend to neglect actual water users' diversity and water rights complexity. They not only perceive 'equals', but also actively construct them in accordance with their own model of equality.

The era of top-down water policy formulation and implementation has surely ended. Concepts of equality, freedom, inclusion, participation and democracy are key to all modern water policies. Moreover, national legislation and international water policies call for the establishment of 'rational, efficient and accountable water uses'. The standards for 'rational' and 'efficient', however, are set by those who govern and not by the governed. Global water policy experts are assumed to know the cross-cultural, cross-continental medicines to cure the 'backwardness' of local water cultures. How is it possible that experts continue to ignore the poor results of their plans and policies for marginalized water-user groups? Ironically, the neoliberal champions of 'accountability' are not themselves held accountable for the dramas that they create among the water user populations whom they label as their 'clients'. The water rights frameworks proposed by these experts are based on a myth that social and legal engineering can create a new and improved order. Yet, in the hands of experts, the force of such ideas is extremely powerful in Latin America

Neoliberal godfather Friedrich Hayek foreshadowed the use of myth to justify policy-making:

> *Although those responsible for a decision may have been guided by no more than prejudice, some guiding principle will have to be stated publicly if the community is not merely passively to submit but actively to support the measure. The need to rationalize the likes and dislikes which, for lack of anything else, must guide the planner in many of his decisions, and the necessity of stating his reasons in a form in which they will appeal to as many people as possible, will force him to construct theories, i.e. assertions, about the connections between facts, which then become an integral part of the governing doctrine. This process of creating a 'myth' to justify his action need not be conscious.* (Hayek, 1944, p116)

The strength of myth, creating its own water world, is shown by the impacts and non-adaptation to local contexts that are characteristic of most mainstream water policies that have been applied in Latin America during the last decades and that continue to be promoted.

This book aims to elucidate a classic conflict of political power, cultures, values, livelihoods and resources. The allocation and use of water is a particularly apt framework for understanding the effects of externally imposed cultural and political values. Water is essential to life, cultures and identities. It is essential for local villages and communities aiming to sustain and strengthen their peasant economies in an adverse physical and political environment. Yet, as in many other places, the dominant cultures in Latin America have imposed qualifications on water rights that clash with local cultures. It is poignantly illustrated as water rights and uses in indigenous and *campesino* or peasant communities[2] meet the norms and policies of national governments (Bustamante, 2006; Guevara-Gil, 2006; compare Getches, 2005; Wilkinson, 2005).

Local water-user collectives typically understand water rights as being claims that encompass both the right to access and withdraw water and the right to determine rules for managing water use systems (Beccar et al, 2002). Therefore, issues of material resource distribution merge with cultural and political recognition (Fraser, 2000; Boelens and Doornbos, 2001; Zwarteveen et al, 2005). In the eyes of local user collectives, these fundamental components of their water rights are under attack through encroachment from dominant players and from legislation and policies. This puts their vital resources and rule-making faculties at risk. *Out of the Mainstream* examines the struggles and claims for socio-economic redistribution and political-cultural justice that arise as local user communities seek access to water and decision-making power over its management.

The conflict of cultures, social structure and divergent economies in water rights disputes is especially well illustrated in the Andean communities of Peru, Chile, Ecuador and Bolivia. These countries vary in their professed respect for indigenous and *campesino* cultures; but all have enacted water laws and pronounced policies that can abrade local cultures (Urteaga and Boelens, 2006).[3] Case studies from local and indigenous water-user communities and river basins in neighbouring countries and the Northern Hemisphere provide comparative evidence. Comparisons also deepen the analysis of water–power–identity interactions by simultaneously challenging and enriching locally used frameworks and concepts.

Despite the reality of multiple cultural identities, an array of stereotypes persists about the moral and economic features of peasant communities and indigenous populations. These images have been influenced by racist impulses to discipline the Indians as well as for motives emerging from progressive, humanistic ideals (Salman and Zoomers, 2003; Getches, 2006; Guevara-Gil, 2006). Throughout history, paternalistic, apparently benign attempts to 'civilize' Indians by bringing them into the mainstream have been endemic to nation states that have encountered different cultures and peoples within their borders (Assies et al, 1998; Albó, 2003; Baud, 2003).

Water bureaucracies in the Andean countries have certainly made efforts to 'recognize local cultures, rights and identities'. Corresponding policies, however, simultaneously entail recognition and official rejection of the great diversity of water management repertoires. Commonly, recognition policies are grounded in *de facto* hierarchical structures and essentialist concepts of ethnicity, community and legality that aggravate the complex tensions between official law and local water rights systems. They presume to include the 'extra legals' as equals in a uniform and externally normalized system that is profoundly based on unequal power structures (Boelens, 2009).

These recognition policies are not in opposition to, but rather facilitate, current modernization policies, including the neoliberalism promoted by transnational agencies and governmental entities. As Assies observes in this book (see Chapter 3), the neoliberal state does not simply 'recognize' the community, civil society or indigenous culture, but rather reconstructs them to fit its ideological concept. This policy differentiates between 'good Indians' and 'bad Indians'. The former present cultural demands that are 'compatible' with the neoliberal project, whereas the latter are the 'radical Indians' who call for redistribution of power and resources. Therefore, although the neoliberal project speaks of decentralization, tolerance and respect for multiculturalism, these values are not allowed to impinge upon the model's foundations. Cultural diversity is accepted and encouraged by 'neo-indigenous' policy so long as it does not interfere with market rationality. Thus, community-based natural resource management norms and forms can be attacked because they are seen as jeopardizing free market operation.

Besides political and economic conflicts, indigenous peoples struggle against the characterizations and misrepresentations of their cultures that the hegemonic dominant society assigns to them (compare Laurie et al, 2002; Getches, 2005). Indigenous societies have maintained significant political and cultural autonomy, while also adapting to changing circumstances. Many communities have affirmed their own historical patterns and employed different political thinking that is often the opposite of the government's approach, and sometimes have been moved to resist government actions through lawsuits, protests and struggles in the streets (Assies and Gundermann, 2007). Similarly, many peasant communities and local water-user collectives engage in fierce struggles to defend their cultural expressions and collective management arrangements that can be the most essential fibres of their community social fabric (Gelles, 2002; Rhoades, 2006). Other communities, on the contrary, have been forced to assimilate new norms and practices and have lost social coherence and autonomy as they have come to depend on outside state-defined water-use hierarchies or market-driven water control (Gelles, 2000; Getches, 2006; Castro, 2006).

Commonly, however, 'adaptation' is merely superficial and, in practice, local rights and organizations persist in disguised forms. Beyond simple obedience and domestication according to the presumed official 'legal modernization' of Andean communities, their actual practices often manifest a bottom-up, grassroots-driven, subtle and strategic resistance. Indigenous and *campesino* communities creatively blend water rules, rights and practices of various legal systems, generating new socio-legal repertoires to regulate their resources, while demonstrating their dynamic capacity for change when they strategically adapt longstanding cultural patterns and identities. As the book's authors illustrate, this can be done with organizational tools such as federations of water users formed to defend their water and confront the new challenges of a globalizing world.

In sum, globalization and the neoliberal project of international institutions impose pressures on developing countries to pursue efficiency in water policies and to homogenize national water law. As a condition of multilateral or bilateral development assistance, countries may be coerced into acting more aggressively to reform water policies and, in turn, to force local communities into the mainstream. But the latter are not defenceless. They can even strategically apply cultural, political and legal pluralist strategies to use mainstream norms in order to stay out of the mainstream.

What is the actual and potential impact of the superimposition of national laws and international water policies on indigenous and *campesino* water users? Can decision-makers, some professing to respect indigenous people and local water cultures, appreciate the consequences of their water policies? What are the visible and invisible strategies to resist local water rights encroachment? These questions and their potential answers reach beyond the issue of water resources distribution alone. They entail struggles over the formulation of the rules and rights; over the authority to make decisions and enforce norms; and over the discourses established to impose, legitimize or defend particular policies. Together, the chapters in this book aim for a better understanding of the struggles that ensue as 'modern' water policies confront local forms of control rooted in the culture and identities of the user groups and their networks.

Water rights, collectives and identity in the Andean countries

> *Attach any branch you like, but attach it to a strong trunk.*
> (Martí, 1891, p18)

The Andean region is home to millions of small-farmer families, mostly living in indigenous and *campesino* communities. Beyond their great diversity, they share a series of social, cultural, political and ecological features originating in the long history of the Andes. These community foundations have not lain static but have generally adapted to new power structures. Even the entity called 'community' has part of its origins in colonial legislation designed to 'reduce' the Andean population into controllable groups (*reducciones*). However, this does not mean that indigenous and other rural communities are historical fossils left over from the 'remnant natives' in the Andean landscape. On the contrary, particularly in the highlands, communities are growing in number and in economic, agro-ecological and political importance. Small-scale subsistence farming by families is the foundation of food security, both locally and nationally. They address the climatic and ecological challenges of the Andes in multiple ways, both within their community territories and through supra-local strategies. Moreover, their huge contribution to preserving biodiversity is just beginning to be recognized. Finally, communities are major political stakeholders in local, regional and national settings (Bebbington, 2001; Mayer, 2002; Guevara-Gil, 2006, 2008).

The norms and practices of user organizations and peasant and indigenous communities play a key role in local water management by incorporating elements of Andean, colonial and republic-period water traditions with 'modern' norms and technologies (Oré, 2005; Perreault, 2006). In this process, 'peasant', 'indigenous' or 'local law' is intertwined with the norms, procedures and organizational forms of official law. Although the rural and indigenous authorities are generally the first level of control in water rights and conflict resolution, the rules and representatives of official institutions are also involved. Thus, users often appeal to combined rules and practices originating in diverse normative sources (Orellana, 2004; Peña, 2004; Guevara-Gil, 2006). In a single small-farmer community, water rights can derive from multiple frameworks that, moreover, interact: official law, peasant or indigenous law (sometimes called customary law), religion-based water control norms (from Catholic, Protestant or Andean religious beliefs), the 'law of development projects', or norms of other origins (von Benda-Beckmann et al, 1998; Roth et al, 2005).

Water rights express the legitimacy of claims to water and to decision-making power over water management. One can talk of 'rights' only when the water use is certified by an authority with legitimacy among users and non-users and a capacity for enforcement (Beccar et al, 2002; Gerbrandy and Hoogendam, 2002). State officials commonly equate 'legal' and 'legitimate' water rights, but local user groups usually differentiate between the two and challenge this conflation. Most Andean areas have more than one authority, each representing different socio-legal systems and often taking divergent positions on the legitimacy of different claims to water use (Boelens and Hoogendam, 2002). These different water rights regimes coexist, complement, overlap or even contradict one another so that legitimate authority in Andean water management is not restricted to state agencies or rights emanating from state law (Roth et al, 2005).

As an integrated part of their production strategies, indigenous and other local water users take advantage of the existence of legal plurality to defend their interests and manage their conflicts. By doing so, they actively produce an inter-legality in which the stakeholders involved in water management inter-relate, reinvent and experiment with rules and rights of diverse origins, ranging from the Andean communities themselves to outside water management systems (Guevara-Gil, 2008). Moreover, some rules that are considered to be local 'community' rules are actually official norms, adapted and internalized by communities (von Benda-Beckmann et al, 1998). This counsels that the societal stakeholders cannot be characterized as the untainted representatives of a given legal system or 'water culture' (Cohen, 1986; Roth et al, 2005).

This is also why an analysis distinguishing among *campesino* or indigenous normative systems and the official legal system, as if they were two completely different worlds, cannot explain the dynamics of inter-legality. This would only support the well-known statement that plurality is a structural fixed property in any multicultural society. On the contrary, water resource management in the Andes is best understood as a product of legal pluralism and interactive coordination and conflict among several socio-legal repertoires. This perspective avoids the error of considering local law as a normative structure consisting solely of a set of 'ancestral customs', 'ancient norms' and 'indigenous traditions'. Of course, customs and traditions may inform current norms, but are neither the sole source nor unmoving referents of today's norms. These are norms that are continually undergoing reformulation and transformation (von Benda-Beckman et al, 1998; Roth et al, 2005; Guevara-Gil, 2006).

It follows that water rights embody social relations among the actors involved and are intimately linked to existing social organization and power relationships. In day-to-day political encounters, diverse interest groups challenge the construction, application and reproduction of rights. Local water-user groups engage in a 'struggle for legitimacy' as they seek formal legal recognition of normative constructs and as they strive to gain legitimacy 'outside the legal domain' in everyday water-use practices and water control struggles (Getches, 2005; Vos, 2006; Verzijl, 2007). Therefore, the struggle for local water rights' legitimacy in the Andes provides insights into the political construction of positive justice and of local concepts of equity.

In sum, any analysis of the construction, reproduction and transformation of Andean water rights systems, beyond focusing on the 'truthful representation' of their sources of origin or the 'academic accuracy' of local *usos y costumbres* constructs, needs to consider their constitution as local–national–global hybrids and, at the same time, focus on the question of their political use and convenience for either intervening agents and supra-local rulers or for local user groups who struggle for livelihood defence and rule-making autonomy.

Despite great socio-legal and cultural diversity, most Andean nation-states (perhaps with the exception of Bolivia) have centralized their water resource laws (Urteaga and Boelens, 2006). In these systems, it is the state's exclusive prerogative to promulgate laws, regulations and procedures and to award water rights. Similarly, enforcement of these laws has always been considered the privilege of governmental institutions and agents (Guevara-Gil, 2006). Even where rural groups are formally recognized as having authority to set water policies, the reality, in practice, falls short of the promise (compare Getches, 2005; Vos et al, 2006). Water laws usually reflect the needs of a minority of large irrigation systems and so-called 'modern' growers. Local norms are generally rejected or judged 'irrational, inefficient and backward'. By marginalizing local rights, denying multiple cultural identities, and rejecting context-based management and organizational systems, the Andean governments have undermined indigenous and rural societies' norms and subsistence strategies.[4]

New water policies would seem to offer more room for local legal systems and water cultures. Water scarcity and competition for water resources are increasing (from indigenous and peasant populations and from mining, export-oriented agribusiness, the urban water supply sector and other stakeholders). The idea that problems should be solved with less control from the top down and more local control and initiatives has spawned proposals for less 'state-centred' policies. Still, market-oriented models reflect a universalizing bias by offering blanket prescriptions for the vast diversity of local contexts (Budds and McGranahan, 2003; Castro, 2006; Bakker, 2010; compare Kay et al, 2007; Kay and Akram-Lodhi, 2008; Zoomers and van der Haar, 2000). Furthermore, the approach of 'normative decentralization' often has a privatizing, individualizing aim, clearly working against collective water management by rural and indigenous communities. As observed by Assies et al (1998), the neoliberal model and its policies are not just economic, but include a cultural programme which is a powerful tool to establish a particular market-based relationship among the state, the market and civil society stakeholders (compare Foucault, 1991).

New neoliberal policies create competition for water and foster speculation by current users and by new economic agents, rather than establishing strategies or policies seeking collective action (Swyngedouw, 2005; Gentes, 2006; Bakker, 2007; see also Chapters 2, 6, 10 and 12 in this volume). Water rights in Andean communities differ from privatized models by featuring collective water rights. Even within these systems individual rights are not to be confused with 'private rights'. As outlined by Beccar et al (2002, pp3–4):

> *... collective water rights can be seen as the demands on water use and control by the organization of users vis-à-vis other parties (individuals or collectives), whose interests may collide with their own. These rights also determine the collective forms and conditions for access to the water source and the prerogatives and burdens assumed as a group versus third parties. Individual water rights, then, are inside each water use system, establishing relations for access to water among the different users and their respective rights, privileges and obligations.*

Unlike government-granted or market-allocated water rights, related mostly to individuals, water rights in community systems are generally granted to families for belonging to a collective. Within webs of complex social relations and multiple identities, Andean collective water-control systems are a bastion against commoditized market exploitation and private water rights. User families constitute important elements of their identity by being dutiful community and water-use system members. Moreover, the rights of each individual are derived from and embedded in the collective rights and duties. So, without romanticizing Andean morals, equity statements or presumed indigenous solidarity,[5] the interests of all users in the system comprise an interdependent network that is capable of self-management even in a harsh physical-geographical and politico-economic environment (Boelens, 2008). Andean communities continue to exist because their collective institutions are indispensable (compare Mayer, 2002; van der Ploeg, 2006, 2008; Rhoades, 2006). Neither peasant families and communities nor their irrigation systems will be able to reproduce themselves amidst exclusively mercantile relationships, and they are well aware of this fact. These institutions, therefore, cannot be supplanted by individual, state or market solutions.

In situations of scarcity and mutual dependence on a shared water resource, collective action is necessary and communities take up this task. They are:

> *... complex, multilayered entities whose members, differentiated by class, gender, status and often by ethnicity, try to construct and reaffirm social and territorial bonds because of this mutual dependence to defend and control vital individual and common resources. Similarly, Andean water control collectives refer to groups of internally differentiated water users who are bound by mutual dependency to develop, use and manage their water sources by a sense of collective (culture-space bound) hydraulic identity, and who are determined to* realize *their interdependence and materialize their collective and individual water rights by engaging in collective action strategies.* (Boelens, 2008, p146)

Despite existing inequities, they have the potential not only to protect the least powerful players, but also to strengthen existing water cultures, preserve the environment and increase stability and food security for the entire country. However, in Latin America, as illustrated by contemporary Andean practices, there is a recurring pattern in the political economy of natural resources: although indigenous and rural communities are fundamental for the nation's economy and well-being, they are not only marginalized, but public policy tends to reinforce their cultural, political and economic subordination.[6]

Contents of the book

In this book, authors have gathered to analyse the issues of legal pluralism, enforcement of local water rights and (non-)recognition of communities' own management systems in light of the local cultural foundations and the national and global power relationships that create tensely unequal multicultural contexts.[7] Water rights are not necessarily limited to formal legal disciplinary constructs. They can be linked to diverse fields of knowledge and interpretation, including the technical-biophysical, organizational, socio-legal, political-economic and cultural-metaphysical. Water users assume that these factors are integrated within their water control systems and rights frameworks.

Authors from different disciplines have studied a diverse array of experiences from different countries while analysing the relationship among official water policies and current mainstream scientific conventions that inform those policies and the community-based forms of water management that prevail in certain areas. This review of history, politics, hegemonic thinking and struggles facilitates a better understanding of the dynamics of legal pluralism, cultural diversity and socio-economic injustice in local Andean contexts.

Out of the Mainstream is organized in four parts. Part I introduces the control of water, the most vital of resources, as the crucible for larger political and cultural interactions. The question is framed by discussing the main threads of neoliberal economic ideology. Globalization affects the production and distribution of goods, but it also impacts upon culture and communication. This was seen in the green revolution of the mid 20th century as the policies promoting irrigation of arid lands were realized. Class, ethnic and gender politics are encountered at the local level as national political power and dominant water stakeholders challenge the ability of communities to sustain their identities and livelihoods.

The introductory chapters use examples from the Andes; but some compare experiences there to situations in other continents in order to illustrate the pervasiveness of the issues in water policy-making. This chapter introduces the concepts of legal and cultural pluralism, the nature of local water rights, the complexities of achieving recognition of local and indigenous communities' own management systems, and the politics of scale and identity that relate to the defence of local water territories. Chapter 2 focuses on water rights and water property relations under neoliberal water policy regimes. Market-oriented water policies and private water rights structures in contexts of unequal power relations work to the detriment of less wealthy smallholders, such as water user communities, urban water collectives and poor neighbourhoods. In order to understand the fatal attraction that neoliberal policies pose to policy-makers and planners, the critique scrutinizes the power of neoliberal utopianism and examines the political-philosophical roots of this positivist belief system.

Chapter 3 examines the process of national identity formation and indigenous struggles for autonomy and territory in light of the contemporary forces of neoliberal globalization. Social movements that combine issues of ethnicity and economic redistribution produce increasingly complex forms of struggles because claims based on 'identity policies' receive ambivalent responses from Latin American states that are pursuing 'recognition politics'. Based on illustrations from Bolivia, Colombia and Mexico, the chapter analyses how combining the neoliberal model with recognition policies can give rise to a new official strategy of 'managed multiculturalism' that celebrates cultural pluralism but fails to materialize it in lasting redistributive effects for oppressed ethnic groups. By contrast with transformative multiculturalism – which aims to redistribute power and resources – this 'top-down' multiculturalism reinforces essentialist concepts of group identities that implicitly ignore locally generated cultural norms, self-management strategies and forms of political representation.

Chapter 4 discusses the gender biases of conventional water policies and water professionals that have the effect of denying the importance of women in critical water resources decision-making. The values and discourses of prevailing modern irrigation design and development schools combine with the power structures in local contexts of project implementation to shape the outcomes of water development and reinforce the process of gender-based economic and political subjugation. The chapter examines the resistance of the water profession to gender insights by clinging to the positivist epistemological tradition that historically has dominated water expertise. In water, 'normal knowing' is typically associated with masculine identities. Challenging mainstream water science, therefore, also means challenging established masculine expert identities and cultures.

Part II of the book develops issues of water and the politics of multiculturalism specific to the Andes. The Andean situation provides the context for analysing the impacts of neoliberalism and globalization and, later in the book, for comparing the consequences to situations in other regions.

Chapter 5 analyses how Andean countries' policies over the last two centuries illustrate a contrast between ostensible legal and discursive equality for indigenous populations and the actual persistence of economic and political injustice. Governments and national elites have tried to balance a political system responsive to the 'terms and requirements of modernity', incorporating and at the same time marginalizing indigenous peoples and *campesino* communities. Faced with attempts by governments to assimilate indigenous people, strident movements have arisen, demanding inclusion on their own terms. Their demands express cultural, political, social and economic demands, which often include claims of collective rights over land, water and territory. These mobilizations have increased the visibility of indigenous populations, and their organizations have become major stakeholders in the local, national and international political arena.

Chapter 6 explains why the cultural relevance of water in the Andes goes beyond its importance as an economic commodity. The indigenous culture and identity of Andean communities are dynamic, deeply ingrained in farming and ranching, and often dependent on irrigation water management. In high mountain rural communities, the collective ownership system of water rights is a key institution. Thus, attempts to privatize or wrest control over water from indigenous communities would affect their cultural identity and their local management systems. Chapter 6 also points out that national laws and public policies usually ignore customary law and the uses and practices of community water management. This trend is an expression of the historically troubled relationship that the state has with peasant communities in the Andes.

This part concludes with a discussion of coping strategies that have emerged in Bolivian communities. Chapter 7, for example, analyses the rationale of most intervention policies that modernizing water delivery and application systems is crucial to improving local livelihoods. Programmes are legitimized by neoliberal proposals to privatize land and water, and even by romantic counter-discourses. But most community members have access only to eroded smallholdings. Population growth and powerful outside demands for land and water have driven such communities beyond their land's carrying capacity. Mainstream approaches do not resolve this situation, whether they are neoliberal policies designed to achieve rational, efficient use of natural resources, or the idealized proposal for achieving community survival through reciprocity relationships and ecological adaptation. Chapter 7 explores the importance of water and land in determining local class relationships and then analyses the way in which these forms of 'natural capital' might help to promote the construction of sustainable livelihoods in order to escape poverty. In many cases, the chapter concludes, the only viable option providing economic security may be to find alternative livelihoods or migrate to the cities.

Part III of the book compares the interactions among local water rights systems and the laws and policies of the dominant society. Chapters show how water bureaucracies in North and South America have often neglected or misunderstood local normative frameworks. Sometimes governments have been forced by local or international demands to try to 'recognize the rights of rural communities and indigenous peoples'. The problem is that such recognition policies have been skewed by attempts to protect the interests of private parties and public policy-makers. In other cases, communities have 'voluntarily' adapted their systems, bargaining to retain a modicum of control. This can create tensions between the systems and, in some cases, assimilation of some norms into a hybrid system. The success of hybridization varies.

Chapter 8 examines water legislation in Andean countries and reflects on why the concept of collective rights is generally antithetical to national policies. Under the paradigm of 'modernity', the official policies have marginalized indigenous and peasant management systems in order to achieve 'resource governability' and acquire coveted development projects. Although Ecuadorian policies now claim to break with their firm attachment to neoliberal adjustment and return to public control, the actual consequences for local water-user collectives are still uncertain. Even Bolivian 'indigenist' laws generate frictions when confronted with on-the-ground realities. In Chile and Peru, the official apparatuses excel in their misunderstanding of collective water rights and methods of adaptation to local management and concentrate attention on large 'modern' systems. This causes *de facto* marginalization of Andean collective water control systems, which are implicitly classified as 'irrational', 'inefficient' and 'backward'.

Chapter 9 provides an account of the social impacts of official water law in a highland valley of Peru, the Achamayo River Basin. The chapter shows how mainstream normative accounts of state water law fail to grasp the way in which peasants and local bureaucracies reinterpret and make sense of official regulations. Through this interpretive enterprise, local water organizations resist and reshape the meaning and enforceability of state law. Using an ethnographic approach, the author in this chapter analyses the reconfiguration of official commands caused by local social forces and power relations. This phenomenon itself is a powerful force in strengthening peasant water organizations.

Chapter 10 examines the social implications of Chile's 1981 Water Code. The code, issued by Pinochet's junta, is based on notions of private property and free market mechanisms with minimal government regulation. It has attracted the attention of international agencies which assume that it would improve water-use efficiency and bring social benefits. The chapter explores the case of the arid Atacama region and argues that the mechanisms embodied within the Water Code have allowed commercial interests to obtain greater access to water, while reducing the ability of indigenous communities to assert and defend their customary water rights. Social outcomes are being mitigated to some extent by protective government programmes and modifications to the Water Code in 2005; but the model itself remains unchallenged, threatening indigenous water communities' material and cultural survival if they have to compete for water on the market.

Chapter 11 provides a comparative perspective to the book's central case by analysing how Indian water rights in the US encounter state and federal water policies. Concentrating on Pyramid Lake in Nevada, US, the chapter traces the effective loss of access to water by an Indian tribe as irrigation systems drained away their water sources, eliminating a historic fishery in the tribe's lake and reallocating water to communities of settlers. This case shows how even a well-intended legal doctrine such as the reserved water rights for indigenous people can be subdued by mainstream social and economic interests. It concludes, however, with the tribe gaining sufficient political influence to recover water for the lake's fishery.

In Chapter 12, another case from the US is examined with strong comparative importance for Latin American policies, offering lessons for policy-makers to respect water institutions grounded in local user collectives. The chapter explains the complex relationship between state-centred water governance in Colorado and New Mexico and the ancient Hispanic *acequia* irrigation institutions. While these *acequias* are based on collective rights concepts and complex, organizational forms rooted in community-based obligations and rights, the state administrative system promotes individual rights concepts and allocates water based on its 'optimal use'. In this chapter the author shows the dangers of weakening experience-based and history-based social and natural resource management in which people, place and identity closely interact. The chapter concludes by noting a recent formal state recognition of *acequia* rights.

Chapter 13 examines the institutionalization of legal pluralism in official law and discusses how local and indigenous groups can strategize to achieve legal protection for their resources. The chapter draws on international examples to distinguish between state 'incorporation' and 'recognition' of local rights. Incorporation domesticates local normative systems in the likeness of dominant law. This approach effectively subordinates local law to official national law. Recognition, by contrast, creates room to manoeuvre, enabling indigenous peoples and rural societies to continue with the inter-legal development of their own community resource ownership systems. In the latter process, the indigenous peoples themselves codify local law 'from the bottom up'. Some critics oppose codifying indigenous, local or customary law, feeling that this will end up freezing and alienating it from local social dynamics. The author, however, asks if this process may promote ethnic reconstruction, reinforcing indigenous organization and reinventing local law.

In Part IV of the book, the authors shed light on possible responses by indigenous, *campesino* and other grassroots groups who want to defend their water rights and decision-making authority to manage their resources. The final chapters concentrate on studying the feasibility of mobilizing water users by applying various legal and networking strategies.

Chapter 14 considers the opportunities that indigenous peoples have to use international norms in order to assert their water rights against states whose practices result in intrusions on their cultures or deprivation of local control. These international laws are not specific to water rights, but purport to protect human rights, environmental quality or cultural practices; yet they may have value in struggles to protect water rights. Although legal precedents relate to other natural resources, the chapter concludes that international law may hold some promise against the most egregious state incursions on indigenous water rights.

Chapter 15 addresses the issue of contemporary 'scalar politics' and grassroots networking strategies for shaping and defending indigenous and *campesino* water rights. The authors focus on the multiple ways in which Andean communities and irrigation organizations mobilize, and the varied federations and networks within which local people might organize to pursue their claims. The chapter shows how local social relations are connected to broader-scale political, economic, cultural and ecological issues. It is based on cases such as the mobilization organized by the Interjuntas federation in Ecuador and the Bolivian 'water wars' in Cochabamba and El Alto.

In Chapter 16, the multi-scalar responses to the contradictions of bureaucratic governance and neoliberal reforms in Andean countries' rural water management are explored. The authors examine the strategies used by indigenous and *campesino* federations to counteract undemocratic state reforms and extractive industries in order to regain control over natural resources. The chapter shows how the indigenous and rural movements have made themselves heard throughout the Andean region. Their networking initiatives have resulted in changes in state public policies and laws. Mobilizations, multi-actor platforms and continuous bottom-up pressure have been critical in materializing their objectives.

The concluding chapter integrates the previous ones, their cases and concepts, and questions the pertinence of the Andean nations' water laws and policies to the great diversity of local realities and distinct forms for managing water resources. Who 'owns' official water policy and laws? Are they truly applicable to the local contexts that they aspire to regulate? Is there a political agenda behind scientific and legal discourse? Conflicts and possible answers addressed in Chapter 17 go beyond the issue of water resources distribution and materialize in different political and normative spheres. The chapter places the issues that arise from the cultural justice and water redistributive struggles in a broader perspective. Rather than vaunting the latest conventional policy proposals, Chapter 17 seeks to build on local communities' own strategies and responses. The chapter concludes by stressing the importance of the issues globally – issues which, after all, are fomented by global trends.

Water rights, water territories and the politics of scale and identity

As shown by the chapters in this book, the modern world's power structures can usurp or threaten the rights of rural and indigenous societies. Since ancient times, ruling groups have continually tried to expropriate and control local community resources such as water by strategic use of cultural politics. Enlisting community labour and regulating water rights, for instance, are part of a process of identity formation. But dominant groups have taken over both resource ownership and such rules for water management, manipulating and confiscating patterns of identification. The quest to shape users' water control norms, beliefs, identities and practices, in order to secure their obedience to, and compliance with, a dominant model for water control and 'normal' water rights, is at the heart of the issue (Boelens, 2009; compare Foucault, 1991).

Most contemporary Latin American nation states have promoted a form of multiculturalism that actually destroys community management of rights when it does not follow the market's rules of play. The latest neoliberal wave in the Andean countries is fostered by banks and international agencies and by states themselves which are reordering their policies to conform to liberal theory and to please international institutions. Thus, globalizing forces strengthen the market-based institutional models that threaten to co-opt local systems and community management forms (compare Hendriks, 1998; Budds and McGranahan, 2003; Swyngedouw, 2005; Bakker, 2007, 2010; Perreault, 2008).

National and international policies do not attempt to adapt to local contexts, but rather seek to transform and control them. It is the *users' universe* that is to be adapted (Boelens, 2008). But such changes require more than enacting new laws. A water-political order becomes institutionalized in a users' society only when it becomes integrated within its economic, moral and ideological structure. Laws cannot act, only societal forces can shape such change (Cohen 1986; von Benda-Beckmann et al, 1998; Roth et al, 2005). Andean user collectives therefore resist externally imposed policies that attempt to impose 'normalization' and to take control of their water rights and management systems. Their resistance includes opposing current distributive inequalities and undemocratic forms of representation, challenging the rules of the game and the very politics of truth and identity themselves.

In addition to the matter of how water is distributed, the dispute is fundamentally about the definition of rights to use water and local autonomy for its control (Gelles, 2002; Oré, 2005; Guevara-Gil, 2006). The multidimensional importance of water rights in indigenous and rural communities of the Andes can be seen in the way in which struggles over water have materialized. The disputes dynamically link four different but connected 'echelons of rights' that are at the root of water conflicts (for a conceptualization of the echelons of rights analysis (ERA), see: Boelens and Zwarteveen, 2005; Boelens, 2008). Together, these echelons give substance to water rights struggles and highlight how they involve the formulation and materialization of water rights in the following fields of contention:

- access, withdrawal and use of water, and related infrastructural, material and financial *resources*;
- formulation of the *rules*: contents of water rights, obligations and management rules, and of mechanisms to acquire rights;
- the legitimate *authority* to make decisions, establish the rules and enforce rights; and
- the *discourses* that challenge, impose, legitimize or defend particular water policies and water-political orders.

In the fourth echelon, discourses are developed and invoked as 'socio-technical, socio-natural organizers and political stabilizers': conjunctions of power and knowledge that aim to create and proliferate the belief that particular policies and water rights orders are correct and self-evident. Discourses strategically connect all water rights domains, from the technical-physical to the cultural and metaphysical, to compose and glue together convenient water rights and truth orders. Society's ruling groups seek to enrol and align humans, nature and thought within a single water governance system that is structured according to their interests to control locality and based on non-local rules, truths and frames of reference. In reaction, local users' collectives commonly resist and strategize to construct their own, alternative systems (Boelens, 2008).

These chapters offer a framework to understand how struggles over water arise at each of those four levels, implicating local norms and forms of control and bringing the culture and the identities of the user groups into conflict with modernist water policies. In these struggles, local user organizations trying to defend their autonomy do not avoid interaction with the state or development institutions. Many communities request collaboration. Through such interaction, the state, the users and third parties all try to achieve their own, frequently contradictory, purposes. Here, commonly, local user groups pursue state resources and international funding without handing over local normative power. This raises the question of power: who controls the activities and resources of whom and how? Each side tries to conscript the other in the action programme which they desire. This interaction is between entities with unequal power but a mutual need for each other's resources, where each makes strategic use of some of the other's techniques, norms and rules.

Cultural politics and the politics of identity, therefore, are inherent to dominant, control-externalizing strategies and to the power tactics of control-localizing communities and water user collectives. Each seeks power over the applicable rules, rights and meanings of social practices. Each tries to shape identity and forms of subjectivity in divergent ways (see also Cohen, 1986; Foucault, 1991; Álvarez et al, 1998). In this struggle, the dynamic assertion of water rights diversity is an expression of resistance. By creating or adopting water rights that are profoundly intertwined with local identity formation, a world of difference is perpetuated below the outer appearances of uniformity. Multilayered collective water-rights systems extend existing forms of legal pluralism, implicitly questioning the exclusiveness and self-evidence of formal state- and market-based water rules.

It follows that we need to examine how water user communities and networks construct new, diverse water rights and collective organizational forms in their struggles against control externalization. This process simultaneously implies the active construction and reconstruction of 'water territory' as a socio-productive and cultural–political 'home base', as a rooted and multilayered 'political water community', as a scheme of mutual belonging that enables the rebirth of collective imagination. Such water territories involve socio-natural webs with landscapes and waterscapes in which people live and make livelihoods and identities, for which people feel responsible, in which they are morally involved. Water territories, in this way, contrast with the essentially placeless irrigation systems and watershed plans that are determined by 'technocracy' and that have social arrangements that are presented as convenient arrangements – neutral, universal, timeless and void of context.

Increasingly, struggles to create and defend alternative water rights repertoires and to embed them within water territories are not limited to the local scale (see also Bebbington, 2001; Swyngedouw, 2004; Perreault, 2006). As several chapters show, control-localizing strategies implicate wider political objectives related to social transformation and reform of laws, discourses, governance frameworks and class, gender and ethnic structures. Local user groups engage in scalar politics and scalar alliances to build territorial alliances and networks that link diverse peoples, identities and places. They use these alliances to contest the processes of de-territorialization of water, the separation of water rights and decision-making powers from local livelihoods, and policies and actions that attack the integrity of their territories in general. Advocates of local water control thus have transcended existing boundaries, manageable units and assigned identities, and they continuously produce new webs and alliances for political, normative and scalar articulation (compare Foucault, 1982).

In the end, the water struggle exemplifies not only inter-user conflicts and community–state contradictions, but also the transnational policies and market forces resulting from globalization. Several chapters of *Out of the Mainstream* seek to examine how locally rooted water-user communities develop water control defence strategies, build systems of water rights and water management, and strengthen the definition of 'home territories' for water use. Some authors also show how action on a broader political scale can advance local efforts. This seems appropriate as many of the challenged institutional arrangements and water governance structures themselves emanate from higher levels such as states, transnational corporations and international water development agencies.

It appears that the power to compose or manipulate patterns of multiple scales is crucial. To meet the needs of local water users it is important to reconfigure the scale of water governance. The political ordering of water control and the contents of formal water rights require connecting them on the ground with local livelihoods and perspectives. *Out of the Mainstream* aims to contribute to understanding why water user collectives resist externally imposed change and why they seek creative reproduction of their own cultural and political norms, as well as different forms of democracy and political representation in formal water governance. Thus, the book looks at how asserting water rights pluralism and water identity-based distinctiveness animates challenges to commensuration and control by the dominant power structure. It analyses how local and indigenous communities reject the categories in which the dominating groups want to enclose them, while at the same time pursuing ways in which formal categories can be used for their own purposes. It analyses how political and legal action can be used to challenge dominant forms of power and to 'localize' water control by simultaneously deepening their own water rights systems and borrowing ideas from the dominant system. Along the way, indigenous people, *campesinos* and other marginalized water users are defending their alternative out-of-the-mainstream ideals as they struggle to protect their water domains.

Notes

1. See, for instance, Mollinga and Bolding (2004); Getches et al (2005); Roth et al (2005); Zwarteveen and Bennett (2005); Meinzen-Dick and Pradhan (2005); Castro (2006); Vos (2006); Ingram (2008); Bakker (2010).
2. As will unfold over the coming chapters, the concepts of 'indigenous/*indígena*', 'peasant/*campesino*' and 'community/*comunidad*', as well as the various social

categories and specific denominators of identity by country and place, are contextualized dynamic constructs. In the Andes region, for instance, the relation between self-definition based on class (peasant) and the identification with an ethnic basis (indigenous) is complex and fluid, depending on who uses what labels during what period, context or place. For instance, 'indigenous/*indígena*' (or even *indio, nativo* or *runa*) can be used pejoratively as a racial slur against an ethnic group but may also be a self-affirming category of pride, recently appropriated by the indigenous movement in their discourse on 'first nations' or 'original peoples'. Similarly, 'peasant/*campesino*' is both a construct imposed on indigenous by reformist governments to deny their cultural identity and a term (often insisted on by indigenous peoples themselves) to strengthen class consciousness and political alliances. In the Andes, rural collectives and indigenous peoples commonly use names linked to their territory and home place to express their identity. This book uses the term 'indigenous' as a reference to Andean cultural identity (hybridized, dynamic and self-referential) (Boelens, 2008; compare Cohen, 1986; Gelles, 2002; Albó, 2003; Baud, 2003; Salman and Zoomers, 2003; Guevara-Gil, 2006).

3 Readers should bear in mind that while Chile and Peru are still applying neoliberal policies, Ecuador and Bolivia are currently dismantling those policies and searching for alternative development models. While the Ecuadorian government seems to follow a national popular path, trying to rebuild the developmental and centralized state, Bolivia has started a whole new political experiment in which indigenous people 'have come to stay' (although see Chapter 16 in this book). Unfortunately, both processes are still based on the rents provided by the exploitation of natural resources. This, in turn, is negatively shaping state–indigenous interactions. The current Ecuadorian administration, for example, is ambiguous with respect to developing large-scale mining even at the expense of peasant and indigenous communities.

4 See, for example, Gelles (2002); Albó (2003); Oré (2005); Gentes (2006); Isch and Gentes (2006); Perreault (2006); Urteaga and Boelens (2006); van der Ploeg (2006); Vos (2006); Assies and Gunderman (2007); Verzijl (2007). Compare with Peña (2004); Getches (2005, 2006).

5 The fact that local *campesino* communities, indigenous populations or 'indigenous' laws and governments are not necessarily more 'equitable' or 'democratic' goes without mentioning. See also Laurie et al (2002); Zwarteveen and Bennett (2005); Vera (2006); and Chapters 4, 7, 15 and 16 in this volume.

6 See Mayer (2002); Baud (2003); Salman and Zoomers (2003); van der Ploeg (2008).

7 Preliminary versions of Chapters 3, 5, 6, 7 and 13 in this volume have been published in Spanish in *Agua y Derecho* (Boelens et al, 2006).

References

Albó, X. (2003) 'Andean ethnicity today: Four Aymara narratives from Bolivia', in T. Salman and A. Zoomers (eds) *Imaging the Andes*, CEDLA, Aksant, Amsterdam, pp226–248

Álvarez, S. E., E. Dagnino and A. Escobar (eds) (1998) *Cultures of Politics, Politics of Cultures: Re-Visioning Latin American Social Movements*, Westview Press, Boulder, CO

Assies, W. and A. Gundermann (eds) (2007) *Movimientos indígenas y gobiernos locales en América Latina*, Universidad Católica del Norte, El Colegio de Michoacán, IWGIA, Santiago de Chile

Assies, W., G. van der Haar and A. Hoekema (eds) (1998) *The Challenge of Diversity: Indigenous Peoples and Reform of the State in Latin America*, Thela Thesis Publishers, Amsterdam

Bakker, K. (2007) 'The "commons" versus the "commodity": Alter-globalization, anti-privatization and the human right to water in the global South, *Antipode* vol 39, no 3, pp430–455

Bakker, K. (2010) *Beyond Privatization: Water, Governance, Citizenship*, Cornell University Press, Ithaca, NY

Baud, M. (2003) *Intelectuales y sus utopías: Indigenismo y la imaginación de América Latina*, CEDLA, Amsterdam

Bebbington, A. (2001) 'Globalized Andes? Landscapes, livelihoods and development', *Ecumene*, vol 8, no 4, pp414–436

Beccar, L., R. Boelens and P. Hoogendam (2002) 'Water rights and collective action in community irrigation', in R. Boelens and P. Hoogendam (eds) *Water Rights and Empowerment*, Van Gorcum, Assen, The Netherlands, pp1–21

Boelens, R. (2008) *The Rules of the Game and the Game of the Rules: Normalization and Resistance in Andean Water Control*, Wageningen University, Wageningen, The Netherlands

Boelens, R. (2009) 'The politics of disciplining water rights', *Development and Change*, vol 40, no 2, pp307–331

Boelens, R. and B. Doornbos (2001) 'The battlefield of water rights: Rule making amidst conflicting normative frameworks in the Ecuadorian Highlands', *Human Organization*, vol 60, no 4, pp343–355

Boelens, R. and P. Hoogendam (eds) (2002) *Water Rights and Empowerment*, Van Gorcum, Assen, The Netherlands

Boelens, R. and M. Zwarteveen (2005) 'Prices and politics in Andean water reforms', *Development and Change*, vol 36, no 4, pp735–758

Boelens, R., D. Getches and A. Guevara-Gil (eds) (2006) *Agua y Derecho: Políticas hídricas, derechos consuetudinarios e identidades locales*, Water Law and Indigenous Rights (WALIR) and Instituto de Estudios Peruanos (IEP), Lima, Peru

Budds, J. and G. McGranaham (2003) 'Are the debates on water privatization missing the point? Experiences from Africa, Asia and Latin America', *Environment and Urbanization*, vol 15, no 2, pp87–113

Bustamante, R. (2006) 'Pluri-, multi-issues in the reform process: Towards new water legislation in Bolivia', in R. Boelens, M. Chiba and D. Nakashima (eds) *Water and Indigenous Peoples*, WALIR and UNESCO, Paris, pp126–143

Castro, J. E. (2006) *Water, Power, and Citizenship: Social Struggle in the Basin of Mexico*, Palgrave Macmillan, Houndmills, Basingstoke and New York

Cohen, A. (1986) *Two Dimensional Man: An Essay on the Anthropology of Power and Symbolism in Complex Society*, University of California Press, Berkeley and Los Angeles, CA

Foucault, M. (1982) 'Space, knowledge and power: Interview with P. Rabinow', *Skyline*, March, pp16–20

Foucault, M. (1991) 'Governmentality', in G. Burchell, C. Gordon and P. Miller (eds) *The Foucault Effect: Studies in Governmentality*, University of Chicago Press, Chicago, IL, pp87–104 (originally published 1978)

Fraser, N. (2000) 'Rethinking recognition', *New Left Review*, vol 3, pp107–120

Galeano, E. (2009) Dedication to the book *Aguas Rebeldes: Imágenes de la lucha por el agua y la justicia en los Andes*, by R. Boelens and R. Parra, Imprefepp, Quito and IEP, Lima, Peru

Gelles, P. H. (2000) *Water and Power in Highland Peru: The Cultural Politics of Irrigation and Development*, Rutgers University Press, New Brunswick, NJ

Gelles, P. H. (2002) 'Cultural politics and local resistance in highland irrigation development', in R. Boelens and P. Hoogendam (eds) *Water Rights and Empowerment*, Van Gorcum, Assen, The Netherlands, pp22–35

Gentes, I. (2006) 'La interacción de los derechos locales e indígenas de agua con la legislación chilena', in R. Boelens, D. Getches and A. Guevara (eds) *Agua y Derecho*, Abya-Yala, Quito and IEP, Lima, Peru, pp255–284

Gerbrandy, G. and P. Hoogendam (2002) 'Materialising rights: Hydraulic property in the extension and rehabilitation of two irrigation systems in Bolivia', in R. Boelens and P. Hoogendam (eds) *Water Rights and Empowerment*, Van Gorcum, Assen, The Netherlands, pp36–51

Getches, D. H. (2005) 'Defending indigenous water rights with the laws of a dominant culture: The case of the United States', in D. Roth, R. Boelens and M. Zwarteveen (eds) *Liquid Relations*, Rutgers University Press, New Brunswick, NJ, and London, pp44–65

Getches, D. H. (2006) *Indigenous Water Rights in the United States Law and under International Norms*, WALIR Studies vol 6, ECLAC and Wageningen University, Wageningen, The Netherlands, and Santiago, Chile

Getches, D. H., C. F. Wilkinson and R. A. Williams (2005) *Cases and Materials on Federal Indian Law*, 5th edition, American Casebook Series, West Group, St Paul, MN

Guevara-Gil, A. (2006) 'Official water law versus indigenous and peasant rights in Peru', in R. Boelens, M. Chiba and D. Nakashima (eds) *Water and Indigenous Peoples*, WALIR, UNESCO, Paris, pp126–143

Guevara-Gil, A. (ed) (2008) *Derechos y Conflictos de Agua en el Perú*, WALIR, Concertación, Pontificia Universidad Católica del Perú, Lima, Peru

Hayek, F. A. (1944) *The Road to Serfdom*, George Routledge, London

Hendriks, J. (1998) 'Water as private property: Notes on the case of Chile', in R. Boelens and G. Dávila (eds) *Searching for Equity*, Van Gorcum, Assen, The Netherlands, pp297–310

Ingram, H. (2008) *Beyond Universal Remedies for Good Water Governance: A Political and Contextual Approach*, University of Arizona, Tucson, and University of California at Irvine, CA, http://rosenberg.ucanr.org/documents/V%20Ingram.pdf

Isch, E. and I. Gentes (eds) (2006) *Agua y Servicios Ambientales. Visiones críticas desde los Andes*, Abya-Yala, Quito

Kay, C. and H. Akram-Lodhi (eds) (2008) *Political Economy, Rural Transformation and the Agrarian Question: Peasants and Globalization*, Routledge, London and New York

Kay, C., S. M. Borras and E. Lahiff (eds) (2007) 'Market Led Agrarian Reform Policies: Trajectories and Contestations', Special issue of *Third World Quarterly*, vol 28, no 8

Laurie, N., Radcliffe, S. and Andolina, R. (2002) 'The excluded "indigenous"?: The implications of multi-ethnic policies for water reform in Bolivia', in R. Sieder (ed) *Multiculturalism in Latin America*, Palgrave, Basingstoke, pp252–276

Martí, J. (1962) *Nuestra América*, Obras Completas, vol 6, Editora Nacional de Cuba, Habana, p18 (originally published in 1891)

Mayer, E. (2002) *The Articulated Peasant: Household Economies in the Andes*, Westview Press, Boulder, CO, and Oxford, UK

Meinzen-Dick, R. and R. Pradhan (2005) 'Analyzing water rights, multiple uses, and intersectoral water transfers', in D. Roth, R. Boelens and M. Zwarteveen (eds) *Liquid Relations*, Rutgers University Press, New Brunswick, NJ, and London, pp237–253

Mollinga, P. and A. Bolding (2004) *The Politics of Irrigation Reform,* Ashgate, Aldershot, UK and Burlington, VT

Oré, M. T. (2005) *Agua, bien común y usos privados. Riego, Estado y conflictos en La Achirana del Inca*, Wageningen University, WALIR, and PUCP, Lima, Peru

Orellana, R. (2004) *Interlegalidad y campos jurídicos*, PhD thesis, Faculty of Law, University of Amsterdam, The Netherlands

Peña, F. (ed) (2004) *Los pueblos indígenas y el agua: Desafíos del siglo XXI*, El Colegio de San Luis, WALIR, IMTA, Obranegra, Mexico City and Bogotá

Perreault, T. (2006) 'Escalas socioespaciales, reestructuración del Estado y la gobernanza neoliberal del agua en Bolivia', in R. Boelens, D. Getches and A. Guevara (eds) *Agua y Derecho*, Abya-Yala, Quito and IEP, Lima, Peru, pp285–320

Perreault, T. (2008) 'Custom and contradiction: Rural water governance and the politics of usos y costumbres in Bolivia's irrigator movement', *Annals of the Association of American Geographers*, vol 98, no 4, pp834–854

Rhoades, R. E. (ed) (2006) *Development with Identity: Community, Culture, and Sustainability in the Andes*, CAB International Publishing, Oxon, UK, and Cambridge, MA

Roth, D., R. Boelens and M. Zwarteveen (eds) (2005) *Liquid Relations: Contested Water Rights and Legal Complexity*, Rutgers University Press, New Brunswick, NJ, and London

Salman, T. and A. Zoomers (eds) (2003) *Imaging the Andes: Shifting Margins of a Marginal World*, CEDLA, Aksant, Amsterdam, The Netherlands

Swyngedouw, E. (2004) 'Globalisation or "glocalisation"? Networks, territories and rescaling', *Cambridge Review of International Affairs*, vol 17, no 1, pp25–48

Swyngedouw, E. (2005) 'Dispossessing H_2O: The contested terrain of water privatization', *Capitalism, Nature, Socialism*, vol 16, no 1, pp81–98

Urteaga, P. and Boelens, R. (eds) (2006) *Derechos Colectivos y Políticas Hídricas en la Región Andina*, Abya-Yala, Quito, and IEP, Lima

van der Ploeg, J. D. (2006) *El Futuro Robado: Tierra, Agua y Lucha Campesina*, WALIR, Abya-Yala, Quito, and IEP, Lima, Peru

van der Ploeg, J. D. (2008) *The New Peasantries. Struggles for Autonomy and Sustainability in an Era of Empire and Globalization*, Earthscan, London

Vera, J. (2006) 'Derechos de agua, etnicidad y sesgos de género', in R. Boelens, D. Getches and A. Guevara (eds) *Agua y Derecho*, Abya-Yala, Quito, and IEP, Lima, Peru, pp389–410

Verzijl, A. (2007) *Derechos de agua y autonomía local: Análisis comparativo de los Andes peruanos y los Alpes suizos*, WALIR, Abya-Yala, Quito, and IEP, Lima, Peru von Benda-Beckmann, F., K. von Benda-Beckmann and J. Spiertz (1998) 'Equity and legal pluralism: Taking customary law into account in natural resource policies', in R. Boelens and G. Dávila (eds) *Searching for Equity: Conceptions of Justice and Equity in Peasant Irrigation*, Van Gorcum, Assen, The Netherlands, pp57–69

Vos, H. de, R. Boelens and R. Bustamante (2006) 'Formal law and local water control in the Andean region: A fiercely contested field', *International Journal of Water Resources Development*, vol 22, no 1, pp37–48

Vos, J. (2006) *Pirámides de Agua: Construcción e impacto de imperios de riego en la Costa Norte de Peru*, WALIR, Abya-Yala, Quito, and IEP, Lima, Peru

Wilkinson, C. F. (2005) *Blood Struggle: The Rise of Modern Indian Nations*, W.W. Norton, New York and London

Zoomers, A. and G. van der Haar (eds) (2000) *Current Land Policy in Latin America: Regulating Land Tenure under Neo-Liberalism,* KIT, Amsterdam, The Netherlands

Zwarteveen, M. and V. Bennett (2005) 'The connection between gender and water management', in V. Bennett, S. Dávila-Poblete and M. N. Rico (eds) *Opposing Currents: The Politics of Water and Gender in Latin America*, University of Pittsburgh Press, Pittsburg, PA, pp13–29

Zwarteveen, M., D. Roth and R. Boelens (2005) 'Water rights and legal pluralism: Beyond analysis and recognition', in D. Roth, R. Boelens and M. Zwarteveen (eds) *Liquid Relations*, Rutgers University Press, New Brunswick, NJ, and London, pp254–268

2

Water Property Relations and Modern Policy Regimes: Neoliberal Utopia and the Disempowerment of Collective Action

Hans Achterhuis, Rutgerd Boelens and Margreet Zwarteveen

Introduction

Water policies and reforms are mostly discussed in seemingly neutral terms, focusing on technical, administrative, legal or institutional means of allocating water most efficiently and effectively. The use of technical policy language makes water reforms appear as politically objective. Our aim with this chapter is to show that this is dangerously inaccurate. We argue that water reforms are, instead, deeply political. As we have argued elsewhere, this is because they entail far-reaching redistributions of water – and of water rights, water power and water-based profits (Boelens and Zwarteveen, 2005). But water reforms are also an intrinsic element of the larger project of neoliberalism. For a critical analysis of water reforms, then, there is merit in seeing water reforms as exponents of this larger project, as driven by the dream and mission to create a free market utopia.

In his book *Utopia's Heritage* (1998) Hans Achterhuis convincingly shows the dangers and risks of using utopian dreams as guides to shaping reality. No century before ours began with such lofty ideals, and no century saw its expectations end in so much bloodshed. These things are, as he shows, inextricably bound by a grim logic (Achterhuis, 1998). This chapter extends this analysis to the link between utopian projects, neoliberalism and water reforms. It first explains the characteristics of utopia, and shows how they always and necessarily create their own dystopia through the violent exclusion of others, deviants and rivals. A utopia requires the active destruction of the 'old society' in order to build the new one, pure and unspoiled. The chapter also shows how neoliberalism belongs to the 'utopian family tradition' and illustrates this by examining the ideas of its ideological founding fathers: Hayek and Friedman. They envisioned a society organized around a free market, in which the state's main function would be to create the necessary conditions for competition among free individuals. Private property was among the most fundamental of these conditions because it was seen as the key to individual freedom. While their influence is global and pervasive in many of world's regions and institutions, Friedman and his fellow thinkers searched and found a suitable laboratory for experimenting with their utopian ideas in Pinochet's Chile. The violent military dictatorship of Pinochet installed the conditions for neoliberal 'success' and was instrumental in reconstructing the country using neoliberal ideals and offering a clear and dramatic illustration of the necessary utopia–dystopia connection.

Yet, neoliberal utopias are not necessarily realized through such visibly violent means, and do not always have such clearly identifiable promoters as, for instance, Pinochet, Reagan or Thatcher. Neoliberal utopian dreams most often conquer the world in a much more disguised fashion, and through more consensual and socialized forms of political control by presenting themselves as the most reasonable and rational alternative, and by subtly appealing to common sense. Ayn Rand's philosophy inspired globally influential policymakers such as Alan Greenspan. Her writings and those of other popular writers of the US in the 20th century show how neoliberal dreams succeeded in capturing the imagination of many. Rand's neoliberal-objectivist philosophy, which she explains in her influential novel *Atlas Shrugged*, originally published in 1957 (Rand, 1992), reached a massive audience and its ideas and terminology have become the standard for much thinking about society, economy and human behaviour, even though its ideological and utopian origins usually remain unnoticed.

The second part of the chapter analyses how water reforms form part of the neoliberal utopian project. This analysis elucidates some of the characteristics and effects of water reforms. In thinking about water, and in water policy-making, neoliberal terms have become the normal currency. Concepts such as efficiency, utility maximization, individual choice, and private property rights have become key terms in current water debates and are used to legitimate far-reaching interventions to turn 'backward' water management institutions into more efficient 'modern' ones. The conception of modernity in current water reforms applied by development agencies is deeply inspired by neoliberalist utopian dreams, although this is rarely acknowledged. Indeed, in most water policies and thinking, a barely disguised form of neoliberalism is presented as the only possible alternative for avoiding crisis. It is ironic that promoters of neoliberal water reforms frequently refer to the Chilean water reforms as a successful example, thereby conveniently forgetting to mention that these reforms could only be achieved by violently silencing dissenting voices through state-organized torture, disappearances and massacre.

Although less visible than in Chile, the dystopian side of water reforms can be seen elsewhere. Neoliberal water reforms often require the active destruction of existing management systems and institutions. As the other chapters in this book show, the realization of a neoliberal water dream in the Andean region has envisioned the exclusion or annihilation of the many existing water rights systems and user collectives.

Utopian dreams and dystopian nightmares

'Being human means having a utopia'. This pronouncement made by the American theologian Paul Tillich has the effect of strongly diluting the concept of 'utopia'. Every expectation for the future, every dream about better times, every ideal is defined as a utopia. As distinct from this diluting of the concept 'utopia', our assumption in this chapter is that not only the concept itself, but also its meaning was born in 1516 when Thomas More wrote *Utopia*. The main characteristics of this blissful island are recognizable as Wittgensteinian family resemblances in the many hundreds of utopias that have been published in the centuries since More wrote his brilliant piece. We shall describe three of them here. First, a utopia is a 'makeable' society, one that can be created to perfection. A utopia has its creators and founders, and their ideas are implemented by the inhabitants. Second, a utopia is a project in which a society is produced. The concept 'utopia' cannot therefore be applied to individual experiments in living, such as the 19th-century American philosopher Thoreau's sojourn at Walden Pond. Third, a utopia consists of a society in its *entirety*. Partial improvements and specific reforms can help society to progress; but such changes do not amount to 'utopia'. A utopia is, rather, a total split with the core of the old society in order to construct a new one. As the philosophers Herbert Marcuse and Karl Popper said during their famous debate last century, the issue is 'Revolution versus Reform'.

We have seen various sorts of utopias arise in the course of history. More's *Utopia* can be described as a social utopia. A new ideal society takes shape because social institutions supervise and lead the actions of its citizens. More than a century after More's book appeared, Francis Bacon wrote *New Atlantis*. The residents of this utopian island achieve happiness thanks to the abundance of technical products. A change in political and social institutions is unnecessary in this case; a higher production is sufficient. *New Atlantis* is the prototype of what one might describe as a technical utopia with technical progress being the only thing needed to create a utopian society. The three Utopian family resemblances can be seen in *New Atlantis*. Technology guarantees perfection in society; the transformation of active citizens into passive consumers amounts to a radical split with the traditional subsistence economy.

Social and technical utopias have been important sources of inspiration in Western society since the 17th century, as socialist ideals and technological promises provided guidance to those in pursuit of progress. However, in the 20th century, just when utopia seemed to be within reach, a fear of actually implementing a utopia developed. This fear manifested itself in a new literary form: dystopia. It continued the story of utopia, except that there was no traveller arriving with tales of the outside world. Instead, an inhabitant of the ideal society acted as the narrator. This voice from within made it clear that a utopian society was both repressive and inhuman to the core. A social utopia was exposed in Orwell's 1984, while a technical one seemed to develop into Huxley's *Brave New World*. In fact, all utopian family resemblances are also present in a dystopia. It is the same sort of society in both cases, but because the view 'from outside' is replaced by one from inside – utopian life and policies as experienced by the inhabitants themselves – those same characteristics get a different meaning and are now employed in criticizing utopianism.

A question explored in this chapter is whether capitalism – especially in its neoliberal form – could be regarded as a third version within the family of utopias. If so, as we believe, it explains the attraction and unshakeable faith of many people in such a system. It then also becomes possible to deploy the criticism of utopia in unmasking and contesting dystopia.

Neoliberal capitalism as a utopia?

This section posits that capitalism is driven by a profound utopian mission and inspiration.

Hayek, Friedman and Pinochet's laboratory

While Milton Friedman became the world's icon of 'hard-line neoliberalism', Susan George was right to characterize philosopher (and economist, psychologist and political scientist) Friedrich Hayek as its global founding father. Hayek, like Friedman a Nobel laureate, fundamentally inspired Friedman's doctrine. Moreover, Reagan and Thatcher referred to him as their leading philosopher and the Pinochet regime frequently justified the paradoxes of its neoliberal dictatorship with the ideas of this 'anti-totalitarian thinker'.

When, in 1931, Hayek moved from Austria to the London School of Economics, he was already one of the world's leading economists. After the *Anschluss* he could not return to Austria, and therefore stayed in London to become the philosophical leader of an exclusive group of passionate scientists who, like himself, dreamed of changing the world: 'the Mont Pelerin Society'. David Rockefeller and Milton Friedman were among his students and followers. His best-known book, the *Road to Serfdom*, published in 1944, argued that socialism necessarily leads to totalitarian societies, and that a planned economy as present in most states was a clear sign that the world was heading for slavery. 'Few are ready to recognize that the rise of fascism and Marxism was not a reaction against the socialist trends of the preceding period, but a necessary outcome of those tendencies' (Hayek, 1944, p3). In the chapter entitled 'The Great Utopia', Hayek explicitly engages with the utopian–dystopian philosophical discussion, disclosing democratic socialism as a dangerous utopia which necessarily breeds its own dystopia.

Although Hayek found his inspiration in the liberal tradition of the 19th century, and fiercely opposed nearly all forms of central planning of both economic and social policies, he certainly did not preach a *laissez-faire* capitalism. He argued that societies need to be organized around free markets, but emphasized that the state is instrumental in enforcing a legal order and installing the institutions and abstract rules that make market competition among free individuals possible:

> *Most people still believe that it must be possible to find some Middle Way between atomistic competition and central direction... Yet mere common sense proves a treacherous guide... Planning and competition can be combined only by planning for competition, but not by planning against competition.* (Hayek, 1944, p31)

Hayek argued fiercely against redistributive policies, claiming that 'equality before the law is incompatible with any activity of the government deliberately aiming at material or substantive equality of different people. Policies aiming at a substantive ideal of distributive justice must lead to the destruction of the rule of law' (Hayek, 1944, p59).

The postulation of a causal link between private property and freedom is central to Hayek's ideas, as outlined in the *Road to Serfdom*:

> *The system of private property is the most important guaranty of freedom, not only for those who own property but scarcely less for those who do not. It is only because controlling the means of production is divided among many people acting independently that nobody has complete power over us, that we as individuals can decide what to do with ourselves.* (Hayek, 1944, p78)

This characteristically neoliberal idea is deepened in his book *The Constitution of Liberty* (Hayek, 1960).

In 1950 Hayek became a professor at the University of Chicago. Twenty years later he joined Friedman in his admiration for the Chilean dictator Pinochet, whom Hayek saw as an advocate who would free the country from state regulation. As Grandin (2006a) explains in his *Empire's Workshop*, Hayek defended the general's massacres by writing that he had 'not been able to find a single person even in much maligned Chile who did not agree that personal freedom was much greater under Pinochet than it had been under Allende'. As Grandin (2006a, p173) continues: 'Of course, the thousands executed and tens of thousands tortured by Pinochet's regime weren't talking.'

During the 1970s and 1980s, the free market ideologues of the Chicago School founded by Hayek and Friedman designed Chilean economic policy and became truly influential in dictating the terms of public policy for most Latin American countries. As discussed below, the water reforms propagated in the last decade of the 20th century throughout the Latin American continent and in much of the rest of the world were part of this larger neoliberal project. The Chicago-trained economists introduced Latin America to what is known as the Friedman doctrine. Elaborated upon in Friedman's book *Capitalism and Freedom*, this doctrine posits that economic freedom is the precondition for political freedom and, further, that companies do not, and should not, have social responsibility to the public, but need only to focus on profits in order to shape a free society (Friedman, 1962, p133; see also Friedman, 1970, 1998).

Friedman himself came to Chile to offer his services to General Pinochet, and many of the 'Chicago Boys' who were trained in a special University of Chicago programme at the Universidad Católica de Chile became ministers and took other leading political and economic posts in the first years of the dictatorship (Catalán Aravena and Frank, 1984). Central to the doctrine is his above-mentioned concept of 'freedom':

> *Economic freedom is an essential requisite for political freedom. By enabling people to cooperate with one another without coercion or central direction, it reduces the area over which political power is exercised. In addition, by dispersing power, the free market provides an offset to whatever concentration of political power may arise. The combination of economic and political power in the same hands is a sure recipe for tyranny.* (M. and R. Friedman, 1990, pp2, 3)

Indeed, the message of capitalism based on freedom and voluntary action is repeated everywhere in his work. 'No external force, no coercion, no violation of freedom is necessary to produce cooperation among individuals all of whom can benefit' (M. and R. Friedman, 1990, p2). It is therefore more than ironic that these ideas could only be turned into reality under Pinochet's 'laboratory conditions', where local communities and their acts of resistance were coercively controlled.

Friedman appeared aware of this in noting that the dictator, responsible for thousands of executions and tens of thousands of people tortured, was 'sympathetically attracted to the idea of a shock treatment' (cited in Grandin, 2006a, p164; see also Klein, 2007). As detailed below, for the poor population groups and peasant and indigenous water-user communities upon whom this book focuses, the new economic programme was, indeed, a shock treatment. The junta offered neoliberal policy model-makers the opportunities for experimentation, meanwhile creating the conditions for the model to succeed. As leading Chicago graduate Cristián Larroulet concluded, under Pinochet's steady direction, Chile became a 'pioneer in the world trend toward forms of government based on a free social order' (cited in Grandin, 2006a, p75). A state monopoly on violence actively supported the construction of the neoliberal dream, and the social costs that this construction entailed were deemed negligible in view of the ultimate prospect for utopia. Historical teleology, or the belief that development and history are processes with a determined end goal (the free market society) thus served as the ideological and rhetorical justification for what are today's suffering and sacrifices.

The experiment went far beyond a mere economic and productive transformation; it included the conscious conquering of people's hearts and minds:

> *Where Friedman made allusions to the superiority of economic freedom over political freedom in his defence of Pinochet, the Chicago group institutionalized such a hierarchy in a 1980 constitution named after Hayek's 1960 treatise* The Constitution of Liberty. *The new charter enshrined economic liberty and political authoritarianism as complementary qualities. They justified the need of a strong executive such as Pinochet not only to bring about a profound transformation of society but to maintain it until there was a* 'change in Chilean mentality' *[our emphasis].* (Grandin 2006a, p173)

This central and all-important role of the state in constructing a liberal economy contradicts the oft-made distinction between (neo)liberal and planned economies, with the former often associated with less state interference. Illustratively, Grandin (2006a) mentions the speech 'The Fragility of Freedom' which Friedman gave in Chile, fiercely attacking the way in which welfare states destroy freedom: 'He praised the general for putting Chile back on the "right track"' (Grandin, 2006a, p166). Friedman simply turned a blind eye to the strong and often violent meddling of the Pinochet regime in people's lives. Citing the Uruguayan writer Eduardo Galeano – '*torturing people so prices could be free*' – Grandin (2006a, p175) argued that the freedom Friedman refers to is either theoretical, a future promise or a freedom to be enjoyed only by dominant groups who strongly benefit from the system, rather than one that is based on the experiences of common people.

Rand and Greenspan: Towards mainstream acceptance and worldwide implementation of neoliberal dreams

The cases of Chile, and also the examples of Reagan and Thatcher, are rather deliberate and forcefully implemented forms of neoliberalism, made possible by 'shocks' or crisis situations (compare Klein, 2007). This suggests that neoliberalism tends to take a relatively easily recognizable shape, with clearly identifiable 'culprits'. However, and unfortunately for those who are critical of neoliberalism, it often takes much less visible forms and it is most often achieved through more seemingly consensual and socialized means. This is because many of the central ideas of neoliberalism form part of a widely accepted ideological doctrine, ideas that have come to occupy a central place in everyday and 'normal' thinking. With that, the beliefs in free market capitalism as the end goal of development and as the means to end poverty and hunger have also powerfully influenced international development agencies. Because this approach has become the norm for international development it is difficult to recognize neoliberalism as inspired by utopian ideals.

That neoliberalism is, nevertheless, deeply utopian can perhaps best be illustrated by referring to Ayn Rand. She was a Russian-American author of fiction and philosophy who developed a philosophy all of her own: Objectivism. However, she was primarily successful as a novelist who wrote a voluminous utopian fictional work: *Atlas Shrugged* in 1957 (Rand, 1992). The title implies that Atlas would be well advised to stop holding the celestial globe on his shoulders. Instead of thus bearing the responsibility for the social well-being of the whole world population, he would do better to start thinking of his own interests and freedom. This book typifies capitalist utopia. Although not well known in Europe, this novel was shown by a survey of American readers to be the second most influential book (the Bible being the first) of the last century.[1]

Ayn Rand's indirect and worldwide influence is considerable. For example, her most important philosophy student was Alan Greenspan, until recently the chairman of the American Federal Reserve System, the monetary policy of which affects literally every person in our globalized world. When, recently, Greenspan appeared as a witness before a committee of the US Congress, the utopian inspiration for his financial politics became entirely clear: 'Those of us who have looked to the self-interest of lending institutions to protect shareholders' equity – myself especially – are in a state of shocked disbelief'. Greenspan admitted that his belief in the self-regulatory powers of the free market were erroneous. His consternation sounded genuine. 'I made a mistake in presuming that the self-interests of organizations, specifically banks and others, were such that they were best capable of protecting their own shareholders and their equity in the firms' (*The Guardian*, 24 October 2008). It was extremely painful for Greenspan to have to bid farewell to a theory he had held for 40 years.

Greenspan's frank acknowledgement makes it interesting to find out more about the theories on which he based his ideas: the utopia he borrowed from Ayn Rand and in which he deeply believed. In his intriguing autobiography *The Age of Turbulence*, Alan Greenspan describes how at the age of 26 he met Ayn Rand while he was starting his career and how he was profoundly influenced by her. He became an enthusiastic devotee of Objectivism, which 'championed *laissez-faire* capitalism as the ideal form of social organization' (Greenspan, 2007, p40). Being a Russian citizen who fled her country after the Bolshevist Revolution, Rand constructed her own ideal society as the complete counterpart to Soviet communism. Greenspan described in detail how Ayn Rand became 'a stabilizing force' (Greenspan, 2007, p51) in his life. Her fellow students of Objectivism became his 'first social circle outside the university and the economics profession' (Greenspan, 2007, p51). He describes himself thus as 'a convert': 'I engaged in the all-night debates [held by the Objectivists] and wrote spirited commentary for her newsletter with the fervour of a young acolyte drawn to a whole new set of ideas' (Greenspan, 2007, pp51–52).

He subsequently toned down these radical ideas to a certain extent, however. For example, for him it was taking things too far to say that a government that levied taxes was immoral because it appropriated private property in such a violent manner. He was unable to believe in voluntary, rationally based contributions to the financing of government, but remained convinced that taxes should be as low as possible and governments as small as possible.

Generally speaking, he remained loyal to Ayn Rand's ideas. Significantly, Rand stood next to him when he took the oath of office in the Oval Office under President Ford. Rand and Greenspan maintained a close relationship right up until her death in 1982. Greenspan fully acknowledged the decisive influence that she had on his life and thinking, and he was 'very grateful' for that influence.

Ayn Rand reappears later on in Greenspan's autobiography, in relation to the economic reforms in Eastern Europe and Russia after 1989 (Greenspan, 2007, pp133–134). In recognizing that these reforms would plunge large groups of people into deep poverty and misery, Greenspan's sympathy prevailed, totally against his intellectual convictions. He wondered whether some kind of safety net for the unemployed would need to be provided. Klaus, the person with whom he was in discussion, 'cut him short. "In your country you can afford such luxuries", he said. "To succeed we need a clean break with the past. The competitive market is the way to produce wealth, and that's where we're going to focus"' (Greenspan, 2007, p134). Greenspan admitted that he almost abandoned his belief at that point: 'this was the first time in my life that I was rebuked for not sufficiently appreciating the power of free markets. It was a singular experience for an admirer of Ayn Rand' (Greenspan, 2007, p134).

Greenspan's *An Age of Turbulence*, published shortly before the outbreak of the financial crisis, held firm to his free market faith. At the end of his book, Greenspan acknowledges that 'even for the most well-informed market players, the colossal worldwide financial system has become incomprehensible'. Financial regulators were powerless to do anything and, by definition, no longer had any control over the system. But as late as 2007, Greenspan believed that not to be a problem:

> *Markets have become too huge, complex and fast-moving to be subject to 20th-century supervision and regulation. No wonder this globalized financial behemoth stretches beyond the full comprehension of even the most sophisticated market participants ... the most promising anti-crisis policies are those that maintain maximum market flexibility – freedom of action for key*

> *market participants such as hedge funds, private equity funds and investment banks. The elimination of financial market inefficiencies enables liquid free markets to address imbalances. The purpose of hedge funds and others is to make money, but their actions extirpate inefficiencies and imbalances, and thereby reduce the waste of scarce savings. These institutions thereby contribute to higher levels of productivity and overall standards of living.* (Greenspan, 2007, p489)

Greenspan was therefore totally against the finance ministers and the central bankers in the G10 attempting to regulate this enormously complex international financial system: 'Regulation, by its nature, inhibits freedom of market action, and that freedom to act expeditiously is what rebalances markets. Undermine this freedom and the whole market-balancing process is put at risk' (Greenspan, 2007, pp489–490). As long as all market players are driven by 'self-interest', there would appear to be nothing wrong. His advice looking toward 2030 was confident: 'We have no sensible choice other than to let markets work. Market failure is the rare exception, and its consequences can be assuaged by a flexible economic and financial system' (Greenspan, 2007, p492). Greenspan thus regarded it as his main task to organize this system in such a way that the market could operate in total freedom.

In the Objectivism of Ayn Rand, 'Selfishness' is the central concept. She presumed she could offer with her Objectivism a new, complete and fully fledged philosophical system that included metaphysics, epistemology, ethics and aesthetics, as well as a political philosophy that drew on psychology and economics. All of these components are dealt with comprehensively in *The Ayn Rand Lexicon: Objectivism from A to Z* (Rand, 1988). She argued that, historically, all ethical systems arose from a tradition of assuming that selfishness is evil and altruism a virtue. She believed the opposite was true.

This was the reason she wanted to break totally with tradition. By means of Objectivism, man could learn once more that he should always act 'with his rational self-interest in mind'. Rand was emphatic in stating that she was not talking about indulging in the irrational desires that she associated with Nietzsche's philosophy. Man usually selects rational self-interest by means of creative production. Anything he creates in this way is his own property and this is sacrosanct. Rand totally rejected the idea of sharing with others; she believed it to be an irrational act. The creator of property (almost always a super-capitalist in Rand's novels) nevertheless supports society because he provides employment and supplies consumer products. However, it should never be his overt aim to contribute to society in this way. If that does happen – if the needs of others begin to play a role instead of one's own self-interest – free society will come to grief.

Ayn Rand's philosophical adversary is Immanuel Kant. 'Kant's philosophy in terms of every fundamental issue is precisely the opposite of Objectivism', she stated. She was of the opinion that Kant's ethics comprised a doctrine of self-denial and self-sacrifice, while his sense of duty was no more than irrational exhortation to lose and to sacrifice oneself. In this way, Kant doomed free, creative and rational producers, creating the ultimate dystopia. 'Kant is the most evil man in the history of the world', said Rand apodictically.

Rand's great influence and reputation are not primarily based on this sort of philosophical exposition, but on her novels. Despite her disputable reflections on other philosophers, *Atlas Shrugged* is an exciting book that propagates her main utopian message with verve. Below we outline the main utopian theme of *Atlas Shrugged*, although it is impossible to do justice to the many characters and story lines in this novel with its 1168 pages.

The title is a good reflection of the main message. Right through history up until today, the burden of human survival has always fallen on the shoulders of free and brilliant producers. They have ensured that there was food for everyone; they taught their fellow citizens to study nature rationally and to develop techniques to govern it. In other words, they ensured that life became increasingly better. However, because of the influence of mystics, priests and the founding fathers of various religions, as well as philosophers (see Kant), these creative but solitary individuals also became convinced that it was essentially their task and duty to efface themselves in favour of their needy fellow citizens.

In the future era hypothesized in Rand's novel, the social system disintegrates. Outside America, in old Europe, people no longer believe in production for a free market. People's republics, where production resources have been nationalized, have arisen everywhere. That leads to disaster and misery. As the last remaining capitalist state in the world, America has to support and feed the Europeans. However, even this last bastion of the free market falters for two reasons. First, the administration in Washington, because of its socialist aspirations, puts increasing pressure on the economy by imposing regulations and allowing more power to the trade unions. Taxes keep rising and cartels (which Rand fervently supported) are forbidden. In other words, as intervention in the free economy continues, production decreases.

Second, industrialists disappear mysteriously. When this happens, their properties – the means of production – are destroyed. Oil wells are set on fire and factories are sabotaged. And as if that is not bad enough, every time a successful capitalist disappears, no one seems to be capable of carrying on the business. Every successor – including managers appointed by the government – who takes over such a company fails hopelessly. The result is that the economy collapses and society slides back into the barbaric times of hunger and poverty. There is no central heating any more, the trains do not run, and food transportation from the western US fails to reach New York and the other large cities in the east. Ayn Rand describes the decline of America as a classical dystopia, similar to Hayek's examination of 'The Great Utopia' in *The Road to Serfdom* (Hayek, 1944). Utopian statements made by Marx contribute to a highly coloured description that shows how a classical communist utopia turns out to be a dystopia when exposed to the hard economic reality.

Meanwhile, in a valley in the Colorado Desert Mountains that has been made invisible by means of various technical applications, a new society is being prepared in secret. This will be the utopia of greed. The super-industrialists, leading bankers and the judge, who wants to defend property rights, have all moved here. This valley is – in their words – their 'Atlantis'. In the world outside, the capitalists have gone on strike, a strike which manages to attain what workers' strikes never could: the wheels of society come completely to a halt because the hand of capitalism demands that. In their own 'Atlantis', the inhabitants meanwhile live their free market capitalism to the fullest. Everyone produces, competes and trades with each other. When Dagny Taggart, the owner of the railway company and female protagonist in *Atlas Shrugged*, accidentally crashes her aeroplane in the valley, she is allowed to remain only if she earns her keep (as housekeeper for one of the capitalists), although everyone had been looking forward to her coming and welcomed her warmly. Nothing is given in friendship or through a feeling of concern; everything has to be earned by working for it. The super-capitalists in 'Atlantis' boast how much they enjoy competing with and swindling each other. After all, that is better for production and the welfare that is growing so rapidly.

As soon as the collapse of society in the outside world becomes unavoidable and the New Yorkers attempt to flee their city and state in despair, the inhabitants of Atlantis are ready to go and save the world. John Galt, their leader, who has worked hard to destroy the old world, pronounces at the end of the novel: 'The road is cleared. We are going back to the world. He raised his hand and over the desolated earth he traced in space the sign of the dollar.'

The capitalist community of Atlantis, described by Rand, is a full-fledged utopia. All family resemblances belonging to the utopian tradition are present – from perfection ('makeability') to purity, from emphasis on upbringing and education to ideas on love and sexuality (see Achterhuis, 1998, 2006, 2008). But let us return to the dollar sign at the end of the book; everything in Atlantis certainly revolves around money, which seems to be the ultimate criterion for determining the value of production, even the value of a human being. The wealthier someone is, the more valuable he appears to be. Rand echoes the ideas of Hayek on the relation between money and 'freedom'. Money is freedom of opportunity and the dream is making free life possible:

> *Strictly speaking, there is no 'economic motive' [since the latter merely means] the desire for general opportunity, the desire for power to achieve unspecified ends... If we strive for money it is because it offers us the widest choice in enjoying the fruits of our efforts ... money is one of the greatest instruments of freedom ever invented by men. It is money which, in existing society, opens an astounding range of choice to the poor man.* (Hayek, 1944, p67)

Atlantis is first and foremost an economic utopia.

Burning down impure society: 'De-patterning' and 're-patterning'

The creation of Rand's utopia requires the complete destruction of the old society. In order to build neoliberal society, 'The world as it is must be erased to make way for their purist invention. Rooted in biblical fantasies of great floods and great fires, it is a logic that leads ineluctably towards violence' (Klein, 2007, p19). In Ayn Rand's book, capitalist utopians strive to destroy the existing society, by all means. When the main character John Galt, at the end of the story, is finally challenged to reanimate the collapsing economy more or less according to his own principles, he refuses. Only when the decline and destruction of the old world is complete can he and his disciples reconstruct a new utopian economy based on free market capitalism. The break with the past has to be complete in order for a utopia to take shape (Rand, 1992).

In her latest book, *The Shock Doctrine*, Naomi Klein (2007) shows how in the real world neoliberalism also does not tolerate rivals. She explains how neoliberalism is a fundamentalist doctrine that 'cannot co-exist with other belief-systems'. Friedman and his Chicago Boys also discovered precisely these tenets of destruction, crisis and a total break with the past. In the US, a democratic society with strong trade unions, Friedman could not initially find an audience for his radical ideas. That is why they were first tried out in other countries suffering from the shock of a serious crisis. In such cases it was possible to force a complete break with the past and start again with a clean slate. Klein argues that such crises are not exceptional, but rather constitute an integrated key element of current day capitalism:

> *An economic system that requires constant growth, while bucking almost all serious attempts at environmental regulation, generates a steady stream of disasters all on its own, whether military, ecological or financial. The appetite for easy, short-term profits offered by purely speculative investment has turned the stock, currency and real estate markets into crisis-creation machines.* (Klein, 2007, p426)

Klein also elaborates upon the Chilean example. The military coup against the democratically elected President Allende, implemented with the knowledge and support of US intelligence and politicians, preceded the important role of the Chicago school. They had already worked out plans for the economic restructuring of Chile when the coup took place. Society and the economy could therefore immediately be reorganized on a large scale. The trade unions were banned, the standard of living decreased sharply, unemployment shot up. A forceful shock programme was necessary to get the Chilean people to swallow the utopian pill. The aim of the excessive violence, mainly manifested in widespread torturing, was, according to Klein, not to fight the left-wing opposition to the coup, as that was suppressed within a few days; rather, the intention was to create a state of shock in which the free market revolution could be enforced without too much resistance.

Klein refers to torture as a metaphor: electro-shocks aim to 'de-pattern' and 're-pattern' individuals into 'obedient, rational human beings' just as societies are de- and re-patterned into modern, rational market model societies:

> *In each case, the experiments have failed, while inflicting lasting and often irreparable damage on those who were subjected to them. But has the free market experiment failed?... An entire society has been destroyed, but the corporations that operate in the ruins are doing rather well. Klein's message, then, seems to be that – at least in its own profit-centred terms – disaster capitalism works.* (Gray, 2007b, p3)

The Shock Doctrine describes the many, often successful, attempts that Friedman and the Chicago Boys made after the Chile of 1973 to get the utopia of free market capitalism to be adopted worldwide. The regimes run by the generals in Argentina and Brazil belong here, as does the fierce capitalism that was introduced into Russia under Yeltsin after 1989. Friedman and his disciples supplied the recipes for a new economy, although the advisers came from Harvard in the case of Russia. After 1989, with the decline of the communist utopia, it seemed as if we had reached the end of history; that is how Fukuyama, quoting Hegel, described it. Capitalism had won for good and, under leaders such as Thatcher and Reagan who had been directly inspired by Friedman, everyone gradually fell under the spell of the free market. Privatization of government services and organizations, increasing deregulation of the economy and elimination of trade unions, where possible, were ingredients also of the European economies. Naomi Klein shows how the neocapitalist utopia increasingly coincided with a general neoliberal and neoconservative ideology. Box 2.1 illustrates how the political-philosophical justification for forcefully intervening in existing societies and destroying deviant collective property arrangements – in combination with a civilization–privatization project – was constructed centuries ago.

The ultra-capitalists in *Atlas Shrugged* built their arguments in praise of neoliberal utopia by fiercely demonizing the socialist dystopia. Hero John Galt (Rand, 1992, p1020) talks of 'the Black Mass' of the worship of government to which his fellow citizens were surrendering. John Gray applies this phenomenon of the Black Mass in his most recent book (with the same title) to all modern utopias. The black mass is a blasphemous ritual in which the Catholic holy mass is celebrated backwards. The black mass of all the utopias that Gray describes demonstrates how they all end up as dystopias, full of violence against deviant believers and destruction of presumed impurity. Gray shows how the project of Friedman and the Chicago Boys was more closely related to utopian thinking than to science: purifying market thinking that was blemished by collectivist ideology. Just as all

Box 2.1 Civilization, private property and the justified war against collective rights systems

The neoliberal utopia was not the first to portray America as the virgin territory that could be used to experiment with new ideas about how best to organize society. In fact, much of the neoliberal argument rings familiar old bells to those acquainted with the history of the Americas and to those aware of the history of political philosophy in Europe. Long before Columbus reached the Americas, an image of the New World already served as a frame of reference for European thinkers and was central in representing and defining the European Self against an imagined Indian Other (see Wolf, 1982; Lemaire, 1986; Boelens and Zwarteveen, 2005). Philosophers and politicians discursively constructed the Americas' indigenous societies and their property relations to clarify and promote their own philosophical-political ideas about European civilization. The debate between Juan Ginés de Sepúlveda and Bartolomé de las Casas in the 16th century about whether Indians were to be considered humans *who could potentially become equal* (de las Casas, 1999, originally published 1552) or, instead, needed to be considered and treated as slaves is a famous example of this thinking. In his *Treaty on the Just Causes of the War against the Indians* (1996, originally published 1550) de Sepúlveda defended that 'the perfect should rule over the imperfect' (de Sepúlveda, 1996, p19). He reasoned that Indians were animal like, among other grounds, because:

> *... they do not have written laws, but barbaric institutions and customs...* They do not even have private property... *How can we doubt that these people – so uncivilized, so barbaric, contaminated with so many impieties and obscenities – have been justly conquered? [our emphasis].* (de Sepúlveda, 1996, pp105–113)

Private property, indeed, assumed a central place in the debate, symbolizing civilization or its opposite: decline and downfall.

In the seminal ideas of John Locke, a direct relation is established between a person's labour and his right to privately appropriate nature:

> *It being by him removed from the common state nature hath placed it in, it hath by this labour something annexed to it, that excludes the common right of other men: for this labour being the unquestionable property of the labourer, no man but he can have a right to what [was once joined and] ... left in common for others.* (Locke, 1970, originally published 1690, para 27)

In *Utopia*, Thomas More already had loosely established the same relationship. When the island of Utopia became overpopulated, a number of people left the country to start a colony elsewhere in a place 'that hasn't been cultivated by the local inhabitants' (More, 1975, originally published 1516, p79). When these inhabitants resisted the occupation of their lands by the Utopians, war was declared and the original inhabitants were 'expelled from the area marked for annexation'. This kind of war is considered perfectly just because of the 'natural right to derive nourishment from any soil which the original owners are not using themselves' (More, 1975, p80).

In the views of More and Locke, the European colonizers could legitimately take the land of the Indians because the latter did not efficiently and productively mix their labour with the land they were living on. Therefore, they could not claim the land as their rightful property. Hence the famous phrase of Locke: 'In the beginning, all the world was America' (Locke, 1970, para 49). He considered America as mainly empty, wasted soil where there was 'much more land than the inhabitants possess and make use of' (Locke, 1970, para 184). Since all resources were considered to be openly accessible and 'virgin', property systems in this new world were considered primitive or non-existing. The huge genocide and massive starvation because of diseases and exploitation that accompanied its colonization helped to preserve this myth. Because of the decimation of the population, natural resources became abundantly available and the continent, indeed, presented itself as open and bare, a garden for utopian experiments. Even the critical philosopher Hannah Arendt (1990, originally published 1963) went along with the portrayal of the Americas as virgin territory. She argued that the 'natural abundance found in America' made the 'social question' of redistribution less relevant as if the abundance itself was not the result of dispossession. According to Flores Galindo (1988, p33): 'Europe creates the ideas, the Americas perfect them by means of their materialization; the territory par excellence for practical utopias'.

It comes as no surprise that neoliberals such as Friedman were happy to contribute to this myth of 'land that was unproductive before' in order to preach the link between 'freedom' and private initiative and property rights:

> *The fecundity of freedom is demonstrated most dramatically and clearly in agriculture... Unquestionably, the main source of the agricultural revolution was private initiative operating in a free market open to all.* (Friedman and Friedman, 1990, p3)

modern fundamentalists, they ended up with a modern parody of the classical economy that has very little to do with Adam Smith. Classical economists 'had a good understanding of the flaws of market societies. Lacking this insight, neo-liberals turned classical economics into a utopian ideology' (Gray, 2007a, p120).

To summarize, we have argued that neoliberalism, indeed, displays all of the characteristics of utopia. It is based on the belief that society is 'plannable' and 'makeable', while neoliberalism has identifiable designers such as Hayek, Friedman and Rand whose ideas are implemented. In addition, neoliberalism is a societal project, which concerns the whole of society. It requires a total break with the past to build the new neoliberal, utopian society. Water reforms are intrinsic to this larger neoliberal utopia.

Understanding water policies as utopian neoliberal projects

The World Bank expresses the profound belief in the beneficial effects of water markets in the following way:

> *Tradable water rights allow the price of water to reflect the value of its alternative uses, which creates incentives to put it to the most productive use... Secure water rights are particularly beneficial for smaller farmers, who have been most vulnerable to reductions in their water allocation over time and who have few other sources of security. Tradable water rights, by empowering existing users, help to reduce the abuses of administrative allocation and give assurance to poor farmers that their water availability will not be reduced.* (World Bank, 1996, pp11–12)

Although the quote reflects the ideas of Hayek and Friedman, few water scholars would invoke them or neoliberalism to justify their ideas. Instead, neoliberal water policies are usually defended on technical grounds, such as better efficiencies or improved accountability. Yet, it is no coincidence that the best-known example of introducing market incentives in water management comes from Chile under Pinochet, and was part of the larger neoliberal plan discussed earlier.

Water reforms in Chile entailed the privatization of water rights, which, in practice, often implied the expropriation or auctioning off of existing customary and indigenous water rights to the highest bidder. These rights were conveniently labelled as 'unused rights'. The many social dramas that this created were not a cause for concern by most water policy analysts, who continued promoting the neoliberal model on the basis of its efficiency and modernity. Studies say water transfers happened 'voluntarily', and assert that 'tradable water rights can benefit the poor and increase user participation in water allocation and investment decisions' (World Bank 1996, pp12, 1). The same studies defend the 'superiority of markets' (World Bank, 1996, pp12, 1) and portray 'Chile's experiences with water markets' as:

> *... very positive. Water users are particularly pleased by the flexibility and control over their water rights... Allowing rights to be traded increases the value of the right and its transfer to more productive purposes increases employment possibilities. As a result, the humanitarian and equity aspects of water allocation are likely to be better under a market regime.* (World Bank, 1996, pp8, 15)

The 'Chilean water policy experiment' (see also Chapter 10 in this volume) has been extensively studied and criticized (Bauer, 1997; Hendriks, 1998; Dourojeanni and Jouravlev, 1999; Castro, 2002; Budds, 2004; Boelens and Zwarteveen, 2005; Gentes, 2006; compare Kay et al, 2007) by scholars who showed the profound social and ecological problems generated by free markets, and there is considerable evidence that it failed even when measured against its own objectives. Moreover, these studies have shown that Chilean water reform was a deeply political project, inspired by the utopia of neoliberalism. Although justified on the basis of less state involvement, it required strong state backing in the form of powerful laws and rules. Implementation of Chile's water reforms could only occur under a violent and repressive dictatorial regime: 'What is presented as an economic system governed by the iron laws of a kind of social nature is, in reality, a political system which can only be set up with the active or passive complicity of the official political powers' (Bourdieu, 1998, p86).

The importance and popularity of the Chilean model was enormous, particularly after the 1980s (see, for example, World Bank, 1995, 1996). As Trawick (2003, p977) rightly observed: 'It is no exaggeration to say that a single draft law, modelled on the 1981 Water Code of Chile, is now being circulated in the Andean countries, throughout most of Latin America, and throughout much of the "developing world."' Furthermore:

> *In most countries water is still regarded as public property... The track record of such administered water allocation systems has not been impressive. Despite growing water scarcity and the high costs of hydraulic infrastructure, water is typically underpriced and used wastefully, the infrastructure is frequently poorly conceived, built and operated, and delivery is often unreliable... The results show that a market-based system of allocation is likely to lead to a more efficient outcome in water-scarce countries than are traditional systems of assigning water rights. Such a system has the potential to increase the productivity of water use, improve service delivery, stimulate private investment and economic growth, reduce water conflicts, and free government resources for activities with a public good content or positive externalities... The Bank could assist in such efforts by raising awareness of the potential benefits of tradable water rights and by providing technical assistance to establish such rights... The findings have been presented at professional workshops on water management in Berlin, Paris and Washington, DC, and in Bank-wide seminars. They have also been presented to senior government officials in Mexico and Peru, members of Peru's congress, a group of non-governmental organizations in Peru, and a convention of water user associations in Peru.* (World Bank, 1996)

During the 1980s and 1990s, many Latin American countries were forced by the World Bank, the International Monetary Fund (IMF) and the Inter-American Development Bank to adopt neoliberal water legislation, copying the Chilean model. Such water reforms were part of larger structural adjustment programmes that made significant decreases in public spending conditional for receiving further World Bank and IMF loans. Investments in public drinking water and irrigation facilities, together with their management, were among the more expensive items on many national budgets, and turning the responsibility to provide water to the private sector or to users therefore seemed a particularly attractive way to cut public expenditure. The banks engaged Chilean water experts to promote the model in all of its neighbouring countries. When, under nationwide popular protests, countries could not adopt this extreme neoliberal proposal, they were threatened with not receiving new bank loans.

Most World Bank texts present the Chilean-inspired water policies and laws as technical, neutral and scientifically based proposals for more efficient and equitable water use, delivery and allocation, rather than as ideologically inspired plans for reallocating water among uses and users and, ultimately, for an alternative organization of society. The need for implementing these plans was, and is, justified on the basis of a looming water crisis. This idea is widespread and figures clearly as the dystopian horizon, inspiring fear that serves to mobilize support for neoliberal shock-based water reform. The dominant supranational institutions in the water sector all agree that scarcity not only limits potential growth, but threatens human existence. The apocalyptic terminology rallies the public around painful decisions that need to be made for humanity to survive the crisis. Scarcity is simultaneously presented as an opportunity to reorganize and improve established rules of conduct, and to unsettle entrenched bad habits. Thus, tough measures are needed to solve a daunting problem (see Ahlers, 2005; Wegerich and Warner, 2009). The tough measures that are proposed come straight from the Hayek and Friedman textbooks, and argue that state regulation and administration should be replaced by free markets that will reallocate resources not only most efficiently, but also most beneficially for all sectors in society. For this to happen market forces should not face obstacles created by state intervention. Economic deregulation and institutional devolution are to create competitive markets.

Mathematical and economic reasoning based on universal and generic neoliberal postulates of freedom, rational choice and individual profit-maximizing assumptions are the foundations for promoting these measures.

Neoliberal water policies are fundamentally based on the belief that formal-legal and market incentive structures determine the activities of water users and managers and thereby largely determine the economic efficiency of water control and resource use. Water users are seen as individual rational decision-makers who aim to generate net economic benefits through resource allocation. The outcome of organizational and political processes in water management are seen as the sum of the rational decisions made by these individuals based on material self-interests that can be defined objectively and universally (Moore, 1989; Bakker, 2007).

The three basic ingredients of neoliberal water reform recipes – decentralized decision-making, private property rights and markets – are all meant to create political and economic behavioural incentives for water actors.[2] Whenever the incentives are right, private motives of profit maximization and accumulation (selfishness and greed) will automatically foster the use of water and public funds as efficiently as possible.

In sum, neoliberal water policies are a clear element of the neoliberal utopia, revolving around the unequivocal celebration of market forces and private ownership. The popularity of this neat neoliberal simplicity among water policy-makers and analysts can perhaps be explained by their traditionally strong preference for large-scale standard policy initiatives and their fondness for 'design principles' for building viable institutions. They apply universal criteria, conditions or principles to design an institutional transformation. They believe that money and water need to follow universal scientific laws and that human beings follow the same rational utility-maximizing aspirations everywhere (Moore, 1989; Zwarteveen, 2006). Such beliefs offer an attractive way to make sense of complex problems while continuing to believe in the need for universal water knowledge and authority. This fits the general neoliberal idea that one single economic and political system will emerge as humanity evolves towards one universal civilization in which everyone shares the same values and institutions (Gray, 2007a).

Water property rights, equalization and the destruction of otherness

Private property rights are at the core of neoliberal policies; they subsume all other rights, and all rights are defined in terms of private property. Not only are individual rights necessary to avoid market externalities, security of property is essential for equality and individual control perpetuates freedom. The well-known 2004 Milton Friedman Prize winner, who deeply inspired World Bank policies on 'people's capitalism', Hernando de Soto, eloquently articulated these ideas in his influential book *The Mystery of Capital* (de Soto, 2000). With a utopian flavour, he emphasizes that neoliberalism is needed to resolve poverty:

> *Everyone will benefit from globalizing capitalism within a country, but the most obvious and largest beneficiary will be the poor... As a result, they will support the agenda of reform enthusiastically. The poor will become the most effective public relation machine for reform.* (de Soto, 2000, pp190–191)

According to de Soto, those who defend their own 'extralegal' rules and institutions must be mal-informed or need to be educated. For him, like for most neoliberal adherents, it is unthinkable that somebody would not share the utopian dream: 'The most striking feature of these institutions throughout the world is their *desire* to be integrated into the formal sector', or, in other words: 'the extralegals *want* to come in from the cold [our emphasis]' (de Soto, 2000, p178).

Water policies echo much of de Soto's reasoning, and much current thinking about water rights is actually inspired by neoliberal belief that water needs to be transferable and marketable for it to be used efficiently, producing the highest possible marginal returns. For markets to succeed, clearly defined, enforceable and private water rights need to be in place. In neoliberal thinking, defining rules for the allocation and use of water resources provides the means for identifying committed water uses. Water rights allow water to be priced per unit consumed, inducing users to waste less water. In addition, water rights provide the basis for allocating maintenance responsibilities among beneficiaries. They also, and importantly, provide secure tenure to users, thus establishing incentives for investments in infrastructure (Boelens and Zwarteveen, 2005).

The utopian character of water reforms is further shown in the assumption that they depend on a complete break with the past. To make the neoliberal water dream possible and enforce successfully private property rights in water, all existing organizational, normative and legal barriers against market rules need to be cleared (see also Box 2.1). In particular, existing water tenure arrangements and collective forms of water management are characterized as 'backward' and 'anomalous', requiring eradication (Boelens and Zwarteveen, 2005). As legal pluralism scholar, Marc Galanter concluded years ago: 'modern law includes techniques for eroding away and suppressing local law by official law… It tolerates no rivals; it dissolves away that which cannot be transformed into modern law and absorbs the remainder' (Galanter, 1966, pp163–164). Similarly, the neoliberal water modernity project, if it is to succeed, cannot respect the existing plurality of water rights, water identities and management modes (e.g. Zwarteveen, 2006; Bakker, 2009; Boelens, 2009). Only those local water institutions that obey the rules of the modern market (i.e. 'managed, neoliberal multiculturalism'; see also Chapter 3 in this volume) are allowed to exist.

This destruction of existing legal and institutional frameworks does not necessarily happen through overtly violent and oppressive means, but often occurs much more subtly through new regulations and water development interventions that aim to 'normalize' and 'equalize' water rules and water actors because water markets demand a stable, uniform playing field.

Existing water-use systems in the rural Andes have evolved through centuries of collective investments in infrastructure and as the dynamic outcomes of extended struggles over rules and principles for water management. Acquiring water rights with locally sanctioned legitimacy is rooted in wider legal and normative systems. In many local water-user communities, individual water rights of households exist within collective agreements and are enforced through local, collectively legitimized authorities. On the whole, the collective features of water rules, rights and duties, their hybrid origins, their enormous diversity and their integration within local community structures makes them difficult to reconcile with the private property ideal of neoliberal water policies. Their multifaceted, deeply rooted and dynamic character makes them intangible and unrecognizable in positivist (bureaucratic and neoliberal) water rights frameworks.

Their consequent discursive demise also implies a political disappearance: they can no longer be the subject of water negotiations, and so they either disappear from water policy agendas or are treated as 'anomalies'. Existing local water rights frameworks, in other words, lose their legitimacy and become either illegal or get to be seen as non-rights. By constructing them as 'non-existent' or disqualifying their legitimacy, neoliberalism succeeds in opening up local communities' water sources to markets.

To explain this process of privatization, Swyngedouw (2005) makes use of Harvey's concept of 'accumulation by dispossession': the introduction of private property rights in water is a legally and institutionally condoned form of theft. 'Of necessity, the process of "privatization" equates a process of "dispossession"' (Swyngedouw, 2005, p1). The idea that this is theft fundamentally challenges the claims that justify privatization reforms: freedom, efficiency, equality and civilized progress. Studies support this characterization as theft by showing how tenure security of smallholder farmers typically worsens under neoliberal regimes (Zwarteveen 1997; Boelens and Zwarteveen, 2005; Bakker, 2007). One example is that the neoliberal model promotes delinking of water rights from territory in order to extend competition and enhance free trade of water rights to the most productive use via the highest bidder. This poses an enormous threat to tenure security of those communities who base their rights on socio-territorial claims, and opens up their water for outside buyers with little or no interest in maintaining collective infrastructures. The individualization and dissolving of water rights by detaching them from collective rule-making tends to generate conflicts while weakening collective means of system operation and maintenance.

Many local communities have secured their rights to water by investing tremendous amounts of labour and other resources to construct the system. According to local views, collective construction of infrastructure is the source of water rights. It is also the basis for collective decision-making in operating and maintaining the system (Boelens and Hoogendam, 2002). Privatization and subsequent market purchase of individual rights forcefully break up these systems of water rights and strike at the heart of Andean water control. In general, through the installation of neoliberal regulations,[3] local property systems become alienated from community relationships and security systems, and collective arrangements are replaced by outside market-controlled rules.

Neoliberal water reforms thus indeed work to destroy the existing social organization of water, causing social dramas and inequalities. Yet, the promises of increased efficiencies made to justify these reforms remain unrealized. Although the utopian horizon moves further away, the neoliberal dream continues to be used as an indicator of its achievements and existing common property rights regimes and collective water use are characterized as backward and stubborn for not responding to the universal logic of the market (see also McCay and Jentoft, 1998; Kay et al, 2007). Measured against the teleological ideal of neoliberalism, traditional communities do not act 'rationally' or 'democratically' and thus need to be adjusted to fit into the model. If not, they are doomed to wither and disappear. Instead of analysing the breakdown of collective water rights systems as a consequence of the aggressive encroachment by free market actors – mining companies, power plants and agribusiness – and instead of showing how neoliberal rules provoke the breakdown of community coherence, collective action and shared management, the model is presented as the only way to solve local water control problems. Paradoxically, the remedy that is prescribed is to increase free market rules in local communities so that private interests can improve management efficiency and enforce water rights, with the result that outsiders take over their resources, rules and authority.

Breaking down collective systems by opening them to private buyers is seen as a necessary step towards realizing the ultimate utopian reality. It is seen not as a failure but as an achievement of the policy model. As Milka Castro shows for water user communities subject to Chilean free market ideologies: 'The dominant vision of a modern political and economic scenario appeals insistently to the need to overcome "old, archaic" traditions and move on to "new, modern" forms, proposing for indigenous "marginalization" to give way to "integration"' (Castro, 2002, p197). The excluded are included and become 'potentially equal, normal water users'. Ironically, Hayek (1944, p114) understood it well when he criticized socialist practice for its disciplining normalization: 'To make a totalitarian system function efficiently it is not enough that everybody should be forced to work for the same ends. It is essential that the people should come to regard them as their own ends.' But he failed to see that his words neatly apply to the neoliberal system that he so much desired.

In neoliberal water policy there is a clear, discursive need to present all water actors as potentially equal; markets, meetings and transactions only work when all participants can interact 'as equals'. Equalization and uniformity of objectives and rules framed in terms of individualized water control, commoditized water rights and users who rationally pursue the economically most beneficial water use is a precondition for the model to work. Clearly, the cultural, political and normative standards for this equality – and, thus, for normalizing the abnormals and equalizing the unequals – are set by neoliberal policy model-makers. And the standards or norms are powerful since they are presented as unavoidable laws of nature and almost religiously based on the requirements of civilization (compare Foucault, 1975). Differences in terms of decision-making power, monetary income, skills, information and education that strongly favour private companies and business actors are neglected or presented as a gap that can be bridged by potential equals when joining the market game.

Neoliberal equality and freedom are 'the two faces of the same basic value', as Friedman argues. It fundamentally refers to the fact that everyone is 'to be free to go into any business, follow any occupation, buy any property, subject only to the agreement of the other parties to the transaction' (Friedman and Friedman, 1990, p133), without government restrictions.[4] Therefore, water user communities are potentially equal and will become equal if they accept the standards; the discourse actively encourages water users to participate as equals in a uniform market playing field. In neoliberal practice, this freedom sometimes has to be imposed on the deviants. As Chicago economist De Castro (finance minister under Pinochet) explained: 'A person's actual freedom can only be ensured through an authoritarian regime that exercises power by implementing equal rules for everyone' (cited by Grandin, 2006a, p174). But although the uniform rules of the model are supposedly made for all people, it seems that the water elites are treated as more equal than others. The creation of water monopolies that obstruct free market operation as a consequence of the water privatization regime is a clear example of allowing 'rights of the extraordinary'. Peasant and indigenous communities suffer from such water monopolies controlling water that often comes from sources they need for subsistence while the companies sometimes hold it mainly for speculation.

However, some water user collectives resist the idea of potential equality because they have different standards of equality than those of the neoliberal model. An increasing number of peasant and indigenous organizations are standing up against privatization efforts and neoliberal water reform programmes. This resistance shows that 'not fitting' the model is often a conscious choice, rather than the result of backwardness or unreasoned stubbornness. The fierce contest highlights the contested nature of water rights, and the fact that much, very much, is at stake.

Conclusions

In his book *Neoliberalism*, David Harvey (2005, p19) argues that we can 'interpret neoliberalization either as a *utopian* project to realize a theoretical design for the reorganization of international capitalism or as a *political* project to re-establish the conditions for capital accumulation and to restore the power of economic elites'. He concludes that the second interpretation is dominant, and that the first is primarily a system of justification and legitimization for the second. Harvey convincingly shows that when neoliberal utopian principles 'clash with the need to restore or sustain elite power, then the principles are either abandoned or become so twisted as to be unrecognizable'. Yet, with his suggestion that the utopian project merely serves as a rationalization of the political project, he overlooks how the two are intimately intertwined. The convincing force and power of the neoliberal political project is rooted in the enormous appeal of its utopian dreams that have become deeply ingrained in belief systems. Indeed, this appeal of neoliberalism has enabled it to succeed in becoming the 'normal' way of looking at the world.

Harvey's suggestion of a 'maligned elite' who use neoliberal utopian ideas to continue business as usual may seem justified when just focusing on the manipulative tactics of neoliberals such as Friedman, Pinochet, Reagan and Thatcher, or of some representatives of 'big business'. Yet, the neoliberal project is much more widespread and its ideas have influenced millions of people. The anti-totalitarian dreams of Rand and Hayek were a source of inspiration to people such as Karl Popper and George Orwell. Hayek understood the neoliberal project as a battle for ideas and truth. Indeed, to understand why the project endures and flourishes despite the failures and disasters it produced, even despite its economic inconsistencies, it is important to grasp its dream-like persuasive properties.

The book *Para Leer al Pato Donald* was published one year before the military coup in Chile (and burned and forbidden directly after that). Its writers Ariel Dorfman and Armand Mattelart show how the seemingly innocent and imaginary children's world of Donald Duck promotes the values of greed, selfishness and competition. The inhabitants of Disneylandia have a profound horizontal orientation; 'people with the same status and power cannot be dominated or dominant. All they have left – since solidarity among peers is forbidden – is to compete to get ahead of the other guy – that is, to get into the club of the dominant, to go one rung up (as a corporal, sergeant or general of Disneylandia) on the scale of market merit' (Dorfman and Mattelart, 1972, p35) (compare Girard, 1961; Achterhuis, 1988). In this dream world, the self-made man is fundamental, promoting the basic myth of freedom, equality and social mobility in the capitalist system: 'Equal opportunities, absolute democracy: each child starts from scratch and builds up whatever they deserve' (Dorfman and Mattelart, 1972, p120). Dorfman and Mattelart argue that most criticisms of neoliberalism have concentrated on the unfair relations and unequal distribution required and caused by the system, and the imposition of US lifestyles and production structures. However, 'the threat is not to become a spokesperson for the American *way of life*, but because it represents the American *dream of life*, the way that the USA dreams of itself, redeems itself. The real danger lies in how the metropolis demands us to represent our own reality to ourselves for their own salvation' (Dorfman and Mattelart, 1972, p151).

The way in which 'Disneylandia' and its metaphors serve as promotional material for the Objectivism of Rand, and its morality of rational self-interest, is an example of how the dreams, values and ideas of neoliberalism have succeeded in becoming part of common sense. In water policies, neoliberalism sets the tone – a strong paradigm of rational choice, based on economic analyses, coupled with the language of efficiency. The strong belief in the superiority of individualistic rational choice, so powerful in current water management theories, partly finds its roots in Hayek-liberal concepts: 'Common action is limited to the instances where individual views coincide; what are called "social ends" are for it merely identical ends of many individuals' (Hayek 1944, p44). In addition, in water, the neoliberal discourse largely derives its strength and legitimacy from presenting and – subtly or violently – imposing itself as self-evident and actively constructing a logic of its own inevitability. Thereby, the neoliberal model does not just assume and desire universal laws, but also actively establishes them. Coexistence of a diversity of rules, rights and obligations is actively discouraged since such diversity would obstruct inter-regional and international transfers and trades that require a uniform legal framework. Locally particular rules and rights that would inhibit transfers are anomalies that stand in the way of investments and profits. Unlike what the *laissez-faire* doctrine suggests, strict central authorities and legislation actively discipline and counteract pluralism of water rights to allow markets to emerge. Communities and collective rights systems that do not fit the neoliberal picture are deemed inefficient and backward. Peasant and indigenous communities have often lost their rights and voice as a result of neoliberal regulation and privatization. Not just their water access rights, but also their management rules and identities are denied and undermined by a model that simply assumes freedom of expression without actually verifying the conditions of its existence.

Finally, it may be the irony of our time that the words of neoliberal founding father Hayek (1944, p8), warning against governance systems that generate poverty, the loss of freedom and the road to serfdom, are to be turned against his free market utopia:

> *We are ready to accept almost any explanation of the present crisis of our civilization except one: that the present state of the world may be the result of genuine error of our own part, and that the pursuit of some of our most cherished ideals have apparently produced results utterly different from those which we expected.*

Notes

1 Moreover, three of Ayn Rand's works were in the list (according to *Time* magazine) of the most important American books of the last century: besides *Atlas Shrugged* (published in 1957), *The Fountainhead* (in 1943) and *Anthem* (in 1938).

2 Neoliberal postulates hold that individual users and/or managers will obstruct efficient, effective water management whenever the incentives are 'perverse': they are the main cause of poor performance and create net costs for system operation.

3 Neoliberal policies pose many other threats (see Boelens and Zwarteveen, 2005, and Chapters 8 to 12 and 16 in this volume). For example, common property systems are being dismantled by concentrating decision-making rights in the hands of the few. In most communal water systems, the 'one-right-holder–one-vote' rule applies. By contrast, World Bank and Inter-American Development Bank proposals for new water legislation in Latin America stipulated that voting rights should be made proportional to the quantity of water use rights each user holds, like shareholders in a joint stock company (compare Hendriks, 1998).

4 Friedman strongly opposes concepts of equality as 'fair shares for all', which according to him 'is the modern slogan that has replaced Karl Marx': 'To each according to his needs, from each according to his ability' (Friedman and Friedman, 1990, p134). 'Equality of outcome is in clear conflict with liberty' (Friedman and Friedman, 1990, p128).

References

Achterhuis, H. (1988) *Het Rijk van de Schaarste: Van Thomas Hobbes tot Michel Foucault* [*The Realm of Scarcity*], Ambo, Amsterdam, The Netherlands

Achterhuis, H. (1998) *De Erfenis van de Utopie* [*Utopia's Heritage*], Ambo, Amsterdam, The Netherlands

Achterhuis, H. (2006) *Utopie*, Ambo, Amsterdam, The Netherlands

Achterhuis, H. (2008) *Met alle geweld* [*Violence: An Overview*], Lemniscaat, Rotterdam, The Netherlands

Ahlers, R. (2005) *Fixing Water to Increase its Mobility: The Neoliberal Transformation of a Mexican Irrigation District*, PhD thesis, Cornell University, Ithaca, NY

Arendt, H. (1990 (1963)) *On Revolution*, Penguin Books, London

Bakker, K. (2007) 'The 'commons' versus the 'commodity': Alter-globalization, anti-privatization and the human right to water in the global South, *Antipode*, vol 39, no 3, pp430–455

Bakker, K. (2009) 'The ambiguity of community: Debating alternatives to private sector provision of urban water supply', *Water Alternatives*, vol 1, no 2, pp236–252

Bauer, C. (1997) 'Bringing water markets down to earth: The political economy of water rights in Chile, 1976–95', *World Development*, vol 25, no 5, pp639–656

Boelens, R. (2009) 'The politics of disciplining water rights', *Development and Change*, vol 40, no 2, pp307–331

Boelens, R. and P. Hoogendam (eds) (2002) *Water Rights and Empowerment*, Van Gorcum, Assen, The Netherlands

Boelens, R. and M. Zwarteveen (2005) 'Prices and politics in Andean water reforms', *Development and Change*, vol 36, no 4, pp735–758

Bourdieu, P. (1998) *Acts of Resistance against the Tyranny of the Market*, The New Press, New York, NY

Budds, J. (2004) 'Power, nature and neoliberalism: The political ecology of water in Chile', *Singapore Journal of Tropical Geography*, vol 25, no 3, pp322–342

Castro, M. (2002) 'Local norms and competition for water in Aymara and Atacama communities, northern Chile', in R. Boelens and P. Hoogendam (eds) *Water Rights and Empowerment*, Van Gorcum, Assen, pp187–201

Catalán Aravena, O. and A. G. Frank (eds) (1984) *Chili onder Pinochet*, SUA, Amsterdam, The Netherlands

de las Casas, B. (1999 (1552)) *Brevísima relación de la destruición de las Indias*, Ed. Consuelo Varela, Editorial Castalia, Madrid

de Sepúlveda, J. G. (1996) *Tratado sobre las justas causas de la guerra contra los indios*, Fondo de Cultura Económica, Mexico (originally published 1550)

de Soto, H. (2000) *The Mystery of Capital: Why Capitalism Triumphs in the West and Fails Everywhere Else*, Basic Books, New York, NY

Dorfman, A. and A. Mattelart (1972) *Para Leer al Pato Donald: Comunicación de masa y colonialismo*, Siglo XXI Editores, Buenos Aires

Dourojeanni, A. and A. Jouravlev (1999) *El Código de Aguas de Chile: Entre la ideología y la realidad*, ECLAC, Santiago, Chile

Flores Galindo, A. (1988) *Buscando un Inca: Identidad y utopia en los Andes*, Editorial Horizonte, Lima, Peru

Foucault, M. (1975) *Discipline and Punish: The Birth of the Prison*, Vintage Books, New York, NY

Friedman, M. (1962) *Capitalism and Freedom*, The University of Chicago Press, Chicago, IL

Friedman, M. (1970) 'The social responsibility of business is to increase its profits', *The New York Times Magazine*, 13 September 1970

Friedman, M. (1998) 'The suicidal impulse of the business community', Luncheon address, 21 November 1998, San José, CA, www.cato.org/events

Friedman, M. and R. Friedman (1990) *Free to Choose*, Harvest Book, Harcourt, New York, NY (originally published 1980)

Galanter, M. (1966) 'The modernization of law', in M. Weiner (ed) *Modernization*, Basic Books, New York, NY, pp153–165

Gentes, I. (2006) 'La interacción de los derechos locales e indígenas de agua con la legislación nacional chilena', in R. Boelens, D. Getches and A. Guevara Gil (eds) *Agua y Derecho*, IEP, Lima, Peru

Girard, R. (1961) *Mensonge Romantique et Vérité Romanesque*, Editions Bernard Grasset, Paris
Grandin, G. (2006a) *Empire's Workshop: Latin America, The United States, and the Rise of the New Imperialism*, Metropolitan Books, New York, NY
Grandin, G. (2006b) 'The road from serfdom: Milton Friedman and the economics of Empire', *Counterpunch*, 17 November, www.counterpunch.org/grandin11172006.html
Gray, J. (2007a) *Black Mass: Apocalyptic Religion and the Death of Utopia*, Allen Lane, London
Gray, J. (2007b) 'The end of the world as we know it', *The Guardian*, 15 September 2007
Greenspan, A. (2007) *The Age of Turbulence*, Penguin Books, London and New York
The Guardian (2008) 'Greenspan – I was wrong about the economy. Sort of', *The Guardian*, 24 October, www.guardian.co.uk/business/2008/oct/24/economics-credit-crunch-federal-reserve-greenspan
Harvey, D. (2005) *Neoliberalism: A Brief History*, Oxford University Press, Oxford
Hayek, F. A. (1944) *The Road to Serfdom*, George Routledge, London
Hayek, F. A. (1960) *The Constitution of Liberty*, University of Chicago Press, Chicago, IL
Hendriks, J. (1998) 'Water as private property: Notes on the case of Chile', in R. Boelens and G. Dávila (eds) *Searching for Equity: Conceptions of Justice and Equity in Peasant Irrigation*, Van Gorcum, Assen, The Netherlands, pp297–310
Kay, C., S. M. Borras and E. Lahiff (eds) (2007) 'Market-Led Agrarian Reform Policies: Trajectories and Contestations', Special issue of *Third World Quarterly*, vol 28, no 8
Klein, N. (2007) *The Shock Doctrine: The Rise of Disaster Capitalism*, Metropolitan Books, New York, NY
Lemaire, T. (1986) *De Indiaan in ons bewustzijn: De ontmoeting van de Oude met de Nieuwe Wereld,* Ambo, Baarn
Locke, J. (1970) *Two Treatises on Government*, Cambridge University Press, Cambridge (originally published 1690)
McCay, B. J. and S. Jentoft (1998) 'Market or community failure? Critical perspectives on common property research', *Human Organization*, vol 57, no 1, pp21–29
Moore, M. (1989) 'The fruits and fallacies of neoliberalism: The case of irrigation policy', *World Development*, vol 17, no 11, pp1733–1750
More, T. (1975) *Utopia*, Penguin Classics, Harmondsworth, Middlesex, UK (originally published 1516)
Rand, A. (1988) *The Ayn Rand Lexicon: Objectivism from A to Z*, Penguin Group USA, New York, NY
Rand, A. (1992) *Atlas Shrugged*, Penguin Group USA, New York, NY (originally published 1957)
Swyngedouw, E. (2005) 'Dispossessing H_2O: The contested terrain of water privatization', *Capitalism, Nature, Socialism*, vol 16, no 1, pp81–98
Trawick, P. (2003) 'Against the privatization of water', *World Development*, vol 31, no 6, pp977–996
Wegerich, K. and J. Warner (2009) *The Politics of Water,* Routledge, London
Wolf, E. (1982) *Europe and the People without History*, University of California Press, Berkeley, CA
World Bank (1995) *Peru: A User-Based Approach to Water Management and Irrigation Development*, Report 13642 –PE, World Bank, Latin America and the Caribbean, Country Department, Washington, DC

World Bank (1996) *Tradable Water Rights: A Property Rights Approach to Resolving Water Shortages and Promoting Investment*, P. Holden and M. Thobani (eds), Policy Research Working Paper 1627, World Bank, Latin America and the Caribbean Technical Department, Washington, DC

Zwarteveen, M. (1997) 'Water: From basic need to commodity – a discussion on gender and water rights in the context of irrigation', *World Development*, vol 28, no 8, pp1335–1349

Zwarteveen, M. (2006) *Wedlock or Deadlock? Feminists' Attempts to Engage Irrigation Engineers*, Wageningen University, Wageningen, The Netherlands

Zwarteveen, M., D. Roth and R. Boelens (2005) 'Water rights and legal pluralism: Beyond analysis and recognition', in D. Roth, R. Boelens and M. Zwarteveen (eds) *Liquid Relations: Contested Water Rights and Legal Complexity*, Rutgers University Press, New Brunswick, NJ, pp254–268

3

The Limits of State Reform and Multiculturalism in Latin America: Contemporary Illustrations

Willem Assies

Three decades of indigenous struggle

In early 2000, the city of Cochabamba, Bolivia, was the scene of a pitched battle known as the Water War. Aguas del Tunari Company, a Bolivian corporation linked to the multinational consortium Bechtel, won water development and distribution rights in Cochabamba through a shady process, and then boosted city water rates, supposedly to help finance improved services. The protest that followed was not just an urban consumer revolt or an isolated event. Irrigators in the region who saw their rights to water threatened were also involved and the Water War ultimately resulted in a substantial amendment to the Drinking Water and Sanitary Sewerage Law, which had been enacted a few months before. Multilateral agencies had promoted the law to privatize water supply in large cities and turn water into a commodity. The opposition considered water to be a fundamental right and a public good. The Cochabamba Water War was the most heated example of the Bolivian people's broad-based opposition to the neoliberal model, and was considered the first victory after a decade and a half of defeats (Assies, 2001).

This confrontation marked the start of a cycle of protests about managing the country's resources under the neoliberal model that was imposed in 1985. These protests eventually led to the flight of President Gonzalo Sánchez de Lozada[1] and the resignation of President Carlos Mesa in June 2005. In January 2005, the people of the city of El Alto had forced the Carlos Mesa government to rescind a contract with Aguas de Illimani (a company that is part of Suez

Lyonnaise des Eaux de France) for its resistance to extending service to the poorer parts of that city.

Conflicts over water and other natural resources were the focus of popular and indigenous movements in Bolivia and other Latin American countries. The new political context inaugurated by the government of President Evo Morales (January 2006) created a historic opportunity to transform protest into state policy and design a new regulatory framework for natural resource management. With the support of rural, indigenous and grassroots organizations for its working agenda, including the promotion of indigenous and peasant 'uses and customs' regulating water management, Bolivia's new government now faces a huge political, ideological and legal challenge to reconfigure water norms and management (see also Chapter 15 in this volume).

For the last few decades, indigenous movements have become major societal and political stakeholders in the Latin American arena. What has been called 'indigenous emergence' began in the late 1960s and expanded throughout the following decade (Bengoa, 2000). As Baud and Gelles detail in this volume (see Chapters 5 and 6), the relationship between ethnicity and class was revisited, and ethnic identity was revalued. The Shuar Federation in Ecuador, the Katarist movement in Bolivia and the Indigenous Regional Council of Cauca Valley (CRIC) in Colombia are a few of the movements considered to have led the way. These movements shared two key characteristics: they rejected concepts of political identity derived exclusively from economic class and they emphasized the importance of ethnicity in defining political identity. The first CRIC training workbook in 1973, entitled *Our Struggles, Yesterday and Today*, taught that 'we are peasants' and 'we are Indians'. Similarly, the Shuar Federation stressed: 'We cannot agree with the conclusions of people who view Indians just as miners, peasants, or exploited labourers, as the case may be. A Shuar and a non-native tenant are not the same (even if the latter is exploited along with the Shuar), nor is a Shuar chauffeur the same as a *cholo* chauffeur (although both have the same job)'. In Bolivia, the Tiawanacu Manifesto stated: 'We feel economically exploited and culturally and politically oppressed' (compare Bonfil Batalla, 1991).

The 1970s and 1980s witnessed the emergence of indigenous movements in Latin America that often built local networks before attempting to organize on the national level. A major landmark was the revision of the 1957 Convention 107 of the International Labour Organization (ILO) in 1989. The revision, now embodied in Convention 169, rejected the assimilationist orientation of the previous version and, despite multiple constraints, proposed rights geared towards indigenous autonomy within nation states. Convention 169 has been ratified by a dozen Latin American countries, which have thereby made the commitment to reconcile their national legislation with the convention's premises. Similarly, beginning in the mid 1980s, the same number of countries in the region amended their constitutions to recognize the multiethnic, pluri-cultural composition of their populations, formally departing from the homogenizing concept of the melting-pot nation state. Indigenous movements picked up momentum when the United Nations declared the Year and Decade of Indigenous Peoples in 1993 and 1994, respectively. Commemoration of the 500th anniversary of the Spanish invasion also gave rise to international gatherings of indigenous movements.

In the closing decades of the 20th century, emergent indigenous movements practised 'identity politics' and the Latin American states responded with 'recognition policies'. Dialectic between identity politics and recognition policies encouraged expectations and raised hopes about a 'friendly liquidation of the past' (Van Cott, 2000). This process took place under the 'dual transition' – towards civil governments and towards a new development model – and was promoted through structural adjustment policies inspired by neoliberalism. This dynamic then spread and deepened under globalization as *market democracies* emerged: the rise of formal, electoral democracy alongside market liberalization.

After several decades of multicultural policies, there are new questions and issues about the future course of indigenous movements and the scope of Latin American recognition policies. Constitutional amendments and ratifications of Convention 169 suggested a new direction in relationships between indigenous peoples and states, including recognition of territorial rights, indigenous jurisdiction within indigenous territories and a degree of political autonomy. These calls for recognition of people's identities and for greater political and regulatory autonomy are related to the struggle for fairer distribution of resources and means of production and – as this book's chapters show – also deeply entwine with the issues of water control, water rights formulation and water policy implementation.

Evaluating the new multiculturalism

Initial optimism about new 'pluri-/multi-coexistence' appears to be fading. Stavenhagen (2002, pp34, 41) has argued that for indigenous movements the 'going will be rough from now on' because 'redress for historical grievances – which is what originally started the movement – is a limited objective in the long run… It has now become clearer that what began as demands for specific rights and compensatory measures has turned into a new view of the nation and the state' (compare Iturralde, 1997).

Some authors have suggested that recognition policies in an era of globalization should be understood in the context of large-scale transformations of states, leaving behind the 19th-century model of nation states and the mode of territoriality that they entail.[2] The 'homogeneous' state is transformed as overlapping local, national and global jurisdictions emerge with different loyalties, creating a new legal and political pluralism. Potentially, or perhaps ideally, such 'disaggregation' of the state could accommodate indigenous demands for autonomy; however, these demands face accusations of 'Balkanizing' the country. Yet, this argument denies actual indigenous claims and reality. Moreover, if the concern were the integrity of the country, it would also seem to militate against those policies and politicians who bargain away resources to transnational corporations with special jurisdiction arrangements over their 'concession areas' under free trade agreements (e.g. international arbitration, application of metropolitan norms for dispute resolution and keeping litigation out of national courts).

The shift toward multicultural policies not only occurred in the context of democratic transitions, but also in the shift from the national development model towards the neoliberal development model. Neoliberalism extends beyond economics to culture, involving a redefinition of 'citizenship'. A neoliberal criticism of nation state development is that it produces 'dependent' citizens expecting the state to solve all of their problems. Neoliberalism argues that states should stop being 'paternalistic' towards the citizenry and give them back responsibility for their own well-being.

Following on my reflections on the 'cultural project' of neoliberalism (Assies, 2000), Hale (2002) has proposed a theoretical framework that could be used to evaluate the scope of recognition policies within the neoliberal framework. In order to distinguish between managed multiculturalism and transformative multiculturalism, he asks to what degree 'neoliberal multiculturalism' truly involves redistribution of power and resources to indigenous people. Managed multiculturalism celebrates cultural plurality, but without concrete lasting effects for the members of the oppressed cultural group. By contrast, transformative multiculturalism truly redistributes power and resources. One model is 'top down' and the other is 'bottom up'; the former reinforces limited, essentialist expressions of group identities whereas the latter would build on key elements from progressive identity politics such as 'heterogeneity' and 'hybridity', as well as redistribution of resources and means of production to groups affirming their identity.

Building on Foucault's reflections on governance and the 'production of subjects', Hale (2002) proposes to evaluate the consciousness of those struggling for cultural rights. Classical liberalism, which sought to 'liberate' individuals from community and corporate bonds, may be distinguished from present-day neoliberalism, which paradoxically applauds non-governmental entities such as community, civic and volunteer organizations, and churches that 'rescue' the individual. The neoliberal state expects its citizen subjects to resolve their own problems, whether daily or epochal. As individuals or members of volunteer associations, they have to assume their 'responsibilities' and this makes them vulnerable to 'top-down' policies seeking to tap, circumscribe and shape this 'participation'. Professionalized non-governmental organizations (NGOs) would be the vehicles *par excellence* to achieve this new governance. Therefore, the neoliberal state does not simply 'recognize' the community, civil society or indigenous culture, but reconstructs them after its own image, making distinctions, for example, between 'good' and 'bad' Indians. A 'good Indian' presents 'cultural' demands compatible with the neoliberal project and thus may become what Hale (2004) has called the 'Indio Permitido', whereas the 'bad Indian' is the 'radical' calling for power and resource redistribution.

Although the neoliberal project speaks of decentralization and respect for multiculturalism, the latter is allowed so long as it does not get in the former's way. Respect for certain cultural traits such as old-fashioned valuing of 'folk' culture is no problem; but if the idea is to manage resources differently than market logic, it is another matter. The struggles over water control and water policies constitute a powerful example. Decentralization and transferring management of water use systems to the users in the neoliberal mode usually do not involve the transfer of autonomy of rule-making regarding water control and water rights to these users according to their own practices, demands and traditions. The paradox of this kind of decentralization is that it allows natural resource management only when it fits centralized guidelines that are promulgated without consulting affected local stakeholders.

Hale also argues that in Latin America, the neoliberal movement faces autonomous, volatile and variable civil society organizations that allow for the creation of transformative projects. He further argues against the choice between frontal rejection and total acceptance of the neoliberal project. According to Hale, there are not many alternatives to accepting the existence of 'neoliberal multiculturalism'. Although there may be room to manoeuvre because of the volatility and variability of civil society organizations in Latin America, the possibilities and potential for struggles 'from within' should not be overestimated. It all depends on a well-articulated strategy to achieve transformative multiculturalism that will transcend neoliberal multiculturalism and its mechanisms for resource and power allocation. Such a strategy also involves the relationship between indigenous 'cultural' demands (including 'radical' demands to redistribute resources and power) and 'popular movements'.

This gives us some background to discuss the achievements and limitations of the new indigenous movements and new recognition policies in terms of affirming legal pluralism and autonomy, or territorial self-management, by indigenous peoples. This will also help to analyse structural tensions that governments such as President Evo Morales's administration experience when pursuing multiculturalism. This analysis applies not only to the government, but also to relations among the Bolivian indigenous movement and the international cooperation agencies and multilateral entities that all work with different definitions of 'good (or allowed) Indians' and 'bad Indians'.

Multicultural policies in Latin America

Although the figures are disputed, the number of indigenous peoples in Latin America is estimated at 34 million to 40 million (i.e. about 8 to 10 per cent of the total population) (Psacharopoulos and Patrinos, 1994). The figures gloss over a great diversity of situations, from sedentary farmers with variable historical relations to the state and market in the Andean and Meso-American region, to relatively isolated hunter-gatherer peoples, some of whom are isolated in the Amazon region and other 'peripheral' areas. Indigenous people can remain 'invisible' until their renewable or non-renewable resources are discovered and they are incorporated within the national arena for economic or geopolitical reasons.

During the 1960s and 1970s, large-scale land reforms aimed to give 'land without people to people without land', but failed to realize any genuine redistribution of land from hacienda owners to indigenous and peasant populations. In many cases, limited redistribution was attempted in order to help 'modernize' large landholdings (the 'Junker way') while seeking to redirect redistribution (the 'peasant way') towards colonizing 'land without people'. Renewable resources were discovered (mostly forests and water), as well as non-renewable resources (oil, gas, minerals, etc.). These geopolitical processes provoked objections to disposal of resources and gave rise to differing notions of 'territory' in indigenous movement discourse.

Although Amazon peoples occupy land differently from Andean peoples, their notion of 'territory' was soon also adopted by Andean peoples. Territory is currently a major theme in indigenous movement discourse, involving space and resources, as well as jurisdiction and political authority as foundations for autonomy. The territorial claim is also supported by International Labour Organization Convention 169, which states in Article 13-2 that use of the term 'land' must include the concept of territories. And the United Nations Declaration on the Rights of Indigenous Peoples, which was finally adopted by the UN General Assembly in September 2007, recognizes indigenous peoples' right to self-determination and autonomy or self-government and contains important stipulations regarding territories and natural resources. Shortly thereafter Bolivia adopted this declaration as national law.

Bolivia: Community land of origin and people's participation

The 1952 revolution in Bolivia triggered agrarian reform. A 1953 Agrarian Law introduced the concept of the land's 'social function'. In the Andean west and the central valley region, this process led to land redistribution. In the tropical Oriente, by contrast, agrarian policies adopted by revolutionary and post-revolutionary governments consolidated and modernized the large plantations and cattle ranches. In the Santa Cruz region, an agro-industrial complex arose, initially grounded in the production of rice, sugar cane and cotton, expanding into soybeans by the late 1960s. In Santa Cruz and in other Oriente regions, cattle-raising played a major role and woodlands were pursued by logging companies.

Beginning in the 1980s, the indigenous peoples of the Oriente, constituting some 30 groups and nearly 200,000 people, organized and created the Confederation of Indigenous Peoples of the Bolivian Oriente (CIDOB) in 1982. In 1990 the indigenous peoples of the Beni region held a March for Territory and Dignity as a protest against logging activities in their territories. This highly publicized march reminded the country that there were indigenous peoples in their lowlands. President Jaime Paz-Zamora went personally to meet with the indigenous protestors, but was unable to persuade them not to continue to La Paz. By late September, Paz-Zamora had signed the first four supreme decrees recognizing indigenous territories, later followed by another four.

After the 1952 revolution two agrarian reform institutions (the National Agrarian Reform Council (CNRA) and the National Colonization Institute) had distributed land and granted land titles, sometimes issuing several overlapping titles for the same land. Moreover, they also turned over huge pieces of land under questionable circumstances during the *de facto* government of Hugo Banzer (1971 to 1978) and the following military regimes. This led to new agrarian legislation being considered in 1994 under the government of Sánchez de Lozada (1993 to 1997) as well as a counter-proposal by indigenous peasant organizations. Negotiations among the parties achieved a consensus that was then repeatedly ignored by the government. This led indigenous and peasant organizations to mobilize and press for enactment of the proposed Law of the National Agrarian Reform Service (SNRA), immediate recognition of indigenous territories and other demands. The government's offer to recognize territories in the lowlands, however, fractured the movement. Moreover, business organizations, supported by the Civil Committee of Santa Cruz, pressured for amendments that would favour their interests.

Finally, in August 1996, the SNRA Law was enacted, which conceded to several indigenous demands. The government and the World Bank had initially foreseen opening up the land market; but the new law maintained the distinction between land's social function and its economic-social function, resulting in some lands being unmarketable. The former category includes two forms of collective property: community lands of origin (TCOs) and community lands, in addition to individual peasant housing plots and small properties. These types of socially important property are non-transferable or only transferable under certain conditions. It was also established that certain territories already recognized by a supreme decree should be granted title under the TCO mode; another 16 demands were covered, while opening up the possibility for new demands to be presented.

Interestingly, the TCOs provide to 'indigenous and original communities and peoples collective ownership of their land, granting them the right to share in the use and sustainable utilization of renewable resources present there'. Under the law, distribution and redistribution for use within TCOs 'shall be governed by community rules, according to their norms and customs' (SNRA Law, Article 3-III). However, use of non-renewable resources in TCOs is governed by the constitution and special norms regulating them. This has made it possible for mining and petroleum concessions to overlap TCOs, resulting in land and water pollution and use of resources according to a market approach that fails to consider the cultural interests of people living in TCOs.

Another issue is that TCOs are not jurisdictional or political entities, but rather fall under the jurisdiction of municipalities or provinces. However, the amended 1995 Bolivian Constitution, while recognizing the multi-ethnic, pluricultural nature of the country in Article 1 stated in Article 171 that:

> *Natural authorities of indigenous and peasant communities may perform administrative functions and apply their own norms as a form of alternative conflict resolution, in keeping with their customs and procedures, providing they do not contradict this constitution or the law. Laws will make these functions compatible with the powers of the branches of the state.*

In this same article, the constitution recognized the legal status of indigenous and rural communities, associations and unions. This expression of multiculturalism, however, was not implemented by the state in good faith.

The People's Participation Law (LPP) of 1994 recognized indigenous authorities. This opened up the possibility of creating indigenous municipal districts (DMIs). In a DMI, the indigenous authority would be recognized as a sub-mayor. Although DMIs are institutionally weak since their jurisdiction is not defined and they depend on the goodwill of the municipal mayor, it would be an interesting arrangement for indigenous territorial reconfiguration strategies.

TCOs were recognized through the agrarian law and not through the LPP because the state sought to prevent any connection between the notions of territory and jurisdiction, alleging that this would lead to 'Balkanizing' the country. In this regard, Calla has suggested that the mid 1990s involved a silent and largely unperceived dual decentralization process:

- the territorial decentralization is driven by TCO demands, which have arisen in the last few years in the highlands; and
- the municipal decentralization process driven by the LPP (Calla, undated).

Indigenous peoples pursue strategies to make the two legal frameworks converge. An example is the community of Raqaypampa in the Department of Cochabamba, municipality of Mizque, which has a Quechua population of about 11,000. By the late 1990s the indigenous population in the Mizque municipality was organized into unions, sub-centrals and a regional central, and in 1997 the Indigenous Municipal District of Raqaypampa was created. In 2001, they prepared regulations for governance of the DMI that were approved by the Mizque Municipal Council. The unions also discussed the SNRA Law and in early 2002 applied to the departmental National Agrarian Reform Institute (INRA) for recognition of the TCO coinciding with the DMI's area.

The next step in this strategy would be to make the DMI into a municipality under the 1999 Municipalities Law. That would effectively integrate the indigenous government within the structure of the Bolivian government. Similar efforts were under way in many other parts of the Andean region, sometimes with clear ethno-nationalistic overtones, such as in certain Aymara areas. The territorial recomposition can extend beyond the local community based on the union organization, as in Raqaypampa, and, in other cases, based on reorganized *ayllus*. In his assessment of the process – which was written years before the election of Evo Morales – Calla pointed out that little thought had been given to the significance of such processes or their possible consequences for Bolivia's political and administrative structure or distribution of governmental resources.

With the election of Evo Morales and the drafting of a new constitution, which was approved in a popular referendum in January 2009, conditions have significantly changed. The new constitution foresees the creation of indigenous autonomies. How this will work out in practice and how further legislation will take shape remains to be seen.

Finally, the debate about a new water law that was suspended after the 2000 Water War is again a hot issue in Bolivia. This is because of controversial attempts to sell water from the Department of Potosí to northern Chile supposedly to benefit urban populations in that region, but also to benefit mining companies thirsty for water for industrial use. Moreover, after a Gas War in October 2003 over developing and exporting natural gas, the natural resource-use issue and the neoliberal model, in general, have come to occupy an even more prominent place in national debate. Indigenous movements are major protagonists in challenging the market model and proposing that water and other resources be treated as public goods and handled according to criteria of equity, social justice and respect for multiculturalism. The new Bolivian government, for example, has created a Water Ministry – an unprecedented decision for Latin America – and also intends to develop participatory decentralized public policy that will go beyond centralized bureaucratic management of water resources. This proposal entails challenging the patterns of 'rationality' imposed by hegemonic discourse and multilateral agencies. A first step in this direction was the approval of regulations of the 2004 Irrigation Law, which clearly depart from the market logic and stipulate that water and water rights are not marketable. The regulations also stipulate the use rights of indigenous–peasant communities and families and deal with their registration. Bolivia's new constitution defines access to water and sewerage as human rights and forbids their concession or privatization.

Colombia: From hope to despair

Colombia, along with Bolivia, was studied by Van Cott in her 2000 book *The Friendly Liquidation of the Past*, a title reflecting the optimism of the time regarding emerging multiculturalism in Latin America. The 1991 constitutional reform in Colombia gave rise to major hope for the country's indigenous peoples. Although indigenous peoples comprise only 2 per cent of Colombia's population (compared to 62 per cent for Bolivia), their participation in the 1991 Colombian constitutional reform process generated great expectations. Not only were their presence and performance noteworthy, but they also succeeded in getting a series of stipulations in the new constitution (Van Cott, 2000). Their *resguardos* (reservations) were recognized as collective inalienable property governed by indigenous councils set up according to local customs. Unlike the Bolivian case, there was no governmental strategy to separate territory and jurisdiction. Theoretically, the rights set forth in the Colombian Constitution are more coherent and solid.

The *resguardos* were created in Colombia during colonial times. During the course of the 19th century, as in other countries, there were attempts to dissolve them and encourage private property, a process that was stopped for a time when Law 89 was enacted in 1890, which momentarily recognized the reservations and their councils. However, pressures to dissolve them continued, leading to a resistance movement, including armed resistance led by Quintín Lame in Cauca Valley from 1910 to 1918. Despite resistance, by 1960 there were only some 70 reservations left. In 1961, under the auspices of the Alliance for Progress, an Agrarian Reform Law was enacted and the Colombian Agrarian Reform Institute (INCORA) was created; but the law was not implemented until the end of the decade when peasants and indigenous people began recovering land.

By 1980, existing reservations, including extensive reserves created in the tropical eastern area since the 1960s, were consolidated and new ones established. After enacting the new 1991 Constitution, these reservations were also granted *resguardo* status, which provides stronger legal protection for lands. The number of reservations increased, reaching 638 in 2001 and covering some 25 per cent of the national territory, much of it in the Amazon region. It is estimated that now almost 90 per cent of the indigenous people live on reservations (Pineda, 2001, pp36–37; Arango and Sánchez, 2002).

The 1991 Constitution consolidated reservations and also formalized indigenous jurisdiction, regulation of land distribution within reservations, design and implementation of development plans, promotion of public investment, and representation for dealing with the national government. It also addressed without clear articulation the concept of territorial entities under state law (municipalities, etc.) and indigenous territorial entities (Van Cott, 2000, pp94–95). The constitution was supposed to create regional autonomy for several reservations and an Organizational Law for Territorial Distribution was intended to regularize self-government and expand indigenous jurisdiction. Secondary legislation was needed to stipulate which reservations qualified to receive a portion of the national budget as municipalities within the state structure.

Indigenous people influenced the 1991 Constitution and achieved public prominence and highly significant moral leadership. Social movements also played a role, along with the Liberal Party and the neoliberals in promoting decentralization for different reasons. Therefore, recognition of indigenous autonomy was part of a broader context of decentralization and attempts to put an end to violent conflicts in Colombia.

In 1994, funds began reaching reservations. They came through non-indigenous mayors and departmental governors on the basis of investment and development plans that had to be presented by reservations. Funding was accompanied by time-consuming training programmes, prompting one indigenous leader to remark: 'Nowadays, you need a lot of time to be indigenous' (Padilla, 1995, p147). Perhaps conforming to the Colombian government's procedures is not merely inconvenient; but, as Padilla suggests, doing so may actually increase the possibilities for state and para-state intervention in the internal affairs of the indigenous peoples through NGOs. Indeed, it appears to be an effort to 'standardize' government procedures, effectively impeding the capacity of indigenous people to govern themselves (Hoekema, 1996; Hale, 2002; see also Chapter 2 in this volume).

While the César Gaviria government showed some affinity for the indigenous cause (Van Cott, 2000), subsequent governments have been much more interested in implementing 'mega-projects' to extract natural resources (oil, minerals, hydropower and African palm plantations). Implementing such projects often affects indigenous populations in several ways, such as by polluting the water, competing for water resources to irrigate mono-crops for export and building hydropower projects.

A notorious case is the struggle by the Embera Katío people in northwestern Colombia against building of the Urrá I and II dams in their territory. The project would benefit agricultural businesses in the region and generate electricity for Bogotá, but would also affect indigenous fishing areas. Recognition of reservations and indigenous rights have been conditioned upon and subordinated to the creation of 'strategic alliances' with the private sector under Plan Colombia, which attempts to pacify the country and eliminate drug traffic by promoting business sectors at the expense of more vulnerable groups. Indigenous territorial rights are restrictively interpreted. Such policies, along with increased violence related to drug wars, jeopardize indigenous peoples' existence. Violence against indigenous peoples and their allies has been the subject of judicial proceedings in Colombia and before the Inter-American Human Rights Commission. The Colombian government, however, has disregarded the commission's recommendations.

Colombia's new 1991 Constitution declared Colombia a 'social state under the rule of law' and recognized indigenous rights. The constitutional process offered hope for pacifying the country. However, some portions of the constitution were ambiguous and the sections on economic issues were heavily influenced by the neoliberal model. The gap between what the constitution said about indigenous rights and actual practice has been growing as guerrilla conflicts and efforts to combat drug trafficking have put pressure on policies advancing indigenous rights. Especially under the government of Álvaro Uribe, systematic efforts have been undertaken to roll back the advances made in 1991. Neoliberal policies were deepened. In 2007 all dialogue between the government and indigenous organizations came to a virtual standstill when it became known that Colombia was the only Latin American country that had abstained from the vote over the United Nations Declaration on the Rights of Indigenous Peoples. Meanwhile, new legislation on rural development and projects for oil and mineral exploitation constitute a serious threat for the *resguardos*, and the share in tax revenues to local governments, among them the *resguardos*, has been reduced.

Mexico: The reform, betrayed

Mexico was one of the countries that incubated 'indigenism', a current of thought that revisited 'indigenous affairs' during the 1920s and 1930s. The concept was institutionalized with the Patzcuaro Congress in 1940, which gave rise to the indigenist institutions, both nationally and at the inter-American level. Indigenism was considered progressive because it criticized racist theories; but it became a means of denying diversity by seeking cultural and racial mixing and assimilation (see also Chapter 5 in this book). During the 1970s the assimilation aspect began to be abandoned, at least in theory.

In Mexico, the 'indigenism of participation' arose, in part, to contain the emergence of independent movements. For each of the 56 peoples officially recognized, a congress was held which was supposed to lead to the creation of a Supreme Council. One example was the famous Indigenous Congress in Chiapas in 1974. The local government put the Archbishop of Chiapas, Samuel Ruiz, in charge of organizing the congress. But contrary to official objectives, the congress escaped government control and became an occasion for the area's indigenous groups to meet and discuss their common problems and ways of addressing them. From the local congresses and supreme councils, a National Council of the Indigenous Peoples was created in 1975. However, this renewed indigenism soon showed its limits and the attempt to incorporate the indigenous movements within the state system largely failed.

When the ILO adopted its new Convention 169, Mexico was eager to appear progressive and became the first Latin American country to ratify it. Within the country, ratification went virtually unnoticed, however. Mexico amended its constitution in early 1992 to implement Convention 169 by recognizing the pluri-cultural composition of the nation. But, at the same time, Article 27 of the constitution, which had provided the framework for agrarian reform after the Mexican revolution (1910 to 1927), was also amended. The land reform process thus was officially ended and the amended Article 27 was at odds with Convention 169 by opening the way for the privatization of community lands.

The Salinas government (1988 to 1994) sought to present Mexico as a 'first world' country by negotiating the North American Free Trade Agreement (NAFTA). While negotiating NAFTA, Mexico agreed to put an end to agrarian redistribution, to open up the land market, to promote individual property titles for collective *ejidos* and agrarian communities, and to remove prohibitions against dissolving (formerly inalienable) *ejidos* and communities. A new agrarian law included reference to the protection of the land of 'indigenous groups' pursuant to the constitution; but the constitution referred to protection pursuant to the law and there was no mention of territories in the agrarian law. Therefore, the issue of indigenous land remained in a legal limbo. Later in

1992 the government also passed a new forestry law and new water legislation. The water law reversed the social orientation of previous legislation which protected irrigated agriculture and accorded priority to *ejidos* and agrarian communities over industrial water use and use for electricity generation. With the new law the government sought to create a market in water rights and strongly reduced the role of the state in water management.

Such neoliberal policies created the climate for the outbreak of the Chiapas rebellion in January 1994 (Harvey, 1998). Several days of combat ended in a ceasefire, and lengthy negotiations began between the Zapatistas and the Zedillo government (1994 to 2000). The dialogue bogged down because of a lack of government interest. To resume the peace process, in 1996 the Concord and Pacification Committee (COCOPA) of representatives from the local and national congresses drafted a proposal to amend the federal constitution. The Zapatistas agreed to this with some reservations. The government also agreed initially, but then presented a series of 'observations' that distorted the proposal and precipitated a crisis. While the constitutional amendment was pending, the Zapatistas moved back into their municipalities and communities to consolidate their autonomous governments and strengthen their ties with national and international 'civil society'.

Chiapas was one of the central issues in the 2000 presidential election. Candidate Vicente Fox, of the National Action Party (PAN), at one point declared that he would settle the matter in '15 minutes' and promised that, if elected, he would send the COCOPA proposal to Congress. The July elections yielded the historic defeat of the Institutional Revolutionary Party (PRI), and once he took office in late 2000, President Vicente Fox did, in fact, send the COCOPA proposal to Congress. In response, during February and March 2001 the Zapatistas organized a caravan that crossed the country and finally reached the federal capital, where they made a historic appearance in Congress to support the COCOPA reform.

However, in April, Congress passed a constitutional amendment that fell short of the COCOPA proposal. The wording recognized the rights of indigenous peoples and communities to self-determination 'in a constitutional framework of autonomy that will assure national unity'. Nonetheless, instead of recognizing indigenous peoples as legal entities, it recognized them as 'public interest entities' – that is, as recipients of state care rather than subjects with rights. The concept of territories was eliminated while maintaining a reference to Article 27. Communities were also allowed to associate with one another, but only within municipalities. An amendment to Article 115 of the constitution on municipal organization was left out so indigenous municipalities cannot be created. Supporters of these provisions cited 'the nation's unity and indivisibility'. Over 300 municipalities challenged the law in the National Supreme Court of Justice, which declared the challenges out of order for technical reasons. Despite widespread rejection by indigenous movements, the revised amendment was ratified by 16 state congresses – the states with the lowest indigenous populations. In August 2001, President Vicente Fox signed the amendment decree.

While this tug of war was going on at the federal level, a number of states amended their constitutions and, in some cases, passed secondary laws. Out of Mexico's 31 states, more than 17 have amended their constitutions to foster multiculturalism, as has the federal district. Various states have enacted secondary legislation, including indigenous rights and culture laws. Except for a few positive features, the reforms have no significant consequences.

Legal protection for territories and natural resources is weak and conflict is common. For example, in 2004, the Community Front to Defend Human Rights and Natural Resources of the Mazahua people in the State of Mexico organized several protests against taking water from the Mazahua communities for the thirsty federal district without compensation. Protests continued because negotiations and promises by government agencies produced no concrete actions to benefit Mazahua communities in such areas as sanitation, drinking water supply and rehabilitation of irrigation systems. These protests continued during the following years, and in March 2006 the Mazahua Women's Army in Defence of Water presented the case at the World Water Forum in Mexico City. It is just one example of indigenous peoples' involvement in conflicts over water resources throughout Mexico.

Conclusions

After this tour of the reform processes in three Latin American countries, we have to conclude that the results of the new multiculturalism have, at best, been mixed. The policies were introduced in a context of structural adjustment measures. Neoliberal multiculturalism or what has been called 'neo-indigenism' (Hernández et al, 2004) supplanted the old indigenism and its openly assimilationist policies in an effort to develop new forms of governance under the guidance of the multilateral development agencies. At the same time, macroeconomic policies contributed to the poverty of vulnerable social groups, including indigenous peoples, and income distribution tended to become more inequitable. Free trade policies and the opening up of the national economies went in tandem with the introduction of market-oriented policies regarding natural resources, including water, and an intensification of natural resource exploitation.

The new multiculturalism fitted into the process of dismantling the state and transforming social policies, which were already deficient, into targeted programmes aiming to 'help the poor help themselves'. Within this framework the new multiculturalism purported to be culturally sensitive; but this sensitivity goes hand in hand with regulating the lives of aid programme beneficiaries.[3] 'Participation', 'human capital' and 'social capital' became key words in the new discourse. Mexican 'neo-indigenism' (Hernández et al, 2004) is emblematic in this sense. The projects promoted by Mexican regional development funds, for example, are often business oriented and seek to transform communities into community enterprises. These projects ignore alternative visions of development based on organic agriculture, notions of territoriality, food security and collective rights to natural resources, as well as local forms of water management.[4] In Colombia the hopes generated by the 1991 constitutional reform have been dashed and indigenous access to territories has increasingly become conditioned upon accepting development projects in cooperation with the private sector. In Bolivia, mining, petroleum and forestry concessions frequently overlapped with the TCOs. In both Colombia and Mexico, mega-projects such as the development of the Pacific Coast or the Puebla-Panama Plan threaten indigenous and other rural peoples and their livelihoods.

The case of Bolivia, however, is suggestive of the shift in policy orientations that is occurring in the region. The Movimiento al Socialismo (MAS) is an exponent of the 'new Latin American left' and the search for alternatives to the hitherto dominant neoliberal model. The new Latin American left is extremely diverse and the search for alternatives is an ongoing process (Barrett et al, 2008). Some elements, however, seem to emerge: a newly enhanced role for the state in the economy and in natural resource management and the distribution of benefits; a renewal of social policies and efforts to create a more beneficial economic climate for the poorer sectors of the population; experiments with forms of democracy that complement electoral democracy and a more beneficial stance towards indigenous peoples that may go beyond neoliberal multiculturalism. This chapter has discussed the example of the Bolivian irrigation legislation. More broadly, one might point to the processes of constitutional reform in countries such as Ecuador or Bolivia. As in Bolivia, the new Ecuador Constitution defines the right to water as a fundamental human right, indispensable for 'living well'. Whether this results in concrete policies remains to be seen – a key issue that several of the following chapters will examine. In the case of Ecuador, for example, the relationship between the indigenous movement and the Rafael Correa government is rather tense, particularly due to disagreements over natural resource management and exploitation.

Notes

1 The departure of President Gonzalo Sánchez de Lozada was also provoked by the Gas War in October 2003.

2 See Assies (2000). Castells (1998) has called this the 'network state'; others have referred to it as the 'new medievalism' (compare Held et al, 1999, p85).

3 Laws on indigenous rights and culture in Mexico, for example, include lengthy sections to circumscribe indigenous jurisdiction and reduce it to cases of 'chicken theft'.

4 An issue that is profoundly examined in, for example, multiple studies and debates by the Water Law and Indigenous Rights programme (see, for example, www.eclac.cl/drni/proyectos/walir or the WALIR book series with IEP, Lima).

References

Arango, R. and E. Sánchez (2002) *Los pueblos indígenas en el umbral del nuevo milenio*, National Planning Department, Bogotá

Assies, W. (2000) 'Indigenous peoples and reform of the state in Latin America', in W. Assies, G. van der Haar and A. Hoekema (eds) *The Challenge of Diversity; Indigenous Peoples and Reform of the State in Latin America*, Thela Thesis, Amsterdam, The Netherlands, pp3–21

Assies, W. (2001) 'David *vs.* Goliat en Cochabamba: Los derechos del agua, el neoliberalismo y la renovación de la protesta social en Bolivia', *T'inkazos, revista boliviana de ciencias sociales*, vol 8, pp106–131

Barrett, P., D. Chavez and C. Rodríguez-Garavito (eds) (2008) *The New Latin American Left: Utopia Reborn*, Pluto Press, London

Bengoa, J. (2000) *La emergencia indígena en América Latina*, Fondo de Cultura Económica, Mexico City and Santiago, Chile

Bonfil Batalla, G. (1991) 'Lo propio y lo ajeno: una aproximación al problema del control cultural', in G. Bonfil Batalla (ed) *Pensar nuestra cultura*, Alianza Editorial, Mexico City, Mexico, D. F., pp49–57

Calla, R. (undated) 'Derechos territoriales de los pueblos indígenas en los Andes de Bolivia, Perú y Ecuador: Una exploración preliminar', in Oxfam (ed) *Avizorando los retos para los pueblos indígenas de América Latina en el nuevo milenio: territorio, economía, política e identidad*, CD-ROM

Castells, M. (1998) *End of Millennium*, Blackwell Publishers, Malden, MA, and Oxford

Hale, C. R. (2002) 'Does multiculturalism menace? Governance, cultural rights and the politics of identity in Guatemala', *Journal of Latin American Studies*, no 34, pp485–524

Hale, C. R. (2004) 'Rethinking indigenous politics in the era of the 'Indio Permitido'', *NACLA Report on the Americas*, September/October 2004, pp16–21

Harvey, N. (1998) *The Chiapas Rebellion: The Struggle for Land and Democracy*, Duke University Press, Durham and London

Held, D., A. McGrew, D. Goldblatt and J. Perraton (1999) *Global Transformations: Politics, Economics and Culture*, Polity Press, Cambridge

Hernández, R. A., S. Paz and M. T. Sierra (eds) (2004) *El Estado y los indígenas en tiempos del PAN: Neoindigenismo, legalidad e identidad*, CIESA, Miguel Ángel Porrúa, Mexico

Hoekema, A. J. (1996) *Autonomy and Self-Government: A Fundamental Debate*, University of Amsterdam, Department of Sociology and Anthropology of Law (internal memo), Amsterdam

Iturralde, D. (1997) 'Demandas indígenas y reforma legal: Retos y paradojas', *Alteridades*, vol 7, no 14, pp81–98

Padilla, G. (1995) *What Encompasses Goodness, the Law and the Indigenous People of Colombia*, Woodrow Wilson Center Ethnic Conflict and Governance in Comparative Perspective, Latin American Program Working Paper Series no 215, Woodrow Wilson Center, Washington, DC, pp139–153

Pineda, R. (2001) 'Colombia y el reto de la construcción de la multiculturalidad en un escenario de conflicto', in M. J. Cepeda Espinosa and T. Fleiner, M. J. (eds) *Multiethnic Nations in Developing Countries*, Helbing and Lichtenhahn (Institut de Fédéralisme Fribourg Suisse), Bâle, Geneva and Munich, pp1–74

Psacharopoulos, G. and H. A. Patrinos (eds) (1994) *Indigenous People and Poverty in Latin America: An Empirical Analysis*, World Bank, Washington, DC

Stavenhagen, R. (2002) 'Indigenous peoples and the state in Latin America: An ongoing debate', in R. Sieder (ed) *Multiculturalism in Latin America: Indigenous Rights, Diversity and Democracy*, Palgrave MacMillan, Houndmills, Basingstoke and Hampshire, pp24–44

Van Cott, D. L. (2000) *The Friendly Liquidation of the Past: The Politics of Diversity in Latin America*, University of Pittsburgh Press, Pittsburgh, PA

Wade, P. (2006) 'Etnicidad, multiculturalismo y políticas sociales en Latinoamérica: Poblaciones afrolatinas (e indígenas)', *Tabula Rasa*, no 4, pp59–81

4

A Masculine Water World: The Politics of Gender and Identity in Irrigation Expert Thinking

Margreet Zwarteveen

Introduction

Many people who are concerned with gender injustices in water management have noted that there is a huge gap between 'paper' recognition of gender issues in water management, policies and projects, and real on-the-ground efforts to address such inequities effectively. At the same time, they point to the lack of meaningful integration of gender questions in mainstream water analyses and discussions. Gender remains very much a side issue or an after-thought and is not seen as belonging to the core of what water management is about. This chapter looks at one possible reason for the resistance of the water profession to considering gender insights: the incompatibilities between water experts and gender experts in how they conceptualize and act upon water realities.

It is arguable that in irrigation engineering, rather than an unwillingness or a persistent bias of individual water professionals, the problem is that traditional ways of thinking about water are deeply inhospitable to the analysis of social relations and gender. An important conclusion is that *thinking* (and acting on) gender in water management requires active efforts to change mainstream ways of 'knowing in water'.

This chapter critically discusses the terms of discursive existence for gender in mainstream water thinking.[1] The exercise assumes that water knowledge, like most knowledge that is systematically produced, possesses regularities and exhibits systems of rules. These rules are, however, seldom formulated by the participants in the knowledge-generating process. They constitute what

Foucault called the 'positive unconscious of knowledge' (cited in Davidson, 1986, p222). The rules are relatively autonomous and anonymous, and they make it possible to assert claims that count as important, relevant or true within the boundaries of a science or discipline. Much of these unspoken rules in irrigation result from the isolation of scientific knowledge and thinking about irrigation from the social environment in which irrigation takes place, and with the positioning of the knower *vis-à-vis* that which is (to be) known.

The language, discursive practices and textual resources that form the heart of water knowledge are part of a body of cultural resources through which water professionals represent and identify themselves, and that contributes to legitimizing professional activities and choices. Thus, knowledge about water realities reflects prevailing professional water cultures and identities (with their configurations of power, status, authority and funds) as much as it reflects realities in the field. This realization undermines claims of objectivity and neutrality of water knowledge and opens the door to reflecting critically on how knowledge was constructed and by whom, and how the identity and social positions of knowledge producers impact upon the type of truth claims they make. The reverse is equally true: the recognition that water expertise and authority have an identity – a colour, gender and ethnicity – opens the door to questioning critically the symbolism and assumptions that are taken for granted in water, and to welcoming other voices and sources of knowledge.

Misrepresenting women and gender in irrigation

The difficulty of understanding the role and position of women and gender relationships in irrigation is most often attributed to the symbolic, discursive and ideological construction of farmers, irrigators and engineers as masculine and to the fact that being recognized as inhabitants of, and actors in, irrigation worlds requires rights, abilities and character traits that are seen as belonging more to men than to women. This chapter suggests that there are a number of less easily recognized but perhaps equally fundamental conceptual problems in irrigation thinking that lead to misrepresenting women, that prevent the questioning and challenging of gender relations and that misrepresent irrigation realities as genderless or gender neutral. Identifying these problems may help to explain why 'gender mainstreaming' in water remains a difficult and slow process and it may contribute to finding new avenues and entry-points for making the water world into a better place for women.

My conceptualization of gender frames the search for ways to 'think' gender in water management. Most important is the realization that gender is a social construction, and that its meaning is contested and negotiated. Thinking gender, therefore, implies not treating it as something that simply exists and can be known and mapped in a positivist sense. What it means to be a man or a woman is not a given and cannot be determined in any ahistorical or transcendental way. As a shifting, dynamic and contextual phenomenon, gender does not denote a substantive trait of character or personality, but is a relative point of convergence among culturally and historically specific sets of relations. Gender roles, identities and relations are not tangible and static, but are matters of continuous reinterpretation both in terms of practices and in terms of ideas, sometimes leading to controversy and debate.

Such a contextualized and constructivist understanding of gender is hard to reconcile with a feminist wish to establish women as a political category. It is also incompatible with the habit of mainstream irrigation and water experts to strip away context and meaning to uncover universal human values in efforts to generate generic lessons about the performance of water systems. Understanding gender therefore generates some important challenges to more conventional ways of thinking about water. These challenges fall into three categories.

The first category relates to some general features of irrigation thinking, and in particular how it cherishes its lack of a critical interpretative tradition as a virtue of modern science. These features are related to the positivist epistemological beliefs that guide much irrigation thinking.

The second category concerns the way in which irrigation systems and realities are ontologically defined and the ways in which conceptual boundaries are drawn between 'what matters' for knowing irrigation and that which can be ignored. This is illustrated by the choice of metaphors used for representing irrigation realities that structure the world in oppositional dichotomies with clear gender contents and implications. It also shows in the ontological separation of the technical and the social, or between 'the system' and its context.

The third category relates to how human beings and human behaviour are conceptualized in irrigation thinking, and the overall bracketing of power and politics in this analysis. This shows in the use of deductive methods and ideal-typical models and in the direct association of much irrigation research with those who 'rule' irrigation systems. It also shows in the methodological individualism that characterizes much irrigation thinking, and in its narrow and rather functionalistic and instrumentalist concept of human agency.

These three categories of conceptual problems are interrelated, and they intrinsically relate to water politics and identity.

Power, perspective and knowledge

Although different in focus and scope, and although drawing on different disciplinary theories, the mainstreams of professional water thinking share a number of characteristics. First, and importantly, their traditional subject matter is 'non-social'. Water knowledge is, or used to be, primarily concerned with 'the resource': water. The physical, biological and chemical characteristics of water together with the engineering knowledge needed to convey water constitute the heart of much water knowledge. Although efforts are increasingly made to include social questions in the analysis of water problems, preferred scientific languages and methods continue to be derived from the natural and engineering sciences. These are not always best suited for understanding the behaviour of human beings and their interactions.

Second, much water knowledge is visibly rooted in a modernization project – a project that associates positivist science (mathematics, in particular) and modern technology with progress and civilization. Although most contemporary irrigation and water professionals no longer have the strong faith in technology as a motor of progress that their colonial predecessors had, many continue to believe in the superiority and universal applicability of scientifically developed irrigation technologies or institutional and economic models (see also Boelens, 2008). In this sense, mainstream water thinking can be seen as firmly anchored in the Enlightenment tradition, a modernist and Western way of thinking. In this tradition, the 'god trick' is pervasive: the assumption that one can see everything from nowhere and that disembodied reason can produce accurate and 'objective' accounts of the world (Haraway, 1991). Enlightenment is marked by a faith in the neutrality of reasoned judgement, in scientific objectivity, in the progressive logic of reason and in science itself. Through the omnipotence of reason, transcendence is possible, allowing the knower to escape the limits of body, time and space (Hartsock, 1998, p206).

Third, normal water thinking is also pervaded with a belief that given the proper technologies, institutions or incentive structures, human beings will display the same water behaviour everywhere. This belief is rooted in the epistemological claim of human universality and homogeneity, a claim that is also associated with an Enlightenment tradition of thinking. This claim posits that, in essence, all human beings are equal and share a common capacity to reason. Differences among people are fundamentally epiphenomenal, making it possible to make generic statements about human nature, truth and other imperial universalities. In such a humanist or liberal understanding of human beings, gender can only be thought of as an attribute of a person who is characterized essentially as a pre-gendered substance or 'core' (called the person). This is useful for some purposes, but not for the purpose of a critical enquiry into the meaning of gender. Such enquiry requires a relational or contextual conceptualization of gender, suggesting that what the person 'is' – and, indeed, what gender 'is' – is always relative to the constructed relations in which it is determined (Scott, 1986; Butler, 1999). As feminist political theorists have pointed out, the referent for conceptualizing humanity and the human 'core' in much theory has been primarily masculine. Indeed, the term man as used in liberal thought – even by those who are willing to concede that he/him means 'all' – is not simply a linguistic device or a generic label but a symbol for a concept reflecting both masculine values and virtues and patriarchal practices (Dietz, 1992).

A fourth and related characteristic of much mainstream irrigation thought is its denial of the importance of power to knowledge. This denial of the connections between power and knowledge, and between the construction of subjectivity and power, is directly linked to the fact that much mainstream irrigation knowledge is written from the perspective of those who are in control: planners, administrators and managers. Produced knowledge is aimed at helping *them* realize *their* objectives, and enables them to speak more authoritatively through the disembodied, transcendent voice of reason. Much early engineering thinking did, in fact, reflect an implicit juridical conception of power: the locus of control was situated with the head engineer, at the head gates. He was the sovereign, and the irrigators were his subjects. While there used to be discussion about the most effective control strategy, the very possibility of controlling and manipulating behaviour of people and of flows of water and money was seldom questioned in irrigation knowledge. It is telling that contemporary theoretical models for irrigation system performance, such as those embodied in technical designs or as proposed by neo-institutionalism, are tested primarily through the deductive method rather that empirically. Outcomes or outputs are measured against the expectations of the formal models; but the operational and behavioural assumptions of the models are seldom validated. Moreover, designers are hardly ever confronted with operational realities at the field level just as knowledge about designs is rarely tested against field-level realities. Consequently, the beliefs in the model and in the effectiveness of planners' control mechanisms are not challenged, nor are the legitimacy of water professionals and their knowledge questioned. Indeed, the persistence of certain basic assumptions in irrigation thinking can be explained as much by success in generating funds and power – and bolstering the egos of irrigation practitioners – as by success in generating valid theses about the determinants of irrigation system performance.

Much water knowledge sees knowledge producers, such as the head engineers or managers, as transcendent rational subjects who exist outside time, space and context. Through irrigation knowledge, those in control of water are provided with agency and subjectivity, a discursive construction that is conditioned upon the simultaneous denial or severe limitation of agency to users, irrigators or farmers. The latter group is created as the 'others', who are restricted in their capacity to act and speak, the ones who need to be controlled and whose behaviour needs to be adjusted to what is deemed appropriate by the 'knowers'. In the Andean context, water expertise thus constructed indigenous peasants as backward, uncivilized and irrational. Indigenous peasants were marked and named by irrigation experts, who themselves remained unseen and whose own identity (gender and ethnicity) did not matter in terms of their authority and knowledge (see Boelens and Zwarteveen, 2005).

This is not to deny that contemporary water knowledge takes farmers and officers much more seriously than in earlier days. The call for more participation by farmers in design and management processes, and the associated increased appreciation of the value of farmers' knowledge, have resulted in questions about the legitimacy of scientific water knowledge. The hegemonic superiority of engineers' knowledge and their exclusive claims to the ability to design irrigation systems have also been challenged. Shah (2003, p22) convincingly argues: 'while inclusion of farmers' knowledge and farmers' choices in the process of "design" is envisaged by the dominant model, the validity of conventional disciplinary – scientific and engineering – knowledge and the context in which this knowledge is generated, is not very frequently questioned'. Issues of power and identity, of location and time, continue to be shielded from scrutiny through appeals to 'the technical', 'the rational' and 'the scientific'. Irrigation knowledge continues to be, as Harding (1986, p76) calls it, 'part of the labour of ruling'.

And although the more recent irrigation management literature is much more positive about farmers' knowledge and abilities, their practical influence remains limited. The usual concept of human agency is that of the utility optimizer and rational decision-maker who weighs the costs and benefits of alternative choices. This leaves little conceptual scope for considering the actions and choices of the various players in irrigation from their own perspectives and in their own frames of reference. Nor are culture, tradition and apparently less rational explanations for behaviour considered. In the Andes, the farmers' own systems of distributing water and of defining and allocating water rights, for instance, tend to be considered as 'anomalous' in mainstream water management literature (see Boelens and Zwarteveen, 2005).

The philosopher Spivak once asked: 'Can men theorize feminism, can whites theorize racism, can the bourgeois theorize revolution'? She maintained that when the former groups theorize, it is crucial that the members of these groups be vigilant about their subject positions (Spivak, 1988, p253). Spivak argued there should be critical reflection on the identities and positions of knowers and how they impact upon the knowledge that they produce. Her question entails a much needed acknowledgement in water that knowledge production and designs of water systems are deeply social processes in which different stakeholders interact. The nature of these processes and the different perceptions, interests and powers of the stakeholders involved shape the knowledge produced, as well as the ultimate design choices and technical characteristics. The importance of this insight is that it enables the questioning of irrigation designs, designers and knowers. As a result, the design or the technology (the 'irrigation machine'), as well as institutional models, stop being the norm, dictating the behaviour of users, operators and managers. And technical engineering or other expert knowledge is no longer granted highest status in conceptualizing irrigation realities (compare Shah, 2003).

As a central part of their project, feminist scholars have challenged the norms of objectivity that have long guided science. In the strong formulation of Catharine MacKinnon: 'Objectivity is the epistemological stance of which objectification is the social process, of which male dominance is the politics, the acted out social practice' (MacKinnon, 1987, p50, cited in Langton, 2000, p135). For Haraway (1991), 'seeing well' is not just a matter of having good eyesight: it is a located activity, cognizant of its particularity and of the accountability requirements that are specific to its location. 'Seeing well' implies the refusal of any subject/object split in the production of knowledge, insisting on 'the critical and interpretative core of all knowledge' (Haraway, 1991, p191). In situated knowledge-making projects, embodied knowers engage with active objects of knowledge, whose agency and unpredictability unsettle any hopes for perfect knowledge and control. Indeed, there are connections and linkages between subjects and objects, and the two can be said to stand in a 'dialogic' relationship with each other (compare Sayer, 1992, pp22–42).

This renders impossible the wish to provide truth claims in the strict positivist tradition. It makes politically dangerous any effort to describe the irrigation world in one consistent all-encompassing discourse. This is because a single description has totalizing and exclusionary effects, and is academically suspect, because it hides the knowers and their identity and power in cloaks of objectivity. In the words of Nicholson (1995, p5): 'Any discursive move which attempts to place itself beyond question automatically invokes suspicion.' Indeed, only from the falsely universalizing perspective of those who are, or think they are, in control and command can 'reality' have 'a' structure. That is, only to the extent that one person or group can dominate the whole can 'reality' appear to be governed by one set of rules or be constituted by one privileged set of social relationships (compare Flax, 1986, cited in Harding, 1986, p193).

Gendered metaphors and dichotomies

The ways in which boundaries are drawn in much mainstream water thinking are informed by a powerful spatial imagery with rather strong gender connotations. For one, irrigation systems and what goes on within them are often seen as 'the work place', a domain or area that is spatially and socially distinct from 'the home'. It is the place where production for the market occurs and where incomes are earned, separate from the place where consumption and production for personal or domestic use happen. Second, the irrigation system is also the place that is labelled as 'public', in implicit contrast to the 'private' location of home and family. For a long time, the public world of work and production tended to be seen, and used to be ideologically constructed, as the world of men. Such construction rested on normative ideas that men should be the breadwinners and principal income earners, whereas women should be caretakers, cleaners and mothers. A widespread and strong ideological connotation of the word 'farmer' and, by analogy, the words 'irrigator' and 'engineer' as symbolizing male identity worked to reinforce this notion. While most irrigation thinkers today would no longer explicitly adhere to such gender ideologies, the conceptual language and methodological tools used continue to be pervaded by the dichotomies of work and home, production and consumption, public and private. What matters to irrigation professionalism is what happens in the former – in the world of work, production and public politics. This world is seen as relatively disconnected from and unrelated to the private world of care, consumption and intimacy. The irrigation world similarly is seen as the domain of reason and logic in implicit contrast to the domains of emotion and affection that characterize the non-irrigation world.

A number of influential images are associated with the use of these dichotomized metaphors. An important one is that of irrigators as industrial workers whose working places are socially distinct and separate from their homes. Depending on the degree of autonomy granted to irrigators, they are either seen as factory workers or as private entrepreneurs. Irrigating and irrigated farming are, as a consequence, seen as the business of one individual whose irrigation behaviour is primarily informed by imperatives related to the irrigation system. Other family members sometimes assist this individual, but he (most often the individual is seen as a man) is the one in charge and makes all the decisions. This view is problematic since smallholder irrigated farming often is not the sole affair of one individual but a family undertaking. It is also problematic because it is implicitly based on a nuclear family household model. Supposedly, the allocation of family labour time between competing uses is determined rationally by the principle of comparative advantage so that each household member specializes in those activities which give the family the highest relative return (compare Kabeer, 1991).

This also would entail the existence of clear-cut boundaries between the sphere of work and that of home. Positing such boundaries places households outside of supposed irrigation realities, and outside of what needs to be explained by irrigation knowledge. Since the household is seen as the domain of women, further thinking about women and gender also becomes unnecessary. In the Andean situation, with many men migrating to cities and with household livelihood strategies consisting of a combination of activities, this public–private metaphor is particularly ill suited to understand water realities. Research in the Andes suggests that what a household is, and who belongs to it, is itself often an intrinsic part of local negotiations about definitions. The boundaries between a household and its environment are not a given, but require 'social and cultural work to affirm its existence' (Mayer, 2002, p8), and such definitions are particularly important in local water management and maintenance activities since they establish which members of households are allowed or obliged to contribute.

The use of gendered dichotomies is also problematic because the 'masculine' pole of these dichotomies tends to be valued much more positively and tends to be attributed more powers and status than the 'feminine' pole. Some feminists and some streams of eco-feminism have therefore argued for a reversal of this hierarchy, and for a revaluation of the feminine. Others, in contrast, have argued in favour of strategies that would facilitate and encourage women's entry into the masculine worlds of production, politics and reason. Both positions, however, tend to neglect the importance of critically questioning the ways in which the poles are defined. The boundaries that separate nature from culture, private from public, work from home, and so on, are not fixed and ahistorical, but are contingent and socially constructed. It is important to question taken-for-granted gender hierarchies and dichotomies. In addition, the positing of these boundaries invites the treatment of each of the respective poles of the dichotomies as analytically separate, whereas they exist because of and through each other.

From the perspective of irrigators and farmers, home and work are often closely interconnected, both in the fact that the first objective of work often is family survival, but also because family circumstances and considerations greatly influence work decisions and behaviour. Indeed, the boundaries between public and private, as well as those between production and subsistence, blur upon closer examination. Moreover, work and gender are not easy to categorize into two distinct domains, nor are these domains necessarily in harmony – or in conflict, as some feminist scholars would argue. Most smallholder farm households display a high degree of interdependence between production and subsistence activities as well as between the household's farm functions and its family functions. Domestic or reproductive labour is characteristic of all household members' activities across agricultural as well as subsistence production, and is not restricted to women's work. Irrigating and farming are not just about production and are not only associated with the activities of men. Irrigation needs, interests and activities are seldom directly gendered or a function of a person's gender. The ways in which gender mediates irrigation realities depends on time and location and is also affected by class, ethnicity and other cultural and socio-economic structures and identities.

Placing the irrigation system in the productive and public sphere, and conceptually separating it from the domestic and subsistence sphere, is not just analytically problematic. It has important political and distributional consequences in guiding plot and water allocation, and through designating specific users and uses of water as legitimate, and qualifying others as less important or even illegal. This question is also important, for instance, when considering the artificiality of the divide between water for productive and for domestic use when both are taken from the same irrigation system (compare Bakker et al, 1999).

The dichotomous metaphors also 'infect' irrigation thinking in a more diffuse way by associating masculinity with all that matters to irrigation, while implicitly linking femininity with all that is less relevant. Hence, water for productive uses tends to be considered as more important than water for domestic uses, crops grown for the market are more important than subsistence crops, and public decisions are more important than intra-household decisions. Economic incentives for behaviour are also considered more 'real' than, and normatively superior to, those based on emotions, solidarity and affection. Work such as cooking and the provision of meals for agricultural labourers is not normally considered part of irrigation work. And the irrigation conflicts and struggles that are most easily observed and named tend to be of the spectacular and violent type, involving stealing, fighting and bribery. The more hidden everyday forms of resistance (compare Scott, 1985), the silences and strategic invisibilities (compare Jackson, 1998) tend to receive less attention. Hence, while often not directly gendered, the conceptual delimitation of what counts and matters in irrigation, of what belongs to the irrigation domain, and the definitions of what is 'good' irrigation behaviour are deeply coloured by gendered images and connotations. Using such delimitations and definitions may have the effect of reinforcing and further legitimizing such gendered divides, rather than questioning them.

A clear Andean example of this comes from an irrigation project in Cuzco in Peru. Here, the self-esteem and confidence of male landholders were boosted through their participation in training and interactions with irrigation project engineers and other technical staff. Project staff also appointed men as the community spokespersons and decision-makers. The systematic prioritization of men as the main stakeholders, experts and decision-makers worked to reproduce or perhaps even to generate a gendered hierarchy in how water tasks, powers and authority were defined and divided. Men increasingly became responsible for dealing with 'the outside world' and women became increasingly responsible for the physical labour of farming and irrigating, in addition to their domestic tasks (Vera Delgado, 2005). The former gradually came to be defined and seen as irrigation, while the latter were considered as 'non-irrigation'.

What all this means is that a proper understanding of gender within irrigation systems depends on thoroughly rethinking the metaphorical and spatial, and sometimes ideological and normative, images used. One must overcome, or at least question critically, the dualistic conceptual framework founded upon an opposition between the economic, rational irrigation world of production and politics, on the one hand, and the affectionate and emotional world of the home and the family, on the other. This can, for instance, be done by recognizing the subsistence and livelihood functions of farms. It can also be done by recognizing that men are not just irrigators and farmers, but also husbands and fathers, or by acknowledging that women's identities are not confined to those of mother and housewife, but also often include those of farmers and decision-makers. It includes allowing for the possibility that important irrigation negotiations occur in the domestic domain. And it requires a critical revisiting of what is recognized and defined as irrigation behaviour and of who are recognized as irrigators because what is included in these definitions may be gendered. Rather than assuming *a priori* the meaning and boundaries of irrigation systems, households and farmers, and the criteria for inclusion in the irrigation world, the following questions should be addressed: how do different water users, managers, politicians and others define 'inside' and 'outside' of the system? Who is seen as 'belonging' to the system, and who are ideologically, politically or physically excluded, and in what ways? Are these terms negotiable, and are definitions and conceptual categories themselves a way of defining and reconfirming ideas about gender, and of distinguishing masculine from feminine identities?

Technical and management systems and boundaries

In much of today's irrigation thinking, the colonial view of farmers as backward and in need of civilization is no longer popular. Yet, much thinking is still pervaded with an implicit normativity regarding what is 'good' and what is 'bad' irrigation behaviour. In fact, much irrigation knowledge is more concerned with creating the conditions and teaching people the skills for functioning as desired than with understanding what is actually going on. Perhaps as a result of this, people tend to 'matter', and thus discursively exist, in irrigation thinking only to the extent that they relate functionally to the irrigation system as conceived in technical designs and management models. In their conceptualization of irrigation performance, Small and Svendsen (1992, p4) explicitly posit that 'farmers are considered in their roles as irrigators, but their parallel roles in other aspects of crop husbandry are excluded'. This distinction, as the authors explain, 'is necessary to establish a clear analytic separation between the irrigation system and the broader agricultural system of which irrigation is a part' (Small and Svendsen, 1992, p4). They do not deny that all individuals in irrigation play many roles simultaneously (Small and Svendsen, 1990, p286); but this rests on the Weberian assumption that individuals can, and do, consciously separate their irrigation roles and behaviour from their other roles. Who farmers are thus only matters as far as their irrigation identities are concerned. Their identities are achieved because of their rational involvement in the system. Therefore, unless irrigation roles are directly gendered (i.e. if being a woman or a man in itself is seen as an irrigation role), gender also ceases to matter.

Such conceptual insulation of the irrigation system from its environment mirrors the attempts of many irrigation engineers to immunize the irrigation system from outside interferences. It can, in fact, be seen as an attempt to achieve what technology scholars call a process of 'closure' (Latour, 1987; Bijker, 1993). Closure is achieved when the possible meaning and use of the technology is no longer contested and its origins are ascribed to the laws of nature. One of its effects is that the authority to make truth claims about irrigation lies with experts. It is also another illustration of how the irrigation system is metaphorically compared to a factory or workplace. The very concept of 'role' as used by Small and Svendsen (1992) portrays irrigation realities as factory-like settings with strongly pronounced normative definitions of expected modes of conduct. The nature of roles is taken as a given, and it is derived from an ideal-typical model of how the irrigation system should function. As Giddens (1984, p84) remarks about the role concept: 'the script is written, the stage set, and actors do the best they can with the parts prepared for them'. Again, who plays these irrigation roles and in what social context does not matter. What people do in the irrigation factory is conceptualized as a function of the factory, and is unrelated to who they are or to their status, position or power outside of the factory. Their gender, as a result, is also inconsequential for the understanding of the functioning of the irrigation system and therefore does not require further investigation or questioning.

There are an increasing number of studies showing that in day-to-day irrigation realities, the boundaries between the system and its environment are not so easy to draw. In actual irrigation life, people cannot easily set aside their non-irrigation-related identities and interests for the sake of the good performance of the irrigation system. People's irrigation decisions also stem from considerations that are not internal to the system. More often, irrigating farmers know each other and relate to each other in many more ways than just through sharing a joint irrigation facility. Irrigation decisions are tied to and influenced by wider choices related to farming, livelihoods and social networks. Some studies about Andean irrigation systems show how intra-household disputes over farming and irrigation may be caused by wider conflicts between family members. One study documents how a woman sold her water rights to prevent her ex-husband from using the plot that she considered hers (Vera Delgado and Zwarteveen, 2007).

Like 'the hardware' of irrigation systems, 'the software' – or water user associations – is often seen as relatively insulated from the social context. Not much thought is usually given to who are or should be the participants of user organizations. Instead, in most writings on participatory irrigation management, the group of farmers or irrigators is referred to as a group that is already existing and easily identifiable: those people who are served by a common irrigation facility. 'Participation' is about participation of this group in the project or system of the engineers or state irrigation bureaucracy. The ultimate concern is to unravel the determinants of 'well-performing irrigation management institutions', while what good performance means is already decided – based on universal laws of human behaviour and nature – and mostly expressed in rather narrow technical, productionist and economic terms. In other words, existing situations are thus described and judged on the basis of whether or to what extent they follow, or can be made to follow, the ideal model. The existing social relations of power and the existing culture and norms are loosely treated as the raw material from which institutions can be 'crafted', 'the institutional resource bank from which arrangements can be drawn which reduce the social overhead costs of cooperation in resource management' (Cleaver, 2000, p365).

Conceiving of the irrigation management domain to include all that irrigation experts consider to belong to the irrigation system, and nothing more, is not conducive to making women and gender visible. To 'see' the social and gender factors in water management requires understanding that what happens 'within' the formal water management domain is shaped and influenced by what happens 'outside' it. It also requires a realization that events and decisions that have to do with water do not just take place within the formally defined water management domain. Insulation of the formal water management domain from its environment is based on the idealized views of experts rather than on-the-ground realities reflected in women's experiences as participants in user organizations. Women do not stop being seen as women and become genderless rational deliberators once they enter the formal public domain. One clear illustration of this is provided by an irrigation system in Peru, where about half of the members were women, and where both women and men attended meetings. Observations during these meetings showed that although male members, on average, talked for approximately 28 minutes, female members only talked for 3.5 minutes. Although 'speaking time' cannot be used as a straightforward measure of influence, women did explain that they felt diffident about articulating their concerns in meetings, and that they were afraid of making mistakes and being ridiculed (Krol, 1994). In an irrigation system in Mexico, only 15 per cent of the female farmers thought that their opinions mattered in meetings, against 73 per cent of the male farmers. Female farmers also displayed little interest in playing more active roles in the organization since they felt that by doing so they would call into question their moral integrity and status as women (Ahlers, 2000). A study in Bolivia likewise documents how women felt ill at ease in meetings, which is why many preferred sending their sons or husbands instead of going themselves.

Although not cited here, there are many more examples that suggest that gender colours deliberation and decision-making, even in the absence of formal entrance barriers. Public interaction and styles of deliberation almost everywhere are gendered in that there are distinct social norms and rules that define what sorts of interaction are permissible for women and which for men, in what contexts, and using which modes of conduct. Fraser (1997) even goes further to suggest that discursive interactions within the public domain are governed by protocols and styles of decorum that are themselves correlates and markers of gender inequality. In the above cited examples, to be outspoken and opinionated can be positive characteristics for men, markers of masculine distinction in Bourdieu's sense – a way of defining and reconfirming masculinity and male superiority.

At the same time, belying the formalistic and functionalist expert view, water management is not actually confined to formal water management institutions. One of the more telling illustrations of this is the story many Andean irrigation professionals tell when reflecting on gender: men participating in water management meetings always require a second meeting (the following day or week) to be able to make a decision. As the story goes, they want and need to consult with their wives at home. There are other anecdotal examples of women who are playing important but non-formalized, and therefore non-recognized, roles in organizations or in carrying out collective action. Juana Vera Delgado, for instance, notes how women play prominent water management roles 'behind the screens' in the traditional *reginas* in the irrigation system of Coporaque in the Colca Valley in Peru. Usually men assume the traditional water leader position (although some women also do); but it is normal practice and implicitly understood that their wives will assist them (Vera Delgado and Zwarteveen, 2007). Krol (1994) notes how one woman almost singlehandedly adapted the irrigation schedule in response to requests from neighbours and friends who did not understand it or who experienced difficulties irrigating at the times designated to them (Krol, 1994).

Indeed, water management can occur in a number of coexisting and partly overlapping 'domains of interaction' (Villarreal, 1994), which are not limited to the ones recognized and designated for water management by policy-makers and managers. The very fact that formal water decision-making is defined as something belonging to the sphere of men may in itself prompt the emergence of alternative ways and networks for managing and dealing with water questions. Because formal water users' organizations have come to be defined as masculine domains and because water expertise and authority have come to be associated with masculinity, becoming accepted as members and leaders is not easy for women. In Coporaque, Peru, for instance, one woman who stood up for herself by attending meetings and speaking to authorities was looked at with some suspicion by other women and men. They referred to her as a *machista*, which was not meant as a compliment (Vera Delgado and Zwarteveen, 2007).

Technical and organizational water systems are embedded in wider social and political relations and hierarchies that are not entirely based on or derived from water. Irrigators belong to wider social, cultural and normative systems, and are informed by locally specific ecological conditions. This recognition of embeddedness opens the conceptual door to the recognition of gender: because all social and political environments are gendered, gender shapes and colours all irrigation interactions and irrigation decisions.[2] The work of Giddens and Long assists in recognizing the social positioning of irrigation actors in power relations, including gender relations. Rather than seeing actors solely in relation to the resource or activity of interest to the knower, Long (1992) suggests perceiving them as complex individuals, partly involved in the projects of others and partly involved in their own. Giddens (1984, pxxiv) argues that a person should be recognized as positioned in multiple ways, with social relations conferred by specific social identities. Such recognition could help to explain that Andean women enter formal water-user organizations on different terms than men precisely because they are women, as in the previous example of Ecuador by Krol (1994); they cannot leave their gender identity behind when dealing with water. It would also help to recognize how women and men can manipulate and strategically use their gendered identities, rather than just accept how they are labelled by outsiders.

The implication of embeddedness is that what the system is and how its boundaries are drawn is importantly constituted by the social, political and ecological context in which it functions. This realization leads to a different ontological definition of irrigation or water systems than the one used in mainstream thought, one that allows the physical/technical and the social to be analysed simultaneously as different but internally related dimensions of a single object (Mollinga, 1998). Notions such as 'socio-technical systems' (Mollinga, 1998), 'waterscapes' (Swyngedouw, 1997) and 'nature cultures' or 'cyborgs' (Haraway, 1991) provide ingredients for such an ontology of irrigation systems that does not isolate the water system from its social, cultural and ecological environment. These notions envision human activity and nature as being interactive, shaping landscapes that are dynamic and continuously contested in a process that is constituted by, and simultaneously constitutes, the political economy of access and control over resources (Haraway, 1991; Swyngedouw, 1997). Importantly, the boundaries of the system are not static but change over time and are the subject of negotiation and struggle. This is why describing and understanding an irrigation system requires what Haraway (2003) refers to as an ontological choreography. It requires explicit inclusion of how different actors define and manipulate the boundaries and constituent elements of the system.

'Seeing' gender in water management, then, not only requires allowing women to enter into the already defined and ideal-typical domains of irrigation decision-making. It also requires rethinking the boundaries and functions of these domains. And it includes a critical enquiry into how drawing boundaries between identified domains serves to maintain or erode existing modes of gendered power and gendered identities. The current association of water authority and expertise with male identities and the perception of water management as a masculine domain may rest on implicit gendered beliefs and ideologies that serve to preserve and strengthen gendered power hierarchies, as well as on the actual division of water rights and powers. For women, entering a masculine domain, and assuming water identities that are associated with men, involves revaluing and redefining female identity and work, and a rejection of rules and regulations that tie them to specific roles. This typically happens through calling into question their sexual integrity and moral virtues. They are, for instance, accused of being 'public women' and risk physical and verbal abuse (Arroyo and Boelens, 1997, 1998). The account of Inés Chapi about the early days of her water leadership in an irrigation system in Ecuador is illustrative:

> *We [the women who got together and organized themselves] were told that our children were not from our husbands, that they were children from 'gringos', and the priest told me that we were Negroes. To our husbands they said: 'Listen, you are dummies, you have to take off your trousers, your wife does such and such things.' In the mass in church, people were told not to associate with me and Rosa, that we were bad women leading bad lives.* (cited in Arroyo and Boelens, 1998, p400)

Inés's comments show that not just women, but also their husbands and men, in general, risk losing respect and authority when women assume identities of experts and water decision-makers that tend to be reserved for men. Husbands risk being considered as 'weak' or 'effeminate', while the job of water manager loses respect and imparts less status when women can also do it.

Conclusions

This chapter explains how mainstream professionals conceptualize water realities and the implications for 'seeing' and misrepresenting women and for understanding gender. Ways of framing – of talking and thinking about irrigation – are an intimate part of the larger projects of maintaining or challenging gendered hierarchies and norms. Professional identities in irrigation use languages and ways of producing truths. As such, irrigation knowledge and discourses are part of a larger range of cultural expressions through which professionals represent themselves. This process of identity formation is deeply gendered, both in how it continuously works to reconfirm the masculinity and, thus, the power, strength and authority, of knowledge producers. One example is the labelling of some activities and domains as masculine and others as feminine. In addition, choosing certain metaphors and drawing the boundaries of subject matter in particular ways allows normal irrigation knowledge to reproduce gendered hierarchies and reconstitute gendered identities.

Water science is a peculiar form of science. Read in a Foucaultian frame of analysis, the construction of irrigation knowledge is tied to the development of particular modern forms of practising irrigated agriculture and to 'disciplining' the practitioners. Detailed prescriptions about how to optimize the use of land and water for the production of crops are concretized in technological and managerial designs. As such, irrigation knowledge can be seen as a project to turn farmers and irrigators into 'docile' bodies whose movements can be controlled in time and space (compare Foucault, 1979). This is not to say that the results of irrigation research are misused or misapplied by governments. However, the irrigation activities and policy agendas that address significant irrigation problems are intertwined so that the values driving irrigation policies also determine policies for much irrigation research. Many irrigation texts have been funded by development agencies and a large number of studies have arisen out of, or were commissioned to inform, specific irrigation programmes. Helping to make irrigation systems perform better is a major objective of much research. This has an important effect on the applicable standards of research competence.

For international experts, familiarity with the international irrigation discourse is often of far greater importance than knowledge of a particular country or water use context. Many studies are produced for consumption by the agencies or universities that sponsored them and are circulated only within a privileged circle of policy-makers or academics. Many studies present quite basic information and are predominantly descriptive, providing evidence to substantiate a selection of key themes. Their thrust, in general, is to provide better irrigation designs or management models, rather than producing sharper analysis. Diverse irrigation realities across the world are reduced to 'key performance indicators' (see Perry, 1996; Molden et al, 1998) which can serve as the basis of comparison to compound 'a screening process for selecting systems that perform relatively well and those that do not' (Molden et al, 1998, p19). Such systematic exclusion of context, or of the specifics of the cultural, social and political environment, allows sustaining the façade of a universal and generic 'water expertise', which can be applied the world over with only minor adjustments. It is an expertise that is intrinsically resistant to seeing and understanding gender because gender is necessarily about context. It is also an expertise that may not be hospitable to critical reflection because that would risk unveiling and threatening the foundations of unequal economic and political relations on which mainstream water knowledge is based and that it helps to sustain.

'Normal' water knowledge often continues to be typically positivist, and much of it continues to be prescriptive: it is concerned with how water realities should be, and possibly with why actual realities are different. It is less concerned with trying to understand the logic and determinants of how such realities actually are. Through prescriptive ways of 'ordering' realities – water systems, organizations, institutions, economies – people themselves are also 'ordered' and 'normalized'. They act on the basis of sets of incentives that are clearly identifiable and known to planners, managers and knowers. The incentives can be manipulated or at least, to some extent, controlled by those in power. Gender, just as other social differences and social relations of power, are 'assumed away'. They do not fit in the rational, logical and scientific organization of the water world either because they are seen as despicable remnants of backward cultures or traditions, or because they are perceived as belonging to the world of the family and the private that supposedly do not matter for understanding what goes on in water.

For making women visible, and providing them with a legitimate existence in mainstream water discourse, there are two distinct rhetoric strategies that are not mutually exclusive. The first is to show that there are women among irrigators, water users and managers and among the inhabitants of the world of reason and work. This strategy posits women as similar to all other irrigators. Its strategic effectiveness in gaining recognition for, and attention to, gender importantly rests on convincingly showing that women, too, are endowed with the gifts of reason and rationality, that they too can irrigate and farm, etc. In other words, they are humans too. It implicitly questions the ideological and symbolic association of productive work and the public domain with masculinity and domestic work and the private domain with femininity. It is a successful strategy to claim rights to water and land for women. Yet, it is not very effective in questioning gender inequities as they relate to water. Because of the way in which irrigation and water systems are perceived, the discursive transformation of women into irrigators and water managers entails screening off their non-irrigation identities. This is how women cease to exist as women. Mainstreaming gender, therefore, implies its disappearance as a theme that can be discussed. Women, just like men, get to be treated as 'universal' subjects who are implicitly modelled on men, and 'the gender question' is reduced to one of exclusion or lack of integration. Gender becomes irrelevant because rational water behaviour is not influenced by the gender category to which one belongs, but conceptually 'bracketed away' and defined by one's function and location in the water system.

The second strategy to show that women 'matter' is to create them discursively as a distinct functional group in relation to the water system. This implicitly argues that they are different. For instance, this can be done by showing that women's water needs are distinct from those of men. Or it can be done by showing that female farmers systematically have different assessments of the water system's outputs, impacts and internal operations compared to male farmers. It entails the establishment of another important category of individuals next to the already existing category of water users and farmers, and claiming a degree of acceptance and 'normalcy' for this group. It may also entail a change in the ideas about what the irrigation system 'produces' or should produce, such as by including providing water provision for domestic uses. And it may further entail shifting the system's boundaries – for example, by including women's homestead gardens in the area that is to be served by the system. It implies, then, the expansion of the water reality with a distinct 'women's domain'.

This second way of making women visible clearly does put them on the water map, and allows thinking about their specific water needs and demands. Yet, it is not without problems. Women are made visible as women, as individuals whose identities and needs are derived from the fact that they belong to the female gender. Their link to the water system and, thus, their existence in water discourses also come to be seen as primarily determined by their gender. What is problematic about this is that women's water existence is linked to their gender, while that of men is simply there and unrelated to their social identity. Masculinity is thus assumed and taken as the norm, while femininity is defined as the 'other' which needs mentioning. Such reasoning dangerously limits the definition of the female subject to gender identity to the exclusion of other identities. An illustration of this is provided by van Koppen et al (2001), who describe the different types of members at the lowest organizational level of the West Gandak irrigation system in Nepal. Next to, for instance, a chairman and a vice-chairman, a woman is mentioned as one type of member.

Such reductionism discursively constructs women in implicit opposition to the construction of irrigators, who are assumed to be men. Women's professional identities as farmers and irrigators become difficult to see and understand, while men's identities as irrigators are overemphasized to the neglect of their other identities, including gender. The two categories are defined as mutually exclusionary and dichotomous. Gender – or at least female identity – then becomes itself a determinant of water behaviour, dividing the water world into 'normal' water users or irrigators and women. This dichotomous conceptualization of gender analytically is also problematic because it leads to the universalization and essentialization of gender differences, and thus risks 'freezing' them rather than questioning and challenging them. It is based on the construction of women as an already constituted coherent group with identical irrigation interests and desires, regardless of class, ethnic or racial location, or contradictions. This group of women exists prior to the process of analysis, and prior to their entry into the arena of social relations or the irrigation system (compare Mohanty, 1991).

Questioning gender requires a social relational approach in which men and women are seen as parties to sets of social relations involving rights, resources, responsibilities and meanings. These relations with other men and women are the vehicles through which what it means to be a woman and a man, in that time and social place, is defined and experienced. Gender operates within social categories rather than through pre-existing bounded groups of men and women. Categories of men and women are to be deconstructed, allowing differences within gender divisions, recognizing male gender interests and identities, and separating actually existing women and men from hegemonic femininity and masculinity (Connell, 1995).

In the end, critical knowledge of water is not just concerned with water realities 'out there', but also, and importantly, by how such water realities are interpreted and understood at different levels of governance and by different actors. Gender importantly colours and influences the construction of knowledge, and the identities of who are recognized as water experts. Gender, therefore, is not just a part of water realities in the field, but also fundamentally colours and structures ways of thinking and making sense of those realities and of how identities are constructed. Struggles over meanings and discourses, about how truth claims are made in water, about expert identities and cultures, are and should therefore assume a much more central place in attempts to mainstream gender in water.

Notes

1 By calling a particular way of knowing in water 'mainstream', I do not mean to imply that it is uniform, static or uncontested. Mainstream irrigation wisdoms have always been contested and continue to be challenged by various civil society groups, as well as by water scholars. I use the word 'mainstream' to denote its widespread acceptance and status of 'normalcy'. Indeed, most produced knowledge about water needs refer to it – whether in agreement or in disagreement – to be counted as knowledge, or to have an influence in debates and policies; see Zwarteveen (2006), which is also the basis for the contents of this chapter.

2 The fact that most accounts of embedded realities of water management hardly mention gender may be due to the fact that many of these studies describe and understand irrigation situations in the terms used by irrigation actors themselves, and uncritically accept their gender connotations. Most studies also uncritically adopt the local or conventional methods for identifying relevant actors. Hence, where farmers, irrigators and water leaders in the local understanding are men, researchers accept rather than question this. A focus on visible and audible conflicts, and on a tacit limitation of observations to the 'public' realm of irrigation (fields, canals, meeting rooms and offices) may further hide gender (and women) from the view of irrigation researchers, at least in situations where women's struggles occur in less open and visible ways and where women are not routinely among those present in recognized public irrigation spaces.

References

Ahlers, R. A. (2000) 'Gender relations in irrigation districts in México', in C. Tortajada (ed) *Gender and Water Management in Latin America*, Oxford University Press, New Delhi, pp203–216

Arroyo, A. and R. Boelens (1997) *Mujer campesina e intervención en el riego andino: Sistemas de riego y relaciones de género, caso Licto, Ecuador*, CAMAREN, Quito, Ecuador

Arroyo, A. and R. Boelens (1998) 'Andean irrigation and gender equity', in R. Boelens and G. Dávila (eds) *Searching for Equity: Conceptions of Justice and Equity in Peasant Irrigation*, Van Gorcum, Assen, The Netherlands

Bakker, M., R. Barker, R. Meinzen-Dick and F. Konradsen (eds) (1999) *Multiple Uses of Water in Irrigated Areas: A Case Study from Sri Lanka*, IWMI, SWIM Report 8, Colombo

Bijker, W. E. (1993) 'Do not despair: There is life after constructivism', *Science, Technology and Human Values,* vol 18, no 1, pp113–138

Boelens, R. A. (2008) *The Rules of the Game and the Game of the Rules: Normalization and Resistance in Andean Water Control*, Wageningen University, Wageningen, The Netherlands

Boelens, R. and M. Zwarteveen (2005) 'Anomalous water rights and the politics of normalization: Collective water control and privatization policies in the Andean region', in D. Roth, R. Boelens and M. Zwarteveen (eds) *Liquid Relations: Contested Water Rights and Legal Complexity*, Rutgers University Press, New Brunswick, NJ, and London, pp97–123

Butler, J. (1999) *Gender Trouble*, Routledge, New York and London

Cleaver, F. (2000) 'Moral ecological rationality, institutions and the management of common property resources', *Development and Change*, no 31, pp361–383

Connell, R. W. (1995) *Masculinities*, University of California Press, Berkeley and Los Angeles, CA

Davidson, A. I. (1986) 'Archeology, genealogy, ethics', in D. C. Hoy (ed) *Foucault: A Critical Reader*, Blackwell, Oxford, and Cambridge, MA

Dietz, M. (1992) 'Context is all: Feminism and theories of citizenship', in C. Mouffe (ed) *Dimensions of Radical Democracy: Pluralism, Citizenship and Community*, Verso, London and New York

Foucault, M. (1979) *Discipline and Punish: The Birth of the Prison*, Vintage, New York, NY

Fraser, N. (1997) *Justice Interruptus: Critical Reflections on the 'Postsocialist' Condition,* Routledge, New York and London

Giddens, A. (1984) *The Constitution of Society: Outline of a Theory of Structuration*, Polity Press, Cambridge, UK

Gutiérrez, Z. and S. Cardona (1998) *Las relaciones de género en la gestión de aguas en sistemas de producción intensiva: Estudios de caso Combuyo, Vinto, Cochabamba*, PEIRAV, Cochabamba, Bolivia

Haraway, D. J. (1991) *Simians, Cyborgs and Women: The Reinvention of Nature*, Free Association Books, London

Haraway, D. (2003) *The Companion Species Manifesto: Dogs, People and Significant Otherness,* Prickly Paradigm Press, Chicago, IL

Harding, S. (1986) *The Science Question in Feminism*, Cornell University Press, Ithaca, NY, and London

Hartsock, C. M. N. (1998) *The Feminist Standpoint Revisited and Other Essays*, Westview Press, Boulder, CO

Jackson, C. (1998) *Social Exclusion and Gender: Does One Size Fit All?*, Mimeo, School of Development Studies, University of East-Anglia, Norwich, UK

Kabeer, N. (1991) *Gender, Production and Well-Being: Rethinking the Household Economy*, IDS Discussion Paper 288, Institute of Development Studies, Brighton, UK

Krol, M. (1994) *Irrigatie is mannenwerk: Genderverhoudingen in een kleinschalig irrigatieprojekt in de Ecuadoriaanse Andes*, MSc thesis, Wageningen University, Wageningen, The Netherlands

Langton, R. (2000) 'Feminism in epistemology: Exclusion and objectification', in M. Fricker and J. Hornsby (eds) *The Cambridge Companion to Feminism in Philosophy*, Cambridge, UK, pp127–145

Latour, B. (1987) *Science in Action: How to Follow Scientists and Engineers through Society*, Harvard University Press, Cambridge, MA

Long, N. (1992) 'From paradigm lost to paradigm regained? The case for an actor-oriented sociology of development', in N. Long and A. Long (eds) *Battlefields of Knowledge: The Interlocking of Theory and Practice in Social Research and Development*, Routledge, London and New York, pp16–43

Mayer, E. (2002) *The Articulated Peasant: Household Economies in the Andes*, Westview Press, Boulder, CO, and Oxford

McCay, B. J. and S. Jentoft (1998) 'Market or community failure? Critical perspectives on common property research', *Human Organization*, vol 57, no 1, pp21–29

Mohanty, C. T. (1991) 'Under Western eyes: Feminist scholarship and colonial discourses', in C. T. Mohanty, A. Russo and L. Torres (eds) *Third World Women and the Politics of Feminism*, Indiana University Press, Bloomington and Indianapolis, IN, pp333–358

Molden, D., R. Sakthivadivel, C. J. Perry, C. de Fraiture and W. H. Kloezen (1998) *Indicators for Comparing Performance of Irrigated Agricultural Systems*, IWMI Research Report no 20, Colombo

Mollinga, P. P. (1998) *On the Waterfront: Water Distribution, Technology and Agrarian Change in a South Indian Canal Irrigation System*, Wageningen University, Wageningen, The Netherlands

Nicholson, L. (1995) 'Introduction', in S. Benhabib, J. Butler, D. Cornell and N. Fraser (eds) *Feminist Contentions: A Philosophical Exchange,* Routledge, New York and London, pp1–16

Perry, C. J. (1996) *Quantification and Measurement of a Minimum Set of Indicators of the Performance of Irrigation Systems*, Mimeo, IWMI, Colombo

Sayer, A. (1992) *Method in Social Science: A Realist Approach*, Routledge, London and New York

Scott, J. (1985) *Weapons of the Weak: Everyday Forms of Peasant Resistance*, Yale University Press, New Haven, CT

Scott, J. (1986) 'Gender: A useful category of historical analysis', *American Historical Review*, vol 91, no 5, pp1053–1075

Shah, E. (2003) *Social Designs: Tank Irrigation Technology and Agrarian Transformation in Karnataka, South India,* Wageningen University Water Resources Series, Wageningen, The Netherlands, Orient Longman, New Delhi

Small, L. E. and M. Svendsen (1990) 'A framework for assessing irrigation performance', *Irrigation and Drainage Systems*, no 4, pp238–312

Small, L. E. and M. Svendsen (1992) *A Framework for Assessing Irrigation Performance*, IFPRI Working Paper on Irrigation Performance, no 1, Washington, DC

Spivak, G. C. (1988) *In Other Worlds: Essays in Cultural Politics*, Routledge, New York and London, p253

Swyngedouw, E. (1997) 'Neither global nor local: 'Glocalization' and the politics of scale', in K. R. Cox (ed) *Spaces of Globalization: Reasserting the Power of the Local*, Guilford Press, New York, NY

van Koppen, B., L. van Etten, P. Bajracharya and A. Tuladhar (2001) *Women Irrigators and Leaders in the West-Gandak Scheme in Nepal*, IWMI Working Paper no 15, Colombo

Vera Delgado, J. (2005) 'Irrigation management, the participatory approach and equity in an Andean community', in V. Bennett, S. Dávila-Poblete and M. N. Rico (eds) *Opposing Currents: The Politics of Water and Gender in Latin America*, Pittsburgh University Press, Pittsburgh, PA, pp109–122

Vera Delgado, J. and M. Zwarteveen (2007) 'The public and private domain of the everyday politics of water: The constructions of gender and water power in the Andes of Peru', *International Journal of Feminist Politics*, vol 9, no 4, pp503–511

Villarreal, M. (1994) *Wielding and Yielding: Power, Subordination and Gender Identity in the Context of a Mexican Development Project*, Wageningen University, Wageningen, The Netherlands

Zwarteveen, M. (2006) *Wedlock or Deadlock? Feminists' Attempts to Engage Irrigation Engineers*, Wageningen University, Wageningen, The Netherlands

Part II

Politics of Identity and Andean Livelihoods

5

Identity Politics and Indigenous Movements in Andean History

Michiel Baud

Introduction

There are certain fundamental patterns in the modern history of the Andes. They are visible, first, in the make-up of Andean rural society. The ecological circumstances and the complex ways in which local populations have adapted to them have made Andean agriculture unique. The specific spatial use of microclimates, and the heritage of indigenous systems of water control and labour organization show so many similarities that we can consider it one historical region. This is not to deny, of course, the great cultural and socio-economic differences in the region (e.g. Salman and Zoomers, 2003). David Lehmann (1982, pp1–2) writes: 'Andean rural society presents a remarkable opportunity for studying the variety of patterns of social and economic organization which can arise even in a region which, over five centuries, has been subjected to common successive forms of political and economic domination' (see also Sempat Assadourian et al, 1980, pp14–17).

In the final part of this quotation, Lehmann identifies political history as a fundamental factor in understanding the contemporary Andes. All indigenous groups in the Andes were confronted by centuries of Spanish colonialism. The colonial emphasis on silver production in Potosí and the ways in which the Spaniards tried to use existing patterns of social and political organization created a colonial structure that stressed regional differences, yet considered all communities to be part of the same time/space framework. These processes continued after independence. The nation states that were created in the first decades of the 19th century had legally rejected the Spanish colonial dualism which divided society into two *repúblicas*, the Republic of Spaniards and the

Republic of Indians, but were, in practical terms, unable and unwilling to accept the legal and social consequences of this merger. The struggle for independence did not end the ideology of exclusion and ethnic (in contemporary vernacular, racial) inequality. The Andean republics during the 19th and 20th centuries were characterized by a continuing political and social exclusion of the majority of the indigenous population by modernizing ethnocentric states.

Nevertheless, it would be wrong to assume that indigenous and black populations were hapless victims of a repressive, authoritarian Creole state. The weakness of governmental institutions of the new republics, the lack of financial means and the inter-elite feuds all precluded implementing policies to subdue and assimilate rural populations. Recent historiography tends to show how indigenous populations in different ways maintained part of their cultural and political autonomy and often managed to adapt quite effectively to changing circumstances (Tandeter, 1993; Larson and Harris, 1995; Larson, 2004). On the one hand, Indian communities maintained separate cultural, social and economic spheres reproducing the colonial idea of separate 'republics'. They pursued their own historical trajectory according to political logics that were different from, and often diametrically opposed to, those of the state. On the other hand, they tried to work within the system by using economic opportunities, taking advantage of legislation and pressing governments to protect their interest by way of petitions, lawsuits and protests.

This chapter provides historical and political background to the encounters between the Andean nation states and local communities and movements for the next chapters' themes of cultural justice and socio-economic distribution in water control. Here, the relationship between indigenous communities and the post-colonial state will be central. One of the crucial issues in Latin America's 19th- and early 20th-century history is the relationship between ethnicity and politics. History has greatly complicated the usage of categories such as 'Indian' or 'indigenous'. While these terms were, and are, customarily used in the daily press, political discourse and even in academic literature, their significance and understanding have been subject to continuous change and negotiation. Thus, these categories should be used in a dynamic historical and non-essentialist manner (Assies, 1999, pp21–55).

The purpose of this chapter is to demonstrate how the history of the indigenous populations in Latin America can only be understood as the consequence of the continuing interaction among the indigenous population, politicians and intellectuals, local government structures and the national state (Albó, 1999). This interaction initially took place in rural areas involving conflicts about land and water, but became more urban and national in the course of the 20th century. It involved not only day-to-day confrontations on practical aspects of politics and economics, but also struggles about perceptions, meanings, and political and moral expectations. The study of Andean regional history is best understood when this permanent, practical and discursive interaction between different historical actors is taken into account.

Equality and exclusion in Latin America

The political history of the Latin American republics is defined by the contrast between the legal and discursive equality and the continuation of forms of economic and political inequality and social apartheid that characterized Latin American societies. During most of the 19th and 20th centuries, politicians and intellectuals tried to come to grips with this specific predicament of Latin American state formation. Of course, the terms of this debate changed over time because of changed circumstances on a national and international scale and the active political engagement on the part of Indians.

During the 19th century the question '*¿Qué hacer con el indio*?' ('What shall we do with Indians?') was almost invariably answered in terms of the racialized social Darwinist language of the day (Demélas, 1981; Irurozqui-Victoriano, 1992). In these visions, the Indian population was considered the racial 'other' that could be included only by force in the sacred project of modernity. However, they were not always or exclusively guided by anti-Indian ideas. In her work on Central America, Marta Elena Casaús has recently drawn attention to the influence of a whole range of spiritualist ideas that contradicted and often opposed the prevailing materialist racism. These ideas were not only imported from Europe, but also tried to express a specific Latin American spiritualism (Gonzales Alvarado, 1996; Casaús-Arzú and García-Giráldes, 2005). These ideas were quite influential in the process of nation-building and the creation of new imagined communities. In her work on Bolivia, Marta Irurozqui-Victoriano (1999, p707) states that 'governments' efforts to eliminate indigenous communities were not driven by an anti-Indian policy, but by a will to homogenize, a belief that cultural diversity is an obstacle to the nation's development, and the wish to have a country full of citizens'. This observation is extremely ambiguous because this 'will to homogenize' has often been a euphemism for racial and colonial political projects; but it leaves room for the analysis of more benevolent projects and policies. Ultimately, it was the practical results of these policies that counted – not only for the populations involved, but also for historical research. Looking at state policies in the 19th and early 20th centuries, it seems important to distinguish between concrete racist policies to repress and even destroy indigenous people and their culture versus the more open-minded attempts to create new forms of coexistence and inclusionary politics.

Despite fundamentally unequal relations, colonial and capitalist societies have always produced the 'sympathetic outsiders' who cared, or at least expressed sympathy for, the downtrodden masses; they were the 'colonizers who refused', to use Memmi's famous term. Memmi (1967, p24) writes: 'Since he has discovered the colonized and their existential character, since the colonized have suddenly become living and suffering humanity, the colonizer refuses to participate in their suppression and decides to come to their assistance.'

From the late 19th century onwards, the changing context of urbanization and the expansion of export-oriented agriculture turned the discussion increasingly towards issues of assimilation and the integration of the Indian masses in the project of modernity. Simultaneously, the encroachment on Indian lands and their traditional water rights and the realization that the colonial *pacto de reciprocidad* between government and indigenous communities was no longer valid provoked a generalized discontent among Indian peoples and led to a great number of indigenous uprisings. These changes expressed themselves in the emergence of various brands of *indigenismo* in which white *mestizo* elites tried to find solutions to the plight of the '*infelices indios*'.[1] Although Memmi's analysis cannot simply be applied to a society with such a different (colonial) history and his psychological analysis tends to ignore the material origins of these processes, his approach fits the position of the *indigenistas* quite well. They were normally middle class, urban, white intellectuals and politicians who understood that the incorporation of the indigenous population could not be done without understanding and, to some extent, respecting parts of indigenous society and culture. Although their approach was often superficial and romanticized, the *indigenistas* attentively observed native ways of managing natural resources, their irrigation traditions, and their production and livelihood strategies. Departing from different social, political and regional viewpoints, they professed sympathy for the dilemma of the Indians and denounced their suppression by the dominant classes.

Often these ideas were grounded in a longer-lasting humanist or inclusionary tendency that had taken root already during the colonial period and which was further developed after independence in the republican period (Forment, 2003). With the mixture of indignation and remorse described by Memmi, the *indigenistas* advocated a concerted endeavour from the state to bring about an improvement in the position of the indigenous masses. But aside from the psychological processes underlined by Memmi, historians can discern more structural societal causes for their ideas. To a certain extent, they were messengers of an incumbent state which had started to consider the controlled incorporation of the Indians as an indispensable element of the modernization of their countries. Using a mix of centralism and virulent anticlericalism, politicians and state officials used the supposed protection of the Indians to curb the power of the landowning elites, the so-called *gamonales* and the priests.

There is no doubt about the frequent ignorance and idealization among *indigenista* intellectuals and politicians of the daily reality in which the Indian population lived. In many cases, *indigenismo* was just an element of specific political projects of the dominant classes. Osmar Gonzales Alvarado (1996, p205) observed of the Peruvian situation: 'local intellectuals put together an image of Indians in order to craft a tradition (the glorious Inca past) and seize legitimacy (either to sever ties to the Spanish metropolis, or to consolidate their role as leaders of national progress)'. We may consider the defence of the Indians on the part of people such as Mariátegui or Haya de la Torre as similar efforts to integrate the Indians within national politics, in their case to use the indigenous cause as the basis for a revolutionary project.

Andrés Guerrero (1994, pp197–252) has convincingly shown how even among the liberal defenders of the Indians a string of stereotyped images existed as to the moral and economic capacities of the Indian population. Many of them were still influenced by racialized or eugenistic ideas directed towards 'civilizing' the Indians; others were inspired by the socialist ideals of equality and justice. Larson (2004, p15) observes that elite articulations of liberalism, nationalism and racism took place in 'messy political contexts of rural struggle, market expansion, and political crisis'. One concrete essential area of modernization that was of growing national economic importance throughout the early 20th century was agro-technology development and encouraging rural areas to enter the market economy. For example, this new mission gradually set up water policies oriented towards government intervention, with large-scale centrally controlled irrigation systems on land belonging to rural and indigenous communities. In such attempts the Andean elites were faced with the problem of how to modernize their countries without endangering their dominance over the indigenous majority. In Larson's (2004, p15) words: 'The quandary for Andean Creole elites was precisely how to build an apparatus of power that simultaneously incorporated and marginalized peasant political cultures and livelihood systems in the forced march to modernity.' In order to illustrate the above, the next section briefly outlines the case of Ecuador.

The 'liberal' revolution in Ecuador, 1895

Ecuador was a particular case in point when subsequent liberal governments after 1895 used the indigenous cause to foster their grand project of centralist modernization (Baud, 1996, 1997). On 18 August 1895, newly elected President Eloy Alfaro abolished the *contribución territorial* and some of the labour services (the so-called *trabajo subsidiario*). The considerations that preceded this decree read: 'That the miserable situation of the Indian race must be improved by public authorities' and 'That the Liberal government has the duty to protect the descendants of the first inhabitants of Ecuadorian territory' (Costales and Costales Samaniego, 1964, p723). It was this rhetoric that gained the new president the nickname 'el indio Alfaro'. The most radical advocate of this form of *indigenista* liberalism was certainly Abelardo Moncayo, the minister of internal affairs and police. In 1895 he published an essay called *El concertaje de indios*.[2] Moncayo (1895, pp287–289) starts his essay calling the *concertaje* system '*uso bárbaro*' and then describes it as follows: 'Absolute abdication of freedom, alienation of will and intelligence, the very death of one's personality, that is *concertaje*... It is the systematized degradation of an immense portion of our brethren... It is the legal condemnation of an entire race to brutishness.' In a characteristic paragraph, he summarizes his analysis of the social condition of the indigenous population: 'Condensation of all possible shadows and miseries, roving vileness, ignorance in its simplest terms, servility to an extreme... Here we have the Indian, the masterpiece of Christian lords, of eternal conservative domination' (Moncayo, 1895, pp287–289).

The implementation of this radical *indigenista* rhetoric becomes clear in a circular of 22 June 1898 to the provincial governors, signed by the same Abelardo Moncayo. In this document the illegal use of Indian labour by state and church authorities was prohibited. It was stated that the authorities were no longer allowed to use Indian labourers 'against their will and without paying them the salary due or without remuneration legally agreed upon'. Moncayo admonished: 'It is necessary to punish, therefore, the mayors who in name of the government or with public works as an excuse recruit labourers'. He concluded:

> *All governors, local political authorities, parish directives and police judges are charged with eradicating ... such hateful customs that manifest the intolerable oppression and prejudice burdening the Indian race: such as raiding their homes; seizing their belongings to force them to work for private parties; binding them with ropes, forcing them to labour in places and climates unlike those for which they were engaged; keeping them in prison for month after month at the owner's whim without any legal reason; obliging their wives or children to pay or work instead of their husband or father, always for no payment; forcibly requiring them to sell their goods at arbitrary prices, etc. – all crimes desecrating the guarantees set forth in the constitution and violating the most basic principles of justice, which we can view with such indifference only because we have become inured.*[3]

Whatever their background [and ventriloquist] abilities (Guerrero, 1994), the ideas of *indigenist* intellectuals made a difference in changing the political and legal perspectives of national society and the place of indigenous populations in it. There is no doubt about the veracity of Guerrero's interpretation; but it is interesting to see how the indigenous population tried to take advantage of existing stereotypes by giving them a reversed meaning. Even ill-informed and elitist ideas such as *indigenismo* eventually found their way into politics and legislation and in the end had far-reaching consequences for indigenous society. They created a discursive and legal space to push for new definitions of citizenship and new forms of nationhood. On the one hand, the indigenous populations were liberated from a number of 'old-fashioned' obligations and were – at least rhetorically – recognized as more or less full-fledged members of the Andean nations. On the other hand, racial and socio-economic differences stayed in place, which rendered Indians incapable of acting independently in political and commercial activities (Guerrero, 1991). This ambiguity gradually dissipated during the course of the 20th century when the Indian population started to take advantage of liberal legislation and changed economic and political circumstances. Through a multiple process of resistance and adaptation they fought their way into the Ecuadorian nation and turned their struggle into a struggle for complete – ethnic, political and economic – citizenship (Clarke and Becker, 2007).

Indigenistas and anthropologists

During the first few decades of the 20th century, the context of the debate on the nation changed. With consolidation of US political and economic hegemony in the region, anti-imperialist ideologies rapidly developed in Latin America. They partly expressed themselves in a growing influence of socialist and anti-imperialist ideas, leading most dramatically to the Mexican Revolution. These anti-imperialist ideas led to new nationalistic ideas, which increasingly called for ideological inclusion of all sectors of the population. This led to new forms of state intervention in areas such as education and rural development and simultaneously created a new interest in indigenous and Afro-American popular culture. This trend consolidated around 1930 when the financial crisis added an economic component to this new Latin American nationalism. The economic nationalism exemplified in the politics of *import substitution industrialization* (ISI) and different brands of populist political projects also stimulated new ideas regarding the relationship between governments and indigenous peoples. While governments tried to accelerate the assimilation of their indigenous and black populations – by both benevolent and repressive means – a new, more radical *indigenismo* came into being, demanding inclusion of the indigenous populations on their own terms.

It is important to remember that the *indigenista* movement created images and representations of the indigenous world that not only had political and legal consequences in the period concerned, but also had great influence on 20th-century anthropology and, more generally, analytical ideas of representing the indigenous world. For instance, Castro-Klarén (2003) has stressed the importance of the material remains of the great indigenous cultures to fostering new national identities in Latin America. This led to renewed archaeological interest in indigenous ruins in the 19th century and the concomitant neo-Inca revivalism. These material remains functioned as a kind of mnemonic device, shaping Peru's national identity during the 19th century. Appropriation of the indigenous heritage would, in the 20th century, lead to the *indigenista* movements and, later, to the insertion of the indigenous past in large-scale commercial tourism. A major example of renewed interest in the Incan past is also the emergence of studies regarding Incan waterworks and irrigation management. Water control and irrigation, often considered the foundation of the world's great cultures and civilizations – as opposed to 'natural', uncivilized societies – has since then constituted inarguable proof of the great Andean historical civilization.

Although the timetable differed and *indigenista* ideas started to influence political debate by the first decades of the 20th century, they took root politically in the nationalist period, especially after the economic crisis of 1929. The protective economic policies represented in the ISI model were culturally mirrored by a new wave of political and cultural initiatives meant to incorporate indigenous populations within the bosom of the nation. In countries such as Peru, Ecuador and Mexico, outstanding *indigenistas* were appointed to high-profile public offices. This enabled *indigenismo* to promote significant legislation in Peru and Ecuador. In Peru, the *Oncenio*, the administration of Augusto Leguía (1919 to 1930), began with the most radical *indigenista* policy of all Latin America. These initiatives triggered feverish activity in many fields of social law-making. More laws were enacted regarding the indigenous population from 1921 to 1924 than during the whole preceding century.

'Official', state-sponsored political *indigenismo* was especially strong in Mexico, where the state had always been important since the Mexican Revolution. The administration of Lázaro Cárdenas (1934 to 1940) witnessed a new trend towards internationalizing *indigenista* thinking. In 1940 the First Inter-American Indigenista Congress in Pátzcuaro led to the founding of the National Indigenista Institute (INI) in 1948. An Inter-American headquarters was established in Mexico City with branches in several Andean countries. In the beginning, this institute was a fruitful meeting place for *indigenista* intellectuals such as Moisés Sáenz, Alfonso Caso and Gonzalo Aguirre-Beltrán, with representatives of numerous Indian groups. This prolific exchange was also reflected in new ethnographic interest in the indigenous population, but with contradictory consequences. It led to static folklore fixation, which the Mexican government used to keep indigenous rural society under control, while resulting in an acceptance of the indigenous peoples' cultural diversity and dynamics, and gave rise to new emancipation thrusts. Mexican indigenist Moisés Saenz wrote important treatises on the situation of the indigenous population in the Andean countries.

In the process, official *indigenismo* in the Andean countries acquired new strength. A new artistic interest in the indigenous cultures and livelihoods emerged. Writers such as Jorge Icaza in Ecuador, José María Arguedas in Peru and Jesús Lara in Bolivia created modern images of Andean indigenous culture and stimulated new social and cultural interpretations of the indigenous past and present. Indian society became a source of inspiration for painters, sculptors and musicians. Intellectuals started to devote their energy to compiling and analysing popular culture. This tendency was clearly visible in the project of regional chauvinism in Cusco, where the local elite used their supposed allegiance to the Indian cause ('¡*Nosotros los indios*!') to resist the centralist tendencies from Lima (Escalante, 1976, pp39–52; de la Cadena, 2000). Peruvian historian José Luis Rénique (1991) has beautifully demonstrated how this appropriation of 'indigenous history' was scarcely connected to the reality of the contemporary indigenous population and ignored their situation. While *indigenista* intellectuals in Cusco were discussing the greatness of the Inca past and romanticized indigenous folk culture as expressed in the religious festivities and the culture of the *chicherías*, they ignored the daily reality of the living

Indians in their own environment. They evidently demonstrated what Renato Rosaldo has labelled in another context as 'imperial nostalgia'. They were often hardly aware of what was happening in the surrounding countryside and talked *for*, not *with* the Indian population. Indians could only present themselves as citizens through the mediation of these culturally privileged intellectuals and politicians. On the other hand, Albó (1999) rightly observes: 'Living Indians were no longer just the objects of pity and curiosity; they also evoked admiration.'

These transformations took place in the context of political processes that were sometimes quite violent and had dramatic consequences for rural society. During the Chaco war against Paraguay (1935 to 1938) many Bolivian Indians were recruited by the army and in this way integrated within the national context (Arze-Aguirre, 1987). When the indigenous soldiers who returned after the war were again denied their citizenship rights, they developed deep frustration. They joined forces with the increasingly militant unions of mineworkers and started to push the government for reforms. They were helped by a number of sympathetic governments, the most famous of which was the short-lived administration of Gualberto Villarroel (1943 to 1946). During this period unpaid labour by Indians was formally abolished. In 1945 the first Indian Congress was organized, for the first time making the 'Indian problem' a national issue. Rural unrest became endemic in this period and the indigenous population became increasingly active, especially in the Cochabamba region. The revolution of 1952 can be seen as the culmination of this rural effervescence and the revolutionary reforms of 1953 effectively destroyed the foundations of traditional rural society.

Large-scale peasant movements also shook Peru in 1958 to 1964. They were no longer limited to one region as they had been in 1920 to 1923, but acquired national dimensions. This process of peasant mobilization undermined the hegemony of the traditional landowning class. In 1969 the progressive military government of General Velasco implemented an extensive agrarian reform. The Bolivian revolutionaries and the Peruvian military government eliminated the word 'Indian' and 'indigenous' from official documents, and replaced these terms with the label *campesino*. By doing so, they tried to implement the Bolivarian dream of race-blind politics in Latin America. However, at the same time they tried to reclaim and confirm the indigenous character of Andean societies. The Peruvian military, for instance, accepted Quechua as an official Peruvian language and redeemed Tupac Amaru as a 'national hero'.

In different ways, non-Indians continued to draw attention to the indigenous heritage of Latin America and the continuing importance of the indigenous perspective. The cultural undercurrent of indigenous Latin America (the Mexican author Guillermo Bonfil Batalla, 1990, called it *México profundo*) was increasingly considered the cultural alternative representing the identity and souls of Latin America. Nevertheless, it was clear that the sympathetic outsiders continued to act as the voices of the indigenous cultural heritage. It was inevitable that in the process they frequently resorted to old and new ethnic stereotypes and simplifications. Whether it concerned racial discrimination or romanticization, the question remained as to the authenticity and societal consequences of the representation.

The emergence of indigenous movements in the late 20th century

The last decades of the 20th century witnessed another transformation of the debate on race and ethnicity in Latin America. Social movements and non-governmental organizations (NGOs) created the breeding ground for movements composed of representatives of the indigenous population itself. This tendency acquired momentum with the disappearance of authoritarian governments in many Latin American countries and was reinforced by the increasing importance of vociferous international institutions that provided the indigenous and black populations with an internationally supported discourse of equality and indigenous rights. Social and political ideas that reached indigenous society from the national and international arena were appropriated by indigenous intellectuals and transformed into movements which no longer spoke *for* the Indians, but through the voice of the indigenous population itself (Brysk, 2000). This development was visible throughout Latin America, but it is clear that some noteworthy differences were apparent. Whereas we see active, influential indigenous movements in Bolivia and Ecuador, for example, in Peru the indigenous movement remained small and fragmented. Further research is needed to explain these differences; but there is no doubt that these developments led to a new political intensity in the debate on the position of Indians and blacks in many Latin American countries. A proliferation of indigenous movements and an invigorated political debate stimulated new discussions about the multicultural nature of Latin American societies and led to important legal and even constitutional reforms in a number of countries (see, for example, Wade, 1997).

The underlying paternalism that characterized *indigenismo* and similar reformist political projects should not close our eyes to the sometimes fundamental transformation they caused in the public views on state, ethnicity and nation. Precisely because they occurred simultaneously with, and partly in reaction to, structural changes within indigenous society, they created the foundation for new viewpoints that started to take the concrete reality of the indigenous population as point of departure and gave priority to the indigenous point of view. Indians migrated increasingly to the cities, which implied a new cultural and political insertion in mainstream society but also influenced their places of origin. Simultaneously, social and economic relations in the countryside also changed dramatically, causing diverse and often contradictory combinations of class and ethnicity. These changes led to new social and economic demands and concomitantly to an increasing visibility of the indigenous population in most Andean countries. Popular mobilization became a normal feature of Andean societies and rebellions and land invasions increased dramatically. Often linked to union activity and initiatives for agrarian reform, structural changes in the state stimulated new ideas and demands by indigenous peoples.

In a parallel process, the role of the state also gradually changed. Besides incorporating many *indigenista* ideas (and *indigenista* intellectuals), governments also resorted to land reforms and social and educational projects to integrate and modernize indigenous populations. Sometimes, when radical political movements managed to obtain control of the state, the role of the state became even more salient. This was visible in Bolivia in 1952, but also during the military 'revolution' in Peru after 1968. Although these radical experiments were usually short lived, they had far-reaching consequences because they managed to break the shackles of a strongly established status quo in the Andean countryside and to create new forms of organizing rural society. According to Albó (1999, pp192–193), the agrarian reform movement can be broken down into three broad phases. The first stage concentrated on improving labour relations within the existing system, without questioning the system itself. The second stage saw the break-up of the traditional hacienda system. This was clearest in Bolivia after 1952, but also received a strong impetus through the Cuban Revolution and the subsequent US programme of the Alliance for Progress. This led to the last phase characterized by 'gradualist' state-led agrarian reforms, such as occurred in Ecuador (1964 and 1973) and Peru (1969). This reformism had a strong technocratic, developmental flavour and focused on technological innovation, credit and associative forms such as co-operatives. In the field of development and water policies, this period not only reinforced the drive for hydraulic modernization, but also nationalized water resources, with a major impact. Ecuador and Peru enacted water laws that put all formal ownership of water – previously in private and community hands – under state control and bureaucratic management.

From the 1960s onwards, these national change processes were accompanied by an international quest for inclusion and multicultural nationhood and the opening up of new political and ideological niches. These new perspectives on citizenship and nationhood gathered force and dramatically changed the political context in which indigenous and Afro-American populations fought for their rights. Everywhere in Latin America, indigenous movements emerged that no longer accepted the intermediation of non-Indian groups to express their views. New indigenous movements and their leaders claimed the right to represent populations which had been exploited and discriminated against for centuries. They fostered movements combining a struggle for ethnic recognition with demands for social and economic rights.

Nowadays, these groups have become major players in the political arena. The Confederation of Indigenous Nationalities of Ecuador (CONAIE), the Aymara and Quechua movements in Bolivia and the Pan-Maya movement in Guatemala are just a few examples of the dynamics and contradictions of this modern indigenous struggle. They have caused human rights to be included as a fundamental part of the political agenda, which has led to comprehensive constitutional reforms in countries such as Bolivia, Ecuador, Colombia and Paraguay. They have also posed new questions about indigenous identity. These groups often represent only part of the population. They are sometimes accused of an alternative form of *indigenismo* or new forms of racism. It is still questionable how well they actually represent the indigenous population and what the role of non-Indian sympathizers is. These discussions suggest that the complex interaction between *indigenista* ideas and Andean societies is far from over and that issues of identity and intellectual ownership are certainly not yet resolved. In the early 21st century the debates around issues of authenticity and identity continue to be central issues in Andean politics. And as shown by this book's authors, they also pervade Andean water policies and politics.

Aside from the developments mentioned in the preceding pages, there are some clear explanatory variables that provide additional reasons for the emergence of these new indigenous movements. The first, and perhaps most important, condition was the process of democratization that occurred in Latin America after the 1970s. It created the political space for social movements to express their ideas and to propose ideas on the democratization of society (Peeler, 2003). The indigenous project of reconstituting relations between indigenous and non-indigenous sectors of society can be considered an alternative strategy to reach the goal of a democratic society. Van Cott (2000, p3) observes:

> *Contemporary indigenous peoples' organizations propose to reconstitute Latin American states upon truly democratic institutions based on the recognition of ethnic diversity and respect for the free will of individuals and pre-constituted groups to form a common political community. This project is framed in terms of international law with respect to the rights of peoples for self-determination.*

In the last sentence, Van Cott points to a second factor playing an important role in the emergence of indigenous movements in Latin America: the international connection. This connection can be seen at different levels. First, there is the level of internationally recognized human rights, especially the right to self-determination. Although these laws were interpreted quite differently and have been the object of much conflict, they were backed and implemented by the activities of various United Nations organizations. This made racist and exclusionary legislation increasingly difficult. International support and advocacy organizations have documented and combated racism and the exclusion of indigenous peoples and created global platforms for indigenous rights issues. This globalization of the struggle for indigenous rights had direct influence on the struggle and articulation of demands of Andean groups.

This international connection was even stronger within Latin America. The 'celebrations' of the 'discovery' of America in 1992 gave a sudden impetus to the inter-American cooperation of indigenous movements. Under the motto '500 years of resistance', the indigenous peoples of Latin America presented themselves as involved in one united struggle against colonialism and racism (Albó, 1999, pp826–827). Many indigenous groups tried to transcend national boundaries. Bolivian Katarismo led to the resurgence of Aymara activism in northern Chile. Meetings among Indians from various countries were held in Cusco, Tiwanaku and Quito and attempts were made to join indigenous groups in one organization. Andrés Guerrero (2000) has analysed how these events shaped the present-day indigenous movements in Ecuador. He shows how social movements and NGOs from the 1950s onwards created a breeding ground in which movements gradually came into existence that were composed of the indigenous population itself, directed by their own indigenous leaders. He demonstrates how social and political discourses reaching indigenous society from the national and international arena were appropriated by indigenous intellectuals and transformed into the groundwork for a movement that no longer spoke for Indians, but through the indigenous population's own voice.

Indigenous movements and contemporary politics

It would be impossible to describe thoroughly the varied history of these new indigenous movements and to analyse their background. Here we can only highlight some general characteristics that may explain why these movements meant a radical transformation but at the same time tended to reproduce some of the crucial problems and tensions visible in the relationships between indigenous people and the state from the early 19th century onwards.

First, it is important to stress that present-day indigenous movements address a variety of issues that far transcend ethnic identity *per se*. Although indigenous identity and culture play a significant role in mobilizing their constituency and legitimizing the political leadership, they also embrace issues of class, the majority of the indigenous people being rural or urban poor. The indigenous movements often embody new struggles for entitlements and redistribution with respect to land or water, prices of basic foodstuffs or social benefits. These movements generally aim at territorial autonomy and respect for customary law, but also the inclusion of the indigenous people in the legal system of civil and basic human rights. Significantly, they demand to be recognized as citizens with all rights belonging to that status. They may also insist on state protection against abuses by the police or the military, racial discrimination or violence by landowners, terrorist movements such as Sendero Luminoso or drug mafias.

Second, it is clear that these movements seek new forms of indigenous identity and authenticity. They look to the indigenous culture and history to find roots for new perceptions or alternative societal organization. This tendency is, at once, inward and outward looking. On the one hand, it leads to a new interest for the preservation or resuscitation of indigenous culture and customs. Cultural and religious leaders articulate discourses which aim at a meaningful exchange with the spiritual world and the ancestors. On the other hand, the new importance attached to indigenous culture aims to convince outsiders of the innate value of indigenous culture and its political and social potential. In a reversal of colonial domination and inequality, these movements provide alternative visions of the political world and legitimate new forms of internal organization and leadership. This element forms one of the main attractions for younger generations, who want to recreate their indigenous identities. Of course, these two tendencies are closely linked and lead to new forms of political expression. Rappaport (2004, p120) observes, for example, that the knowledge of religious leaders in Colombia 'provides a potent language for the politicized construction of cultural forms', which aims at destroying ideological colonization by external forces. This combined struggle for cultural, political and socio-economic rights may well be one of the essential characteristics of these movements, leading to hybrid forms of political and cultural activism with widely varying results. In Bolivia, this made for the radical nationalism of the Movimiento al Socialismo (MAS) led by Evo Morales, which stands in contrast to the ethnocentrism of the movement by Felipe Quispe or Pan-Mayan activism in Guatemala. Different as these movements may be, in all of them there is increased political visibility and a variety of interactions with traditional politics.

Third, these indigenous movements entered into new relations with the Latin American states. Sometimes they were supported by the state, as in the above countries, each of which established constitutional reforms and moved towards a multicultural political system. In other situations, the movements emerged in the vacuum left by a receding state or in reaction to the dismantling of state programmes. Yashar (1998), for instance, links the emergence of indigenous movements to the disappearance of government support for rural areas. All in all, there is no doubt that the relationship between the (national) state and the indigenous movement is always ambiguous and contradictory. There is, on the one hand, a strong historically grounded distrust between the different groups which goes back to ages of racial exclusion, repression and ignorance on the part of the state and national elites. On the other hand, no ready-made models are available on which interaction with indigenous populations can be based, and they must often be invented and negotiated from scratch. State policies constantly change and so does the international economic and political context. Indigenous movements are often fragmented and pushed forward by their constituencies. Their cohesion is often fragile and political success may lead to fragmentation and internal strife. This was clear in the case of Ecuador, where access into the formal political sphere caused havoc in the indigenous movement and led to severe fragmentation and weakening of its political leverage. In other cases, such as in Bolivia, the movement's participation in formal politics is continually challenged not only by racist elites and power groups, but also by indigenous and *campesino* groups clamouring for their rights. The problems that most indigenous movements face resemble the class-based movements of the first part of the 20th century. Only by making alliances and compromises will it be possible to obtain sufficient political leverage to change society. In doing so, leaders of these movements run the risk of losing the support of their indigenous constituency.

This phenomenon leads to the fourth element. Recent political processes in the Andean countries suggest that political success may already embody the germs of failure. Alliances with other political groups and rising expectations of indigenous constituencies make these movements vulnerable to encapsulation within mainstream politics and disillusionment among their followers. This problem may explain why issues of entitlements and rights, mentioned in the first point, have become so important in the Latin American indigenous movements. They allow them to pursue issues of indigenous identity while demonstrating that they are addressing some of the more pressing needs of their followers, needs that often have to do not only with issues of identity, but also with income, land and water rights, security and other basic issues.

Conclusions

It is not easy to make general remarks on the objectives and strategies of these new indigenous movements and their relations with sympathetic outsiders or the state. What this chapter has attempted to make clear is that all parties involved in the 'ethnic debate' in Latin America have constantly defined the terms in which this discussion has been pursued.

First of all, interaction between *indigenismo* and new Indian movements shows how hazardous it is to think about schematic dichotomies, in which the white or Western world is basically separated from indigenous culture. Simple contrasts (Indian/non-Indian; city/rural; modern/traditional) fall short in explaining complex historical processes of change and mutual influence. They tend to oversimplify the fragmentary, vague nature of cultural change processes. It will continue to be impossible to interpret indigenous society if we continue to view it as an isolated vestige of a faraway ethnic past.

Second, it is necessary to take into account the life cycle of these movements. Many movements initially relied on outside support and gradually developed into more or less autonomous social and political actors. Almost all of them started out as one-issue movements based upon their indigenous constituency and focusing on issues of indigenous identity and culture. In the course of their existence, they felt the need to extend rapidly their political and social programme. It was nearly impossible to advance the indigenous cause without touching on other themes. Inevitably, therefore, the indigenous movements came to concentrate on broader issues of rights, redistribution and entitlements, with a special interest in urgent issues of land, water, territoriality and collective rights over natural resources. Indigenous identities may have remained important markers for mobilization and political strategy; but social and economic demands were no less important. The resulting interaction between indigenous and class-based political action produces complex and sometimes contradictory forms of struggle.

Third, it is obvious that in the 21st century, national states and their political elites have lost their discursive and political monopoly with respect to the position of the indigenous population in the Andes. A number of issues existent during the 19th and 20th centuries remain unsolved and relevant to present-day analysis regarding the situation of indigenous populations in the Andean region. One of them is the definition of indigenousness, determined both by the indigenous populations and external forces such as the state, international agencies, tourist operators, and the national and international academic community. While it is now generally recognized that indigenous identities are flexible and non-essentialist categories, the struggle for meaning continues. The continuing discussion around 'authenticity' has only become more intense and urgent with the increasing political importance of the indigenous movements.

This also shows that the analysis of the relationship between indigenous populations and sympathetic outsiders continues to be relevant. Although it may be that the paternalistic *indigenismo* of the past has become extinct, the influence of images and political projects of non-indigenous outsiders remains important. Sometimes they can be helpful to indigenous groups in the sense that they provide financial means, technical support or political leverage; but this relationship is always ambiguous for both sides.

The principal ambiguity faced by the present-day indigenous movements appears to be the tension between their cultural identity-based demands and their focus on social and economic rights and increasing political participation. On the one hand, the indigenous movements have stressed their indigenous heritage and, based on that, proposed alternative ways of looking at the world. Different indigenous groups continue to look for new ways to represent their way of life and *cosmovisión*. The great question is how their culture and its social, cultural and ethnic elements have been fundamentally changed through this process. In other words, in what ways did the use and appropriation of *indigenista* ideas cause profound transformations in indigenous culture? It is probably impossible to find a single answer to these questions, but they are a key part of studying the indigenous world in Latin American consciousness.

On the other hand, the indigenous movement included a vast array of social and economic demands and objectives in their political programmes. Their search for secure rights to land and water may be the clearest example of this attitude; but their interventions in national politics, such as in Ecuador and Bolivia, are also signs of this tendency. Indigenous movements will no longer let themselves be isolated in the realm of worldview and identity. They wish to be accepted as full-fledged players in the political and socio-economic arena. They have been able to take part in the various projects of decentralization and *participación popular* introduced into the region during the 1990s. This has enabled many groups to bring forward their demands and to acquire local positions of some power. Simultaneously, they have managed to acquire a presence at the level of national politics, especially in Ecuador and Bolivia. The essential question faced by the present-day indigenous movements – and their supporters – will be whether these movements wish to be accepted as 'normal' political actors or whether they will continue to enter the political playing field as 'others', as political actors with different views, rights and obligations.

Notes

1 On *indigenista* thinking, see Davies (1970); Kristal (1987); Flores-Galindo (1988); Lauer (1997); Baud (2003).
2 Published in Marchán-Romero (1986). The article is taken from a publication from 1923, is dated 1912, but according to Guerrero, was first published in 1895. See Guerrero (1991, pp49, 55).
3 Circular no 26 by Ministerio del Interior y Policía [Ministry of the Interior and Police], 22 June 1898, ANH/C, 10895, Gob. Ad., F.8.

References

Albó, X. (1999) 'Andean people in the twentieth century', in F. Salomon and S. B. Schwartz (eds) *The Cambridge History of the Native Peoples of the Americas, Vol III, 'South America, Part 2'*, Cambridge University Press, Cambridge, pp765–871

Arze-Aguirre, R. D. (1987) *Guerra y conflictos sociales: El caso rural boliviano durante la campaña del Chaco*, CERES, La Paz, Bolivia

Assies, W. (1999) 'Pueblos indígenas y reforma del estado en América Latina', in W. Assies, G. van der Haar and A. Hoekema (eds) *El reto de la diversidad: Pueblos indígenas y reforma del Estado en América Latina,* Colegio de Michoacán, Zamora, Mexico, pp21–55

Baud, M. (1996) 'Liberales e indígenas en el Ecuador a finales del siglo XIX', in H. Bonilla and A. A. Guerrero-Rincón (eds) *Pueblos campesinos de las Americas: Etnicidad, cultura e historia en el siglo XIX*, Universidad de Santander, Bucaramanga, Colombia, pp239–257

Baud, M. (1997) 'La huelga de los indígenas de Cuenca, Ecuador, 1920–1921: Perspectivas comparativas', in L. Reina (ed) *La reindianización de América, siglo XIX*, Siglo XXI Editores, Mexico City, Mexico, pp223–250

Baud, M. (2003) *Intelectuales y sus utopías: Indigenismo y la imaginación de América Latina*, CEDLA, Amsterdam, The Netherlands

Bonfil Batalla, G. (1990) *México profundo: Una civilización negada*, Grijalbo, Mexico City, Mexico (originally published 1987)

Brysk, A. (2000) *From Tribal Village to Global Village: Indian Rights and International Relations in Latin America,* Stanford University Press, Stanford, CA

Casaús-Arzú, M. E. and T. García-Giráldes (2005) *Las redes intelectuales centroamericanas: Un siglo de imaginarios nacionales (1820–1920)*, F&G Editores, Guatemala

Castro-Klarén, S. (2003) 'The nation in ruins: Archaeology and the rise of the nation', in S. Castro-Klarén and J. C. Chasteen (eds) *Beyond Imagined Communities: Reading and Writing the Nation in Nineteenth-Century Latin America*, Johns Hopkins University Press, Baltimore and London, pp161–196

Clarke, A. K. and M. Becker (eds) (2007) *Highland Indians and the State in Modern Ecuador*, University of Pittsburgh Press, Pittsburgh, PA

Costales, P. P. de and A. Costales Samaniego (1964) 'Recopilación de las leyes indígenas de 1830 a 1918', in *Historia Social del Ecuador, Tomo III,* Instituto Ecuatoriano de Antropología y Geografía, Quito

Davies, T. M. (1970) *Indian Integration in Peru (A Half Century of Experience, 1900–1948)*, University of Nebraska Press, Lincoln, NE

de la Cadena, M. (2000) *Indigenous Mestizos: The Politics of Race and Culture in Cuzco, Peru, 1919–1991*, Duke University Press, Durham, NC, and London

Demélas, M. D. (1981) 'Darwinismo a la criolla: El darwinismo social en Bolivia, 1880–1910', *Historia boliviana*, vol 1, no 2, pp55–82

Escalante, J. A. (1976) 'Nosotros, los indios…', in M. Aquézolo (ed) *La Polémica del indigenismo*, Mosca Azul, Lima, Peru

Flores-Galindo, A. (1988) *Buscando un Inca: Identidad y utopía en los Andes*, Editorial Horizonte, Lima, Peru

Forment, C. (2003) *Democracy in Latin America, 1760–1900. Vol 1: Civic Selfhood and Public Life in Mexico and Peru*, University of Chicago Press, Chicago, IL, and London

Gonzales, O. (1996) 'Entre dos incertidumbres: La apropiación del indio en el discurso criollo', *Allpanchis*, vol 28, no 47, pp179–208

Guerrero, A. (1991) *La semántica de la dominación: El concertaje de indios*, Libri Mundi, Quito, Ecuador

Guerrero, A. (1994) 'Una imagen ventrílocua: El discurso liberal de la "desgraciada raza indígena" a fines del siglo XIX', in B. Muratorio (ed) *Imágenes e Imagineros: Representaciones de los indígenas ecuatorianos, Siglos XIX y XX*, FLACSO, Quito, Ecuador

Guerrero, A. (2000) 'El levantamiento indígena nacional de 1994: Discurso y representación política (Ecuador)', *Boletín Americanista*, vol 50, pp123–151

Irurozqui-Victoriano, M. (1992) '"¿Qué hacer con el indio?" Un análisis de las obras de Franz Tamayo y Alcides Arguedas', *Revista de Indias*, vol 52, nos 195–196, pp559–588

Irurozqui-Victoriano, M. (1999) 'Las paradojas de la tributación: Ciudadanía y política estatal indígena en Bolivia, 1825–1900', *Revista de Indias*, vol 59, no 217, pp705–740

Kristal, E. (1987) *The Andes Viewed from the City: Literacy and Political Discourse on the Indian in Peru, 1848–1930*, Peter Lang, New York, NY

Larson, B. (2004) *Trials of Nation Making: Liberalism, Race, and Ethnicity in the Andes, 1810–1910*, Cambridge University Press, Cambridge, UK

Larson, B. and O. Harris with E. Tandeter (1995) *Ethnicity, Markets, and Migration in the Andes: At the Crossroads of History and Anthropology*, Duke University Press, Durham, NC, and London

Lauer, M. (1997) *Andes imaginarios: Discursos del indigenismo 2*, CBC/SUR, Cusco, Peru

Lehmann, D. (1982) 'Introduction: Andean societies and the theory of peasant economy,' in D. Lehmann (ed) *Ecology and Exchange in the Andes*, Cambridge University Press, Cambridge, UK, pp1–26

Marchán-Romero, C. (ed) (1986) *Pensamiento Agrario Ecuatoriano*, Banco Central del Ecuador, Quito, Ecuador

Memmi, A. (1967) *The Colonizer and the Colonized*, Beacon Press, Boston, MA (French original published 1957)

Ministerio del Interior y Policía (1898) Circular no 26, 22 de junio de 1898, in ANH/C, 10895, Gob. Ad., F.8, Quito, Ecuador

Moncayo, A. (1986) 'El concertaje de indios', in C. Marchán-Romero (ed) *Pensamiento Agrario Ecuatoriano*, Banco Central del Ecuador, Quito, Ecuador,

pp287–326 (originally published 1895)
Peeler, J. A. (2003) 'Social justice and the new indigenous politics: An analysis of Guatemala, the Central Andes, and Chiapas', in S. E. Eckstein and T. P. Wickham-Crowley (eds) *What Justice? Whose Justice? Fighting for Fairness in Latin America*, University of California Press, Berkeley, CA, pp257–284
Rappaport, J. (2004) 'Between sovereignty and culture: Who is an indigenous intellectual in Colombia?', in M. Baud and R. Rutten (eds) *Popular Intellectuals and Social Movements: Framing Protest in Asia, Africa, and Latin America*, Cambridge University Press, Cambridge, UK, and New York, NY, pp111–132
Rénique, J. L. (1991) *Los sueños de la tierra: Cusco en el Siglo XX*, CEPES, Lima, Peru
Salman, T. and A. Zoomers (eds) (2003) *Imaging the Andes: Shifting Margins of a Marginal World,* CEDLA/Aksant, Amsterdam, The Netherlands
Sempat Assadourian, C., H. Bonilla, A. Mitre and T. Platt (1980) *Minería y espacio económico en los Andes, siglos XVI–XX,* IEP, Lima, Peru
Tandeter, E. (1993) *Coercion and Market: Silver Mining in Colonial Potosí, 1692–1826*, University of New Mexico Press, Albuquerque, NM
Van Cott, D. L. (2000) *The Friendly Liquidation of the Past: The Politics of Diversity in Latin America*, University of Pittsburgh Press, Pittsburgh, PA
Wade, P. (1997) *Race and Ethnicity in Latin America,* Pluto Press, London
Yashar, D. J. (1998) 'Contesting citizenship: Indigenous movements and democracy in Latin America', *Comparative Politics*, vol 31, no 1, pp23–42

6

Cultural Identity and Indigenous Water Rights in the Andean Highlands

Paul H. Gelles

Introduction: Water and identity in the Central Andes

In the Andes, irrigation water is an important component of production, a source of both conflict and cooperation, and a key element of cultural identity. These different dimensions of water must be viewed together. This chapter argues that indigenous Andean culture and cultural identity continue to be key elements of local social relationships in rural peasant communities, and that these cultural orientations and ethnic identities are intimately linked to agricultural and pastoral production, in general, and irrigation water, in particular. The reasons that in the Andean countries most national laws and politics of state intervention in management of highland hydraulic resources deny or ignore the existence of customary law and the traditional uses and management of water by indigenous communities have to be understood within a larger historical and cultural context. Indeed, water laws and policies imposed by outsiders using external criteria that ignore the Andean context are linked to the region's colonial past and to contemporary cultural politics.[1]

Providing a historical and anthropological understanding of the customary 'traditional' uses of irrigation water by peasant and indigenous communities in the Andean countries, this chapter argues for the need to recognize, politically and legally, these norms, rights and customary practices. Policies and laws should be flexible, respecting local cultural norms and autonomous control over hydraulic resources, as well as the tremendous variation found in traditional uses of water. To that end, we need to understand these norms and uses better.

Water use has been symbolically charged and highly ritualized for centuries in Andean society. Today, in the thousands of highland communities that have irrigation systems and that control vast territories in Peru, Bolivia, Ecuador and Chile, irrigation-based production continues to help constitute Andean cultural identities and orientations. Irrigation water in most highland communities is also generally a 'commons' – that is, a common property resource managed by a local community. Attempts to privatize water or otherwise alienate the control of water from indigenous communities thus challenge the communal control of water that exists in many Andean communities, as well as the cultural identity of millions of highland indigenous peoples.

In fact, the Andes are home to the largest indigenous peasantry on the continent: one out of every three indigenous people of the Americas is Andean. Apart from the many millions of Quechua and Aymara indigenous peoples, there are an enormous number of monolingual Spanish speakers who follow indigenous cultural orientations. It is no exaggeration to state that indigenous highlanders and their urban relatives constitute cultural majorities in most Andean nations (see, for example, Murra, 1982). Therefore, a dynamic and processual understanding of 'indigenous' identity, 'Andean culture' and the indigenous peasant community has a role to play in the formation of national water laws and irrigation bureaucracies in the Andean countries.

This chapter provides a historical and contemporary cultural-political overview of indigenous people, their water-based beliefs, practices and rights, and their organizations; of the way in which different normative frameworks exist in the Andes today; and of a key institution – the highland indigenous peasant community (and its different property regimes) – that informs and complements the particular case studies of each country. The following section reviews some basic definitions (indigenous, peasant, Andean culture) and the politics of representation in Andean studies. It compares two different ways of constructing ethnic identities in Andean society, and discusses the historical importance of water within Andean civilization and cultural identity. The third section is an in-depth look at the highland peasant community, its history and material basis in irrigation water, as well as different institutions and commonly held beliefs that are essential to its understanding. Building on these general understandings of Andean culture and society in relation to water, the fourth section addresses different facets of cultural politics and indigenous mobilization in the countries under study, and the relationship of these to the nation state and its irrigation policies and bureaucracies. The chapter concludes by suggesting strategies for achieving more respect and better recognition of local Andean water rights and norms in state water institutions.

Indigenous people, Andean culture and the politics of representation

The Quechua, Aymara and Spanish-speaking people living in the 'Central Andes' – that is, the highlands of Ecuador, Peru and Bolivia, are firmly tied to, and greatly affected by, national and international political and economic forces. Yet, they also transcend many national and international borders and participate in diverse social and cultural worlds. At the same time, members of this cultural majority have religious beliefs and rituals that are distinctly Andean, and which are tied to fundamental notions of community and ethnic identity, providing important meaning and identity for their lives. Native Andean religious beliefs and ritual practices have long been a fundamental component of local systems of agricultural and pastoral production – activities that sustain life. These beliefs and practices, many of which were forged in an indigenous colonial context (of the Inca), or even in earlier indigenous societies, have been under siege since the extirpation of idolatries in the early Spanish colonial period (see, for example, Salomon, 1991). It is this colonial legacy and its influence in the formation of the nation state in the Andean countries that partly explains why these beliefs and rituals continue to be ignored or denigrated by dominant cultural discourses and state-biased and neoliberal policy-making in the Andean nations.

Of course, the politics of water and indigenous identity in the Andes are part and parcel of the larger shared history of colonialism and the contemporary cultural and political oppression experienced by indigenous peoples in the Americas. Throughout the Americas, indigenous people and their cultural orientations are objects of great discrimination. National and popular discourses throughout Latin America, Canada and the US denigrate, marginalize or erase indigenous identities altogether. Sometimes the latter are appropriated for state building and 'national pulling together' (Urban and Sherzer, 1991), or they are exoticized to generate tourism. Such discourses and practices usually bring little benefit to the indigenous peoples themselves. Until recently, native peoples were infantilized, treated as legal minors and denied voting rights. The words 'Indian' or '*indio*' were routinely used as an insult, usually meaning mean, backward and stupid, and indigenous civilizations, cultures and contributions to contemporary society were consistently erased by national histories or reified as belonging to the past with no relevance to today's world.

This 'crisis of belonging' is especially pronounced in Peru, Bolivia and Ecuador, although the relatively small Aymara communities of northern Chile share many of the same cultural orientations and political processes. Not seldom, indigenous languages and cultures are marginalized from popular and national discourses. Despite new currents, the latter present the Spanish-speaking, white West-facing minority as the model of modernity, the embodiment of legitimate national culture, and the key to these nations' futures.

We must inevitably select and use some problematic terms to frame our discussion, such as Columbus's misnomer: 'Indian'. But the terms '*indígena*' (indigenous) and '*campesino*' (peasant) are used by different political regimes and opposition groups to designate rural dwellers in the Andes in fundamentally different ways and with varying consequences. In many national contexts today, as in the case of the indigenous '*resguardos*' in Colombia,[2] 'indigenous rights can be claimed only by those who authenticate their differentiated ethnic status in order to qualify for special laws and judicial mechanisms' (Guevara-Gil et al, 2002, p47).

As far as 'indigenous' goes, we find quite often that many approaches – from neoliberal to postmodern to Marxist – effectively deny the existence of indigenous peoples altogether (as former US ambassador to the United Nations Jeanne Kirkpatrick put it: 'whoever *they* are'). The United Nations defines 'indigenous people' in the following way:

> *Indigenous communities, peoples and nations are those which, having a historical continuity with pre-invasion and pre-colonial societies that developed on their own territories, consider themselves distinct from other sectors of the societies now prevailing in those territories, or parts of them. They form, at present, non-dominant sectors of society and are determined to preserve, develop and transmit to future generations their ancestral territories, and their ethnic identity, as the basis of their continued existence as peoples, in accordance with their own cultural patterns, social institutions and legal systems.* (Van Cott, 1994, p23; see ILO Convention No 169)

While not all indigenous peoples are peasants and certainly not all peasants are indigenous people, the Andean indigenous majorities of Bolivia, Ecuador and Peru are made up mostly of peasants. Indigenous peasant households send surplus labour to urban centres, and migration is one of many processes linking peasant communities to larger political and economic forces. While indigenous peasants are not static but, rather, 'on the move', and not remote but very much connected to larger economic, political and cultural processes, the term 'peasant' does link identity with the key form of livelihood (land husbandry) and its economic domination by relatively powerful outsiders.

It is crucial that we define and simultaneously problematize the notion of indigenous 'Andean' culture. This is necessary for a number of reasons: Andean culture remains a strong structuring force in highland society and the traditional uses of water; Andean peoples coexist with Amazonian indigenous people, who have a very different history of relations with the state in Peru, Bolivia and Ecuador; and Andean culture has been romanticized elsewhere in the development and anti-development literature, which often portrays Andean society as harmonious, communitarian and egalitarian, as well as timeless, remote and 'utopically' in touch with nature.

The concept of Andean culture, '*lo andino*', has recently come under attack from different quarters. As Abercrombie (1991, p97) puts it:

> *... to suggest the existence of* a *rural/indigenous culture in the Andes, what is often called in the literature 'the Andean' is usually to fall victim to non-Indians' essentializing stereotype of*

> *'the Indian'. In other words, the 'Andean' is only rightly studied as a [usually utopian] image projected by various urban groups.*

Questioning 'Andeanism' has been salutary for the field, forcing anthropologists and others to examine the dynamic movement and plural identities of highland inhabitants (see, for example, Starn, 1991; Albó, 1994; Kearney, 1996). So, too, greater attention has been directed to the ways in which the indigenous peoples of highland Ecuador, Peru and Bolivia are firmly tied to, and greatly affected by, national and international political and economic forces. Moreover, there is great variation within each national context: in some parts of each of the Andean nations, indigenous cultural orientations and ritual life are a vibrant part of everyday life; in other regions, they may be altogether absent.

Yet, the critique and subsequent devaluation of all things 'Andean' can also play into the dominant cultural discourses that effectively deny the validity of highland life-ways. This critique also ignores the many gains of ethnicity-based mobilization in Ecuador and Bolivia, and the potential political uses of '*lo andino*' in Peru. The fact is that while they participate in, and are affected by, diverse social, political, economic and cultural worlds, most indigenous highlanders also have similar beliefs and ritual practices that are distinctly Andean and, as shown below, which are tied to longstanding notions of community and ethnic identity.

As explored elsewhere (Gelles, 2000), Andean culture is best viewed as having been created from a hybrid mix of local mores with the political forms and ideological forces of hegemonic states, both indigenous and Iberian. Some native institutions are with us today, albeit in a thoroughly revised form, because they were appropriated and used as a means of extracting goods and labour by Spanish colonial authorities and those of the republican states after independence; others were used to resist colonial and post-colonial regimes. These institutions, reproduced and transformed through everyday practices, today vary greatly from one locality to the next. Andean cultural production is dynamic and adaptable, providing orientations and identity for villagers, as well as for migrants who transit different national and international frontiers.

Major cultural changes are taking place in the communities and cities of the highlands as they are transformed by transnational migration, the introduction of new information technologies, tourism and neoliberal economic reforms. However, Andean cultural orientations, social mores and communal politics condition these transnational and international processes in important ways. For instance, in a phenomenon such as transnational migration, communally based forms of understanding and Andean cultural orientations continue to exert a strong hold over transmigrants, even when abroad.[3] Contrary to what one might expect, recently introduced technologies such as television, VCRs and even 'telematic networks' are not just being used to spread the cultural images and products of the dominant society, but also to reproduce Andean culture. Certain cultural frameworks that are hundreds of years old –

as found in both water rituals and patron saint fiesta celebrations – still have a strong sway over community life, and these are dutifully recorded and diffused through the latest technology.

Competing ethnic frameworks: Water, mountains and the Earth versus the colonial categories of indio, mestizo *and* blanco

Water occupies an important position within Andean culture.[4] Irrigation water, in particular, which is central to production in many highland agricultural systems today, has long been one of the most culturally and ritually elaborated resources in Andean society and civilization; it has been used as a means of establishing identity for centuries. As Sherbondy (1982, p24) puts it: 'ancient Andean peoples did not only possess practical knowledge concerning subterranean hydrology, developing techniques for using these subterranean waters, but also elaborated a cosmology based on this knowledge which was useful for expressing concepts about ethnicity and political units'.

Indigenous peoples in the Andes, past and present, often trace their origins to sacred mountains, lakes, springs and rivers. This explains why the Inca nobility claimed the largest body of water in the Andes, Lake Titicaca, as the source of their imperial origins (see also Salomon, 1991). Irrigation water is generally conceived of as an extension of subterranean waters, the latter uniting highland lakes, rivers and mountains with the 'mother lake' (*mama qucha*) – that is, the ocean (Sherbondy, 1982, 1998). Together these form a hydraulic network through which the gods and ancestors travel, and from which the human world originated.

Together with the Earth Mother (*Pacha Mama*), mountains are conceptually linked and often ritually celebrated together with irrigation water; indeed, the importance of mountains within this symbology of power also needs to be stressed:

> *Whatever the ultimate source of water, the mountain deities at the local level were usually the controllers of it. They thus were the ones who directly affected the fertility of crop plants, animals, and – in the end – people.* (Reinhard, 1985, p418)

In sum, this Andean cultural logic and definition of ethnic identity, intimately tied to water and mountains, posits a strong bond between a 'place deity' (Salomon, 1991), the territory it controls, and a dependent social group.[5]

Andean ethnicity was irreversibly changed with the Spanish invasion in 1532, when the Spaniards introduced the category 'Indian' (*indio*), gathering under one term a great number of ethnically distinct peoples; the term was used to define the conquered, those to pay tribute and service. The colonial categories of *indio* (Indian), *español* or *blanco* (Spaniard or white, respectively) and *mestizo* (mixed race) have been fundamental to Spanish and republican political economies, and they still exert strong power throughout the Andes.

This kind of ethnic identity formation, as in many European colonial contexts, finds its origins 'in the asymmetric incorporation of structurally dissimilar groupings into a single political economy' (Comaroff and Comaroff, 1992, p54).

The established form of differentiating and ritually marking social groups through native Andean religion continued during the Spanish colonial period, although the colonialists had their own racial categories and ethnic identities. Ethnic identity, then, often became reformulated as a community-based identity, and this was tied to the sacred geography and the patron saints of each town. Among the most explicit expressions of this community-based identity are the myriad public rituals dedicated to water. Some irrigation-based communities in the Andes have little, if any, ritual life; yet there are many who do. Extensively documented throughout many parts of the Andes are the large, communally sponsored, week-long ritualized canal cleanings, known as the Water Fiesta or 'canal scraping' (*Fiesta de Agua*, or *champería* in Spanish, *yarqa aspiy* in Quechua). During the seven or eight days of the Water Fiesta, canals and reservoirs are cleaned and ritually cleansed; offerings are made to the ancestors and local mountains, springs and lakes; ceremonial divisions surface; and ritual competition in dancing, song-writing and feasting takes place between the divisions (Gelles, 1984, 1998). This longstanding ritual complex is widespread and is found in Ecuador, Peru and Bolivia.[6]

Water ritual in the Andes takes many forms. Sometimes irrigation water is ritualized throughout the year or water cycle rather than concentrated in a single week-long celebration. The symbolic elaboration and metaphorical uses of water, while sharing many common elements across regions, can also vary considerably. For example, water is gendered by indigenous highlanders and sexual symbols pervade irrigation ritual. In some regions water is conceptualized as a male force, likened to blood or semen, which fertilizes the mother Earth, while in others, water is conceived of as feminine, coming from 'mother mountains' and likened to mother's milk (Gelles, 2000). Despite regional differences, the symbolic and ritual elaboration of water is a strong structuring force in highland society in many parts of the Andes. Indeed, irrigation rituals constitute a powerful medium for transmitting and reproducing beliefs about fertility, disease, authority and ethnic identity. In Andean society, where the 'communication of deeply held values and of ideas about the order of things is accomplished by preference in ritual acts – that is, in symbolic action and not verbally' (Rasnake, 1988, p220), ritual holds special importance. In accordance with Tambiah's (1985) observations, irrigation ritual in the Andes, then, actively produces practices and policies that constitute social reality.

The foregoing discussion identifies two distinct processes of identity formation that affect irrigation practices and politics in the Andean nations. The first process has to do with an understanding of community, ethnic categorization and production in terms of the larger political economy of the colonial period and the contemporary cultural politics of the state. Indigenous mobilization and the politics of water operate today in the context of racialized identities, discrimination against 'Indians', and differential incorporation by Andean cultural majorities.

Ethnic stratification and the politics of cultural pluralism are, of course, intimately linked. The colonial categories and racist attitudes that were instituted in the countries under study during the colonial period, and which were intimately linked to tribute exactions, have survived with virulence. Today, the term *indio* is part of a racial idiom used to express class and ethnic differences among individuals, and the resonance of whiteness with power is strong throughout the Andean nations (see, for example, Weismantel, 1989; Valderrama et al, 1996). As Abercrombie (1991, p96) says: 'Given their advantage in force, it is not surprising that aspects of the colonizer's value systems have become hegemonic, so that the stigma attached long ago by Europeans to "Indianness" has worked its way into "Indian" self-consciousness as well.' Only during the last decades has the content of the word 'Indian' been inverted and self-proclaimed by indigenous groups who have reappropriated the term for their own strategic purposes.

The second process of identity formation, which is much older and which has to do with finding identity in water and mountains, shows that the politics of identity and ethnicity in the Andes cannot be reduced to the differentially positioned usages of *indio*, *cholo*, *mestizo* and *criollo*. On the one hand, because of the ways in which native Andean religion and ritual practice atomizes power into thousands of mountains (and water sources), each of which has a subject population dependent upon it (and a patron saint) for fertility and prosperity, there is an almost endless differentiation of ethnic identity among social groups in the Andes. On the other hand, it is the shared conviction that mountains and the Earth possess spiritual properties that along with language, dress, diet, etc. differentiates indigenous peoples from mainstream *criollo* society, a society that denies the validity of these Andean cultural orientations – the term *criollo*, which was originally used to designate people of Spanish descent born in the Americas, refers to the 'social unit defined in cultural and class terms ... that has directed state operations since the birth of the republic' (Turino, 1991, p260).

Because indigenous people are denied opportunities within these countries, many individuals choose to sacrifice their cultural orientations and their ethnic identity to achieve greater prosperity and status. The system of stratification racializes class and ethnicity, and is quite fluid. As Fuenzalida (1971, p20) puts it: 'The higher one is on the social ladder, the whiter one appears; the lower one is, the darker.' While the change to *mestizo* status may bring about economic and social progress for the individual, it helps to reproduce the structures that marginalize and stigmatize Andean culture and identity. These same hegemonic structures affect state, regional and local irrigation practices and politics, where issues of cultural identity and agricultural production intersect in the community.

The highland community and water

To understand indigenous rights over water and its relationship to Andean cultural production requires us to understand the highland community, 'the principal source for the reproduction of Andean identity' (Ossio, 1992, p249). This is not to say that Andean cultural production does not take place in the cities of the Andean nations; it most certainly does. But the relationships, social and spiritual, which have come to define *lo andino* in relation to water, are mostly generated in rural community settings.

In Peru alone there are 5700 officially recognized peasant communities, which control over 18 million hectares of land. Indeed, as Mayer (2002, p37) put it, 'communities are an important and growing component of rural settlement patterns and tenure arrangements, and not, as is often expressed, a declining left-over of traditional backward rural systems or an empty shell of colonialism that should be abolished'. Alternatively called 'indigenous communities' (*comunidades indígenas*), 'peasant communities' (*comunidades campesinas*), *resguardos* or *comunas*, officially recognized communities usually have legal personhood that allows them to challenge powerful insiders or outsiders who threaten their resources, including irrigation.

Many of the communities which dot the landscapes of the Andean nations, as well as the beliefs and rituals that link local identity to production, must be understood in terms of irrigation development, which has a long history in the Andes. Together with camelid herding and the vertical control of different ecological niches, terracing and irrigation facilitated the development of the Inca Empire and other pre-Columbian states in the rugged environment of the Central Andes. Covering countless thousands of mountain slopes, the canals and terraces that indigenous states and local polities built are truly monumental structures and represent millions of human days of labour. Unlike the pyramids, palaces and fortresses that pre-Columbian polities also built, the congealed labour found in terraces and canals has reproduced their investment in this form of 'humanized nature'.

Indeed, the political economy of extraction in the Andes has long been bolstered by an extensive and longstanding agricultural infrastructure; the surplus production provided by irrigation helped to fuel state power after the Spanish invasion and after independence. In contrast to the neighbouring Amazon Basin, extractive industries in the Andes and on the coast, such as those based on silver, guano, oil and other resources, have drawn on highland communities that possess well-established territories and relatively stable systems of agricultural and pastoral production. These productive systems have been the material basis for maintaining a large peasantry and cheap labour force, and for extracting foodstuffs, tribute and taxes over the last 500 years.

At the same time, it is important to understand that this irrigation infrastructure has been producing nowhere near its potential; the productivity of these systems diminished considerably with the great population decline that followed the Spanish invasion, when close to three-quarters of the pre-Columbian terraces were abandoned (Masson, 1987), as were most of the irrigation canals. Over the last century, the rapidly growing highland population has put pressure on communal resources and has engendered attempts to recover some of the lost infrastructure.

While the irrigation infrastructure constitutes a key element of the material basis for many communities and villages, we must also understand these rural settlements historically and culturally. As mentioned earlier, Andean cultural orientation and identity formation link people, place and production within particular communities. Salomon, for example, shows how in the early colonial period the use of the term *llaqta*, which can be glossed as 'town', expressed a strong bond between what he calls a 'place-deity' or 'deity-locale', the territory it was imagined as controlling, and the group of people who depended on this territory and who were favoured by the local deity (Salomon, 1991). Such an identity was reconfigured through the colonial period and through the fusion of Andean and Iberian beliefs, practices and institutions. Today, towns are represented by a patron saint and other lesser saints, and these play an important role in defining personal and communal identities in the highlands. As providers of fertility and life for crops, livestock and people, as well as of disease, death and destruction, these different protector spirits and emblems of communal identity must be placated by ritual offerings, libations and religious celebrations (e.g. fiestas). The prosperity of each family, village and community is to a large degree seen as depending on frequent 'gifts' to local mountain deities, the Earth mother and assorted figures in the Catholic pantheon. This is a key feature of life in the Andes, defining ritual practice and social life, as well as cultural and ethnic identity.

At the same time, as was observed above, the Andean community is clearly the product of the 'colonial matrix' (Fuenzalida, 1970) that brought together indigenous Andean and European social, cultural and political forms to constitute a new and unique entity. Many of the highland communities were established in the late 16th century as locally dispersed populations were resettled by the Spaniards into nucleated settlements (*reducciones*) for the purposes of tribute assessment, social control and religious indoctrination. As Abercrombie (1991, pp95, 96) says:

> *... centuries of colonial domination (and resistance to it) have produced many hundreds of small, community-sized 'ethnic groups', centred on 'county seats', towns in which pre-Columbian populations were forced to settle... [They] generally define themselves as members of a local group, coterminous with town-territory, and beyond it, as citizens of the province and department defined by the nation state to which they also pertain.*

Since independence in the early 19th century, successive national governments have see-sawed back and forth between two positions:

1 'emancipating' (i.e. severing) indigenous populations from their communal identities and collective forms of organization; and
2 providing official recognition and legal protection to communities.

During the early Republican period, 'progressive' thought in Latin America held that 'Indians' should be incorporated as citizens within the country, and intermediary forms of association – such as ethnic groups and communal identities – should be done away with. By the same token, communal ownership and organization were seen as obstacles to progress. This mindset led to reforms that often took away the only safeguards that peasants had against predatory haciendas, mines, etc. Such reforms were often later offset by official recognition of communities and their rights over their resources (including water), as well as by land reforms that both strengthened the legal recognition and status of some rural communities while, at the same time, imposing top-down water, land and territorial rule-making.

Today, however, with the neoliberal reforms and privatization policies that are sweeping Andean nations such as Peru, Chile and other parts of Latin America, we see the pendulum swinging back to the policies of the early Republican period, with these economic reforms threatening again to bring back the concentration of land, the removal of communal safeguards, and attacks by different industries, private agents and government agencies on indigenous highland communities, their common property regimes, their rights over water and their cultural identity.

Reciprocity and the commons: Families, community, state

In order to understand indigenous water management and the highland community, it is necessary to understand two key features of highland life: reciprocity and communal control over the individual. It is also important to mention that there are different kinds of property regimes in highland communities. While families and individuals generally hold land privately, and it can be bought, sold, rented or sharecropped (although sometimes only to other members of the same community), some lands are communally held and cultivated in many highland communities.

More important for our purposes here, irrigation water throughout the Andes is a form of common property. The latter, understood as a type of property relationship in which an identifiable community of interdependent users controls a particular resource, usually excludes outsiders while regulating use by members of the local community (see, for example, McCay and Acheson, 1990). In this sense, irrigation water and its regulation has much in common with rangelands, shellfish beds, fisheries, forests and other resources managed by communities.

The key player in the commons game is the Andean household: the constituent unit and building block of highland social organization. The Andean household functions as an economic institution. Peasant agricultural production, 'though linked to and affected by world capitalism, in some ways remains outside of its main transformative thrust' (Mayer, 2002, pxiv). The household can be understood in terms of several ideal models: an economic institution with a certain internal unity and division of labour, as well as boundaries through which goods and services flow in and out; as a set of culturally prescribed social roles associated with kinship and gender; as a productive process; or as a site where household members exchange food, affection, work and other culturally prescribed items.

Andean households have a great capacity to work collectively. At the same time, individual agency operates within this process: 'the socialized institution sets the stage as to how the rules of the game are to be played by giving the mode of transaction a cultural content, whether it is a reciprocal labour exchange, or a barter partnership, or a sale; but it leaves room for the pursuit of personal interest in each case' (Mayer, 2002, p37). Reciprocal exchanges also take place between the family and the community and are perhaps the key process within 'the commons'.

While conflict is as much a part of community life as cooperation, reciprocity – understood as a relationship, sometimes symmetrical and other times asymmetrical, between families and between families and the community – is central to life in the highlands (see, for example, Alberti and Mayer, 1974; Mayer, 2002). Symmetrical forms of reciprocity in the Peruvian Andes include *ayni*, which refers to exchange labour or the equal exchange of goods or services between individuals or households and some forms of *minka*, as when several individuals help a person to build or roof a house (this is sometimes composed of several *aynis*). Sometimes these 'symmetrical' forms of reciprocity are, in fact, manipulated by relatively powerful individuals in asymmetrical ways that lead to unbalanced, extractive relationships (Mayer, 2002).

A clearer form of asymmetrical reciprocity is found in *faena* and *corvée* labour. Communal *faenas*, or obligatory work parties directed by communal authorities, are found in many non-industrial societies, past and present. It is, in a sense, a labour tax levied by rural communities on their members. In effect, in return for complying with *faena* labour in works that supposedly benefit the collective good (such as for maintaining irrigation infrastructure, building roads, schools, etc.) and with *cargo* obligations (a type of civic duty in which townspeople are appointed or volunteer to fulfil a community office, which is a rotating position of authority), each household gains rights to the common property resources of the community, such as the medicinal plants, firewood and pasture lands of the high reaches, as well as irrigation water.

The relationship is asymmetrical in that the community has power over the individual and can constrain his or her behaviour through negative incentives and sanctions, including fines, jail, removal of water rights (Boelens and Hoogendam, 2002) and banishment from the community. It can also provide positive incentives such as fiesta participation or by offering food and drink in festive or ritualized *faenas*. The relationship becomes even more asymmetrical when *faena* labour is used to benefit powerful individuals in the community or when communal labour is used by the state, thus becoming *corvée* labour (as in road-building).

The principle of the commons is manifest in relation to canal cleanings and irrigation-related *cargos* that irrigators must fulfil in order to gain access to irrigation water. Indeed, the 'hydraulic *faenas*' enforced by communal authorities are one of the most productive means by which old irrigation infrastructure is maintained and new infrastructure is created (Boelens and Gelles, 2005). This kind of labour investment is commonly linked to the generation of water rights as well as the creation of water rights when new systems come into being (Gelles, 1986). In short, participating in the maintenance of irrigation infrastructure has long been the key to having the right to irrigate (Boelens and Doornbos, 2001).

As a common property resource, individual access to irrigation water is usually managed and controlled by the community directly or by a village-wide water users' association. These associations, usually but not always made up of men, not only oversee canal cleanings and the distribution of water, but often determine the planting schedule and the cycle for other agricultural activities, such as preparation of the soil, weeding and buttressing of the plants. The activities of the individual, then, are often constrained by the irrigation calendar and decisions established by communal authorities. This is even more the case in sectoral fallowing systems, which sometimes highly constrain individual behaviour and decision-making.

The other side of the commons and community control is that these communities have legal personhood that allows them to challenge not just powerful insiders, but outsiders (mining companies, haciendas, etc.) who threaten their resources, including irrigation water. Indeed, while communal institutions are, at times, subject to abuse by powerful families within the community, these institutions are generally quite effective in allowing local community members and water users to regulate their resources in a relatively democratic and sustainable fashion.

Indigenous mobilization and the cultural politics of water

The common historical, political and cultural processes discussed above relate to an indigenous empire that further developed an already extensive infrastructure, the loss of much of this infrastructure and resettlement of rural populations in *reducciones* during the Spanish colonial period, the 'Andean' type of identity that conceptually links people, place and production to local deities (both native Andean and Catholic), and common property with power of the community over the interests of powerful insiders and outsiders. Overlying this are the separate nation states or 'imagined communities' (Anderson, 1983) of Peru, Bolivia, Ecuador and Chile that developed since independence. The history of indigenous mobilization in Latin America also has implications for Andean nations (see also Chapters 3, 5, 15 and 16 in this volume).

In Latin America, the indigenous rights movement includes a variety of groups working for change in the status and condition of indigenous peoples, claiming for both socio-economic redistribution and their recognition as a distinct cultural group (Van Cott, 1994). While the winds of change were sown during the 1940s, the early 1970s saw the first large wave of indigenous organizations, as well as the emergence of international advocacy groups. The 1980s was generally a time of transition from authoritarian to 'democratic' rule, and in many countries indigenous peoples gained voting rights. According to Van Cott, the fall of socialism and the decline of the Latin American left cleared a space among popular organizations for ethnicity-based mobilization. Brysk (1994) emphasizes the panoply of emerging transnational social movements – such as those concerning human rights, women's rights and the environment – of which the indigenous rights movement has been one of the most successful.

There are conflicting goals and ambivalent ties between ethnicity-centred indigenous movements and peasant organizations, labour unions and armed revolutionaries. Class-centred organizations often trivialize the cultural orientations and ethnicity-based demands of indigenous groups. Nevertheless, liberation theology, peasant mobilization and other left political and intellectual movements were crucial influences on indigenous militancy. Although political reform is a shared goal, indigenous organizations often have a different set of priorities and demands, such as self-determination, cultural legitimacy in the public sphere, and rights over land, water and other territorial resources.

As Brysk (1994, p33) shows, the 'diverse needs of a pan-American movement eventually coalesced around the concepts of self-determination and "ethnodevelopment": informed self-management of cultural and social change... But states tend to interpret the social movement's call for self-determination as a threat to sovereignty rather than a deepening of democratization.' Because indigenous organizing often opposes not just state policies but the elite economic interests that they support, such movements are often met with violent repression by the army. Yet, the state does not have a monopoly on armed forces, and hundreds of indigenous leaders and tens of thousands of indigenous peasants have been killed by private militias and in the crossfire among revolutionary movements, drug traffickers and the army.[7]

Besides constraining factors at the national level, there are legal and institutional obstacles facing self-determination initiatives at the international level.[8] While indigenous groups often have faced major constraints when trying to materialize their interests at venues such as the United Nations and the Organization for American States, they have forged important alliances with non-governmental organizations (NGOs) and the 'transnational regimes' that develop between these (Brysk, 1994; Getches, 2006).

Andean rights movements

In Ecuador and Bolivia, indigenous peoples from both the Andes and the Amazon have made significant political gains. In the case of Bolivia, the 1994 constitution declared Bolivia a 'multi-ethnic and pluri-cultural' country. Still, the tension between neoliberal and indigenous political interests is great and there has been a wide breach between official rhetoric and practice.[9] As Gustafson (2002) argued in his review of Bolivian 'neoliberal inter-culturalism' and 'liberal indigenism', neoliberal economic measures over the last decade have often been accompanied by legislative measures that ostensibly recognize indigenous languages, identities and organizations. Yet, despite the important gains by popular and indigenous organizations in bilingual and inter-cultural curriculum reform, municipal decentralization, collectively entitled indigenous territory and other areas, elitist and conservative political discourses routinely disparage indigenous leaders and their movements in racist terms. For instance, in the different forms of social mobilization and 'water wars' that erupted in 2000:[10]

> *... urban commentators called for government action, and the press adopted a 'we' the urban, modern Spanish-speaking Bolivia against 'them', the anti-systemic, anti-modern, violent racial other... Urban non-indigenous ideas about the 'Indian' are still mired in colonialist discourses, seemingly incapable of processing the reality of new indigenous protagonisms.* (Gustafson, 2002, p288)

Thus, even when Aymara militants were demanding university education and tractors, their protests were characterized as a 'violent rejection of modernizing processes' (Gustafson, 2002, p289). Elements of such racism come powerfully to the fore in the current conflicts among the indigenous highland provinces and the white and *mestizo* lowland provinces, penetrating the debates on regional autonomy, the distribution of national resources and the process of developing a new constitution.

Despite these prevailing stereotypes and the way in which reforms such as municipal decentralization have the potential to sap the strength of indigenous movements in Bolivia, the latter remain well rooted and are expanding their influence. In addition, the 2002 elections brought an overwhelming number of indigenous and peasant leaders into congress. And in 2005, with support of the indigenous movement, the first indigenous president, Evo Morales, was elected. His re-election in 2009 confirmed indigenous and popular support. Significantly, new irrigation legislation was enacted that (though with differing results) aims to foster the water rights of the indigenous water-user communities.

In Ecuador, like Bolivia, indigenous movements have made a political space for themselves, at first receiving impetus from their class-based counterparts, but eventually formulating their own unified demands. Since 1980, a national federation of indigenous people united the ethnic groups. With its nation-paralysing strikes, and demands for bilingual education and state and legal reform, since the early 1990s, the indigenous movement in Ecuador has become one of the most organized and institutionalized of any movement in Latin America (MacDonald, 2002, p176). For many years, the Confederation of Indigenous Nationalities of Ecuador (CONAIE) has effectively linked indigenous peoples from all regions of the country and it maintains an extraordinary capacity to mobilize the indigenous population. Mobilizations paralysed much of the country throughout the last decade, opposing neoliberal agrarian and economic policies and privatization-oriented land and water reforms, while struggling for land and water rights, cultural recognition and a new constitution.

CONAIE and other indigenous organizations are fighting for a 'plurinational' state, one that moves beyond multiculturalism or respect for cultural differences to a federated political mosaic – that is, a state made up of constituent 'nations'. In 2003 the party of the indigenous movement swept into power, installing indigenous individuals in high-ranking positions, such as the minister of agriculture and the minister of foreign affairs. But this formal indigenous state government participation was short lived; the existing power structures and the populist and neoliberal policy reality soon made the movement's government participation end, and even left the indigenous seriously injured, now to recover 'outside the realm of official state power'. The 1998 constitution and especially the 2008 constitution are major advances for the recognition of indigenous rights, although actual changes in everyday society and in the prevailing power structures clearly show that reform is not just a matter of 'changing laws'.

The gains in Bolivia and Ecuador are mitigated by the fact that indigenous resources, leaders and organizations are still under attack and weakened by internal conflicts. But the way in which activists have organized along ethnic-based lines in these nations is virtually inconceivable in Peru (see, for example, Remy, 1994). In Peru, indigenous movements and organizations are mainly found in the Amazon region, less in the Andean highlands. Indigenous organizations there have only recently begun to have some impact upon national politics, and most of these represent the interests of Amazonian 'native' communities (Dean, 2002), as opposed to the large concentration of indigenous people found in Andean 'peasant' communities. Although the Peruvian state has generated organizations and policies for indigenous peoples, 'most of the sectors to which such policies are directed do not identify, organize or mobilize as indigenous peoples, nor do they raise ethnic grievances' (Remy, 1994, p108). Compared to Bolivia and Ecuador, indigenous movements in Peru lack articulation at the national political level, leading one researcher to state that the idea of an 'indigenous Peruvian' is an oxymoron (Dean, 2002).

Unlike Ecuador, peasant communities in Peru have not organized into ethnic-based indigenous federations, but rather peasant federations (see Boelens et al, 2005). The reasons for this have in great part to do with the fact that the 'military radicals' of the Velasco regime (1968 to 1975) largely 'peasantized' the highlands of Peru (see, for example, Mayer, 1994) – that is, made the class-based term of *campesino* the predominant idiom for discussing rural dwellers; the regime replaced an ethnic designation, *indígena* (as in *comunidades indígenas*) with a socio-economic one, *campesina* (as in *comunidades campesinas*). This is part of the mixed legacy of the Velasco regime, which (as with the military of Ecuador mentioned above) brought about important reforms that benefited many indigenous peoples. The regime carried out massive land reform, made Quechua an 'official' language, and ruptured Peru's colonial past. Indeed, the 'democratically' elected regimes since that time have rolled back many of the gains of the early Velasco years, and there has been great political upheaval and a steady decline in the standard of living.

In 1980, the change from military to 'democratic' rule was accompanied by the rise of the Shining Path subversive movement. The latest figures reveal that about 70,000 people, mostly Andean peasants, died in the brutal war between Peru's military and Shining Path. Shining Path, while using Quechua as a language of recruitment and with a rank and file largely made up of individuals from an indigenous peasant background, also eschewed indigenous politics and used the peasantist Chinese model and class-based language to characterize their revolution and the cause of rural dwellers.

From 1990 until 2000, an authoritarian state, Fujimori's civil-military regime, brought about drastic privatization of state-run industries, facilitated in part by 'the forging of an alliance with a Limeño business-military caste that appropriated the state apparatus for its own material and ideological ends' (Dean, 2002, p204). Human rights violations, pervasive corruption and militarization had extremely disruptive consequences for indigenous groups. So, too, the neoliberal legislation introduced by the Fujimori regime 'countermanded constitutional and statutory protection of indigenous peoples' land rights established under previous governments ... private and commercial interests with sufficient resources and political influence have been very successful at subverting the spirit of those laws designed to defend the rights of indigenous communities' (Dean, 2002, p213). Fundamentally, this was perpetuated by the Toledo government and now, in particular, by the current neoliberal government of Alan García that continues to proclaim its eminent domain over resources in order to provide legal security of concessionaires who are granted utilization rights. The lack of indigenous input in these concessions and policy-making, in general, continues.

Marginalization, Andean cultural rights and water policy

While Andean nations have different histories and factors affecting indigenous mobilization, the fights for Andean cultural rights and the legal recognition of community and territory affect any discussion of indigenous rights and water law. The exclusion of indigenous communal and cultural rights to use and manage irrigation water has to be seen in terms of a larger cultural-political dynamic. Rights are denied when a dominant cultural minority deploys its worldview through its educational system, civic ceremonies, language, vision of development and water policies. Water policy and the institutional cultures generated within irrigation bureaucracies are part of this same cultural-political dynamic. The particular kind of imagined community or nation state within which they develop inevitably shapes water law and policy-making. In the Andean countries, it is *criollo* culture (and, to a degree, *mestizo* culture) that holds centre stage in nation-building.

Cultural politics must also be understood in terms of the different normative frameworks that simultaneously exist in water management in the Andes today. Officially, all water in the Andean nations belongs to the state, which ostensibly has the right not only to decide the uses and allocation of water, but also the organizational models by which it is managed. Highland communities dispute the state's control over 'its' water and often refuse to allow the state to determine local irrigation practices. As in the case of Peru, conflict between local and state models of resource management is found in a great number of highland communities (Gelles, 1998, 2000). These models embody fundamentally different historical processes and represent extremely different ways of conceptualizing and implementing water management. One takes a secular, bureaucratic, and increasingly individualized and monetized view of water management. Local models (such as those of the water fiesta discussed above) are focused on ritual assurance and view water as part of a larger social and symbolic universe that is part of larger Andean cultural orientations.[11] These differing state and local normative frameworks and organizational models for distributing water may be used in the same community. Sometimes state models successfully supplant local models of irrigation; but typically the displacement is only partial. In many cases, local models of irrigation successfully resist state intervention altogether.

The adherence to local models of irrigation in the Andes exists within the cultural politics of the nation state, where state policies reflect and deploy the political will and cultural hegemony of a dominant ethnic group. As Herzfeld (1992, p3), in a different context, puts it: 'Nationalist ideologies usually lay claim to some kind of constructed "national character". Their bureaucracies have the task of calibrating personal and local identity to this construct.' As such, bureaucracies are directly tied 'to long-established forms of social, cultural and racial exclusion in everyday life' (Herzfeld, 1992, p13). They treat particular individuals and groups differentially, depending on whether they are seen as sharing the bureaucrats' social world and cultural orientations. This is often expressed with metaphors of race and blood lines.

The ideology of nation-building in these countries and throughout the Americas in irrigation matters and otherwise is thus shot through with racism and is built upon the assumption that indigenous people and their cultural orientations (including their languages, irrigation rituals, agricultural systems, as well as general life-ways involving food, music, etc.) are backward and that *criollo* culture is 'modern' and thus more rational, efficient and superior. One of the key tenets of modernism is that Western science constantly produces more efficient technologies and ways of organizing, and that 'cultural differences will fade away as people discover the effectiveness of rational Western culture' (Norgaard, 1994, p7). Irrigation politics and the imposition of state models of development in highland communities must thus be viewed in terms of a 'long tradition of Western thought which holds that ethnic attachments are irrational and archaic and ought therefore to evaporate as the world moves towards greater modernization and rationality in the conduct of its affairs' (Maybury-Lewis, 1982, p220). In short, the ongoing clash between state and local models of irrigation is conditioned by a racist colonial legacy and by the larger cultural politics in the Andean nations today.

But state officials and models of resource management have a hard time displacing local models, tied as they are not just to resource management and production, but also to Andean cultural frameworks that inform local and ethnic identities, and to the diverse political-organizational levels that link Andean water-user families and communities, from the local community to the national federations and international rights movements.

Conclusions

In the highland community of the Andes, irrigation and water development are linked to ethnic identity formation and are affected by the cultural politics of the Andean nations. The symbolic and ritual importance of water creates a strong structuring force in highland society and is tied to Andean communal and ethnic identities. Yet, the racialized categories of people and colonial legacy in Bolivia, Peru and Ecuador have contributed to a 'crisis of belonging' that frustrates understanding or recognition of indigenous water rights and uses. The local community is the key site for Andean cultural production and the management of irrigation water is essential. Water provides a strong material and spiritual basis for Andean life-ways and cultural orientations, one that has a firm foundation in ancient infrastructure and in well-developed understandings that join sacred landscapes to production, community, the commons and cultural identity. The massive irrigation infrastructure (including terraces) provided the material foundation for the development of Andean civilization; but today the common property systems and the kind of Andean identity that links people, place and production found in highland communities are under assault.

There are major differences in the cultural-political landscapes of Peru, Bolivia and Ecuador. But all of them have a long history of coercive policies that seek to assimilate indigenous peoples or break down their identities and collective landholdings and resources. Yet, today in Bolivia and Ecuador, it is clear that with indigenous ethnicity-based mobilization and state recognition of these countries as multi-ethnic and pluri-cultural, there are emerging options for thinking outside of the *criollo* and neoliberal box in state building, for empowering indigenous cultural rights and for safeguarding indigenous communities, resources and identities. These experiences should be considered in attempting to reinforce indigenous rights over water in varied national contexts. In doing so, it is important to have a deep and contextual understanding of the particular national histories and cultural politics of each country.

An underlying theme of this chapter has been that there are several 'indigenous' 'Andean' cultural orientations that are fundamental to rural life and water management in the highlands. On the one hand, the religious beliefs and practices tied to the customary beliefs and practices of irrigation are just one piece of the Andean community's cultural mosaic: Andean people have at their disposition many different cultural frameworks drawn from communities, towns and cities in their own and other countries. But, on the other hand, the notion of the commons and the cultural orientations found in irrigation remain important and vital elements of highland society. While not all highland communities share these orientations, they remain important to the lives of millions of people in the Andean nations. In many communities, irrigation is the productive domain that receives the greatest symbolic and ritual elaboration.

The perspective advanced here challenges those who would trivialize indigenous life-ways. At the same time, it defies the romanticization and simplified representation of Andean cultures, and recognizes the problems these cultural frameworks face as they confront larger political and economic forces. Implicit in most ideologies of national development in the Andean nations and throughout the Americas is the assumption that indigenous peoples must renounce their cultural orientations and identities in order to progress. Secular, bureaucratic, monetary and supposedly more rational and efficient state models of water management – which increasingly work hand in hand with neoliberalism and the privatization of land and water – claim to provide universal benefits, while in fact extending state control and the cultural orientations of national and international power holders. The cultural politics of the nation state in the Andean countries views the human and natural resources of the Andean highlands as inferior to those found in urban *criollo* society. Such an ideology often blinds state officials to the positive features of the commons and the productive features of customary irrigation practice.

Therefore, there is a need to direct anthropological attention not only to indigenous peoples and their so-called 'traditional' cultures, but also to the irrigation bureaucracies and the dominant national cultures of the countries under study. They are not inevitable or necessarily more rational and equitable, but rather are historically produced and represent the power of a dominant ethnic group. It is urgent to make explicit the discriminatory ways in which water politics operate today. In this same regard, it is necessary to educate the dominant society so that the terms 'Indian' and 'indigenous people' are no longer constraining. This can be done by showing that their customary uses of water and their cultural distinctiveness is compatible with 'modernity', urban spaces, transnational migration and social mobility.

Water in the Andes is not just an economic good, but also a cultural resource. Longstanding and culturally elaborate beliefs and rituals are intimately linked to water management by local communities. They are part of highland communities' cultural patrimony, and indigenous peasants should have rights over how they manage their irrigation systems. Indigenous peoples should have the rights to define water culturally and to have their definitions and customary uses respected by the policies and bureaucracies of the nation state. Yet, indigenous communities are challenged to maintain local control and 'traditional uses' – that is, to continue to define water materially, socially and culturally, when water laws and policies now essentially import neoliberal recipes.

In this regard it is clear that those who aim for more democratic and equitable water policies must find an effective way to interface and build coalitions with national indigenous and other organizations, as well as with local universities and other social actors, to move forward changes in water law and policy in a way that also strengthens indigenous communities, organizations and cultures *vis-à-vis* the state and the predominant 'national' culture. Only in this way can theory and our growing knowledge of indigenous water management systems be converted into sustainable practice in the Andean countries.

Notes

1 This chapter draws on my work in Peru as an ethnographer and anthropologist; in general terms, I argue for an interpretive and historical understanding of irrigation and the cultural politics of indigenous identity (see Gelles, 2000, 2002a, 2002b).

2 Rappaport (1994) shows that in Colombia, indigenous communities must submit historical evidence and assert a distinctly 'Indian' identity to lay claim to the land, political authority and communal protection of the *resguardos*, or indigenous communities.

3 For example, the film *Transnational Fiesta: 1992* (Gelles and Martínez, 1993) shows that during the Patron Saint Fiesta in the highland community of Cabanaconde in southern Peru, migrants return from Peruvian cities and from the US to visit their homeland. At the core of local identity formation, the fiesta has now become the core ritual in the annual reproduction of this transnational community.

4 Many anthropologists, following Kosok (1965) and Wittfogel (1957), have explored the role that irrigation played in the development of pre-Columbian social formations (see, for example, Sherbondy, 1982; Zuidema, 1986). Others have focused on the symbolic and ritual dimensions of water (e.g. Arguedas, 1964; Isbell, 1974; Ossio, 1976; Sherbondy, 1986, 1998; Valderrama and Escalante, 1988). Still others have studied the ecological, social and political facets of contemporary irrigation (e.g. Lynch, 1988; Guillet, 1992; Gelles, 1993, 1998, 2000; Bolin, 1994; Gandarillas et al, 1994; Gose, 1994; Treacy, 1994; Boelens and Dávila, 1998; Gutiérrez and Gerbrandy, 1998; Oré, 1998, 2005; Trawick, 2003; Boelens and Gelles, 2005; Guevara-Gil, 2006). See also the 27th and 28th issues of *Allpanchis* (1986).

5 Ethnicity, then:

... describes both a set of relations and a mode of consciousness; moreover, its meanings and practical salience varies for different social groupings according to their position in the social order. But as a form of consciousness ... [it is] produced as particular historical structures impinge themselves on human experience and condition social action. (Comaroff and Comaroff, 1992, p54)

Yet, 'while ethnicity is the product of specific historical processes, it tends to take on the 'natural' appearance of an autonomous force' (Comaroff and Comaroff, 1992, p60).

6 See, for example, Isbell (1974); Ossio (1976); Arguedas (1985); Valderrama and Escalante (1988); and Salomon and Urioste (1991).

7 At the same time, it is important to realize that indigenous movements vary tremendously within and between different national contexts, and often compete for legitimacy and power within the public and political spheres. Indeed, 'indigenous movements are sophisticated multilayered actors simultaneously engaging an array of international, nation and regional processes' (Gustafson, 2002, p293).

8 See, for example, Chapters 13 and 14 in this volume.

9 See Albó (1994); Bustamante (2002); and Gustafson (2002); see Chapters 3 and 5 in this volume.

10 See Bustamante (2002) and Chapters 3 and 15 in this volume.

11 Irrigation bureaucracies and water policies formulated in the Andean nations often exclude not only those with different cultural orientations, but women as well, both in the field and in the irrigation bureaucracies themselves (Lynch, 1991; Boelens and Zwarteveen, 2002).

References

Abercrombie, T. (1991) 'To be Indian, to be Bolivian', in G. Urban and J. Sherzer (eds) *Nation-States and Indians in Latin America*, University of Texas Press, Austin, TX, pp95–130

Alberti, G. and E. Mayer (1974) *Reciprocidad e intercambio en los Andes peruanos*, Instituto de Estudios Peruanos, Lima, Peru

Albó, X. (1994) 'And from Kataristas to MNRistas? The surprising and bold alliance between Aymaras and neoliberals in Bolivia', in D. L. Van Cott (ed) *Indigenous Peoples and Democracy in Latin America*, St Martin's Press, New York, NY, pp55–82

Anderson, B. (1983) *Imagined Communities: Reflections on the Origin and Spread of Nationalism*, Verso, London

Arguedas, J. M. (1964) *Puquio, una cultura en proceso de cambio*, UNMSM, Lima, Peru

Arguedas, J. M. (1985) *Yawar Fiesta*, University of Texas Press, Austin, TX

Boelens, R. and G. Dávila (eds) (1998) *Searching for Equity: Conceptions of Justice and Equity in Peasant Irrigation*, Van Gorcum, Assen, The Netherlands

Boelens, R. and B. Doornbos (2001) 'The battlefield of water rights: Rule-making amidst conflicting normative frameworks in the Ecuadorian highlands', *Human Organization* vol 60, no 4, pp343–355

Boelens, R. and P. H. Gelles (2005) 'Cultural politics, communal resistance, and identity in Andean irrigation development', *Bulletin of Latin American Research*, vol 24, no 3, pp311–327

Boelens, R. and P. Hoogendam (eds) (2002) *Water Rights and Empowerment*, Van Gorcum, Assen, The Netherlands

Boelens, R. and M. Zwarteveen (2002) 'Gender dimensions of water control in Andean irrigation', in R. Boelens and P. Hoogendam (eds) *Water Rights and Empowerment*, Van Gorcum, Assen, The Netherlands, pp75–109

Boelens, R., I. Gentes, A. Guevara and P. Urteaga (2005) 'Special law: Recognition and denial of diversity in Andean water control', in D. Roth, R. Boelens and M. Zwarteveen (eds) *Liquid Relations: Contested Water Rights and Legal Complexity*, Rutgers University Press, New Brunswick, NJ, and London, pp144–171

Bolin, I. (1994) 'Levels of autonomy in the organization of irrigation in the highlands of Peru', in W. P. Mitchell and D. Guillet (eds) *Irrigation at High Altitudes: The Social Organization of Water Control in the Andes*, Society for Latin American Anthropology and the American Anthropological Association, Washington, DC, pp141–166

Brysk, A. (1994) 'Acting globally: Indian rights and international politics in Latin America', in D. L. Van Cott (ed) *Indigenous Peoples and Democracy in Latin America*, St Martin's Press, New York, NY, pp29–54

Bustamante, R. (2002) 'Legislación del agua en Bolivia', in R. Boelens (ed) *Indigenous Water Rights, Local Water Management, and National Legislation*, WALIR Studies vol 2, Wageningen University/IWE and United Nations/CEPAL, The Netherlands, pp182–263

Comaroff, J. and J. Comaroff (1992) *Ethnography and the Historical Imagination*, Westview Press, Boulder, CO

Dean, B. (2002) 'State power and indigenous peoples in Peruvian Amazonia: A lost decade, 1990–2000', in D. Maybury-Lewis (ed) *The Politics of Ethnicity: Indigenous Peoples in Latin American States*, Harvard University Press, Cambridge, MA, pp199–238

Fuenzalida, F. (1970) 'La matriz colonial', *Revista del Museo Nacional*, vol 35, pp91–123

Fuenzalida, F. (ed) (1971) 'Poder, etnia y estratificación en Perú rural', in *Perú: Hoy*, Siglo XXI, Madrid, pp8–86

Gandarillas, H., L. Salazar, L. Sánchez, L. C. Sánchez and P. de Zutter (eds) (1994) *Dios da el Agua ¿Qué Hacen los Proyectos?: Manejo de Agua y Organización Campesina*, PRIV, Bolivia

Gelles, P. H. (1984) 'Agua, faenas y organización comunal: San Pedro de Casta', *Anthropologica*, vol 2, no 2, pp305–334

Gelles, P. H. (1986) 'Sociedades hidráulicas en los Andes: Algunas perspectivas desde Huarochirí', *Allpanchis* vol 27, pp99–147

Gelles, P. H. (1993) 'Irrigation as a cultural system: Introductory remarks', in *Proceedings of the Chacmool Conference*, University of Calgary Archeological Association, Calgary, Alberta, pp329–332

Gelles, P. H. (1998) 'Competing cultural logics: State and "indigenous" models in conflict', in R. Boelens and G. Dávila (eds) *Searching for Equity: Conceptions of Justice and Equity in Peasant Irrigation*, Van Gorcum, Assen, The Netherlands, pp256–267

Gelles, P. H. (2000) *Water and Power in Highland Peru: The Cultural Politics of Irrigation and Development*, Rutgers University Press, New Brunswick, NJ (Spanish translation: (2002) *Agua y poder en la sierra peruana: La historia y política cultural de riego, rito y desarrollo*, Fondo Editorial de la Pontificia Universidad Católica del Perú, Lima, Peru)

Gelles, P. H. (2002a) 'Andean culture, indigenous identity, and the state in Peru', in D. Maybury-Lewis (ed) *The Politics of Ethnicity: Indigenous Peoples in Latin American States*, Harvard University Press, Cambridge, MA, pp278–301

Gelles, P. H. (2002b) 'Cultural politics and local resistance in highland irrigation development', in R. Boelens and P. Hoogendam (eds) *Water Rights and Empowerment*, Van Gorcum, Assen, The Netherlands, pp22–35

Gelles, P. H. and W. Martínez (1993) *Transnational Fiesta: 1992*, Center for Media and Independent Learning, Berkeley, CA

Getches, D. H. (2006) *Indigenous Rights and Interests in Water in the United States and under International Law*, WALIR Research vol 6, University of Colorado at Boulder, CO, UN/CEPAL and Wageningen University, Wageningen, The Netherlands

Gose, P. (1994) *Deathly Waters and Hungry Mountains: Agrarian Ritual and Class Formation in an Andean Town*, University of Toronto Press, Toronto, Ontario

Guevara-Gil, A. (2006) 'Official water law versus peasant and indigenous rights in Peru', in R. Boelens, M. Chiba and D. Nakashima (eds) *Water and Indigenous Peoples*, WALIR-UNESCO, UNESCO, Paris

Guevara-Gil, A., I. Vera-Dávila, P. Urteaga-Crovetto and G. Zambrano (2002) 'Water legislation and indigenous water management in Peru', in R. Boelens (ed) *Indigenous Water Rights, Local Water Management, and National Legislation*, WALIR studies, vol 2, Wageningen University/IWE and United Nations/CEPAL, Wageningen, The Netherlands, pp44–49

Guillet, D. (1992) *Covering Ground: Communal Water Management and the State in the Peruvian Highlands*, University of Michigan Press, Ann Arbor, MI

Gustafson, B. (2002) 'Paradoxes of liberal indigenism: Indigenous movements, state processes, and intercultural reform in Bolivia', in D. Maybury-Lewis (ed) *The Politics of Ethnicity: Indigenous Peoples in Latin American States*, Harvard University Press, Cambridge, MA, pp267–306

Gutiérrez, Z. and G. Gerbrandy (1998) 'Water distribution, social organization and equity in the Andean vision', in R. Boelens and G. Dávila (eds) *Searching for Equity: Conceptions of Justice and Equity in Peasant Irrigation*, Van Gorcum, Assen, The Netherlands, pp242–249

Herzfeld, M. (1992) *The Social Production of Indifference: Exploring the Symbolic Roots of Western Bureaucracy*, University of Chicago Press, Chicago, IL

Isbell, B. J. (1974) 'Kuyaq: Those who love me: An analysis of Andean kinship and reciprocity within a ritual', in G. Alberti and E. Mayer (eds) *Reciprocidad e intercambio en los Andes peruanos*, Instituto de Estudios Peruanos, Lima, Peru

Kearney, M. (1996) *Reconceptualizing the Peasantry: Anthropology in Global Perspective*, Westview Press, Boulder, CO

Kosok, P. (1965) *Life, Land, and Water in Ancient Peru*, Long Island University Press, New York, NY

Lynch, B. (1988) *The Bureaucratic Transition: Peruvian Government Intervention in Sierra Small Scale Irrigation*, PhD thesis, Cornell University, Ithaca, NY

Lynch, B. (1991) 'Women and irrigation in highland Peru', *Society and Natural Resources*, vol 4, pp37–52

MacDonald, Jr. T. (2002) 'Ecuador's Indian movement: Pawn in a short game or agent in state configuration?', in D. Maybury-Lewis (ed) *The Politics of Ethnicity: Indigenous Peoples in Latin American States*, Harvard University Press, Cambridge, MA, pp169–198

Masson, L. (1987) 'La ocupación de andenes en Perú', *Pensamiento Iberoamericano* vol 12, pp179–181

Maybury-Lewis, D. (1982) 'Living in Leviathan: Ethnic groups and the state', in D. Maybury-Lewis (ed) *The Prospects for Plural Societies*, American Ethnological Society, Washington, DC, pp220–232

Mayer, E. (1994) 'Recursos naturales, medio ambiente, tecnología y desarrollo', in O. Dancourt, E. Mayer and C. Monge (eds) *Perú: El problema agrario en debate*, Sepia V, Sepia, Lima, Peru, pp479–534

Mayer, E. (2002) *The Articulated Peasant: Household Economies in the Andes*, Westview Press, Boulder, CO

McCay, B. J. and J. M. Acheson (1990) *The Question of the Commons: The Culture and Ecology of Communal Resources*, University of Arizona Press, Tucson, AZ

Murra, J. (1982) 'The cultural future of the Andean majority', in D. Maybury-Lewis (ed) *The Prospects for Plural Societies*, American Ethnological Society, Washington, DC

Norgaard, R. B. (1994) *Development Betrayed: The End of Progress and the Coevolutionary Revisioning of the Future,* Routledge, New York, NY

Oré, M. T. (1998) 'From agrarian reform to privatisation of land and water: The case of the Peruvian coast', in R. Boelens and G. Dávila (eds) *Searching for Equity: Conceptions of Justice and Equity in Peasant Irrigation*, Van Gorcum, Assen, The Netherlands, pp268–278

Oré, M. T. (2005) *Agua, bien común y usos privados*, Fondo Editorial Pontificia Universidad Católica, Lima, Peru

Ossio, J. M. (1976) *El simbolismo del agua en la representación del tiempo y el espacio en la fiesta de la acequia en Andamarca*, Mimeo, Pontificia Universidad Católica, Lima, Peru

Ossio, J. M. (1992) *Los Indios del Perú*, S.A. Editorial MAPFRE, Madrid, Spain

Rappaport, J. (1994) *Cumbe Reborn: An Andean Ethnography of History*, University of Chicago Press, Chicago, IL

Rasnake, R. (1988) *Domination and Cultural Resistance: Authority and Power among an Andean People*, Duke University Press, Durham, NC

Reinhard, J. (1985) 'Chavin and Tiahuanaco: A new look at two Andean ceremonial centers', *National Geographic Research Reports*, vol 1, no 3, pp395–422

Remy, M. I. (1994) 'The indigenous population and the construction of democracy in Peru', in D. L. Van Cott (ed) *Indigenous Peoples and Democracy in Latin America*, St Martin's Press, New York, NY, pp107–132

Salomon, F. (1991) 'Introduction: Native lords of Quito in the age of the Incas', in F. Salomon and G. Urioste (eds) *The Huarochirí Manuscript: A Testament of Ancient and Colonial Andean Religion*, University of Texas Press, Austin, TX, pp1–38

Salomon, F. and G. Urioste (1991) *The Huarochirí Manuscript: A Testament of Ancient and Colonial Andean Religion*, University of Texas Press, Austin, TX

Sherbondy, J. E. (1982) 'El regadío, los lagos y los mitos de origen', *Allpanchis*, vol 17, no 2, pp3–32

Sherbondy, J. E. (1986) 'Los ceques: Código de canales en el Cusco Incaico', *Allpanchis,* vol 27, pp39–74

Sherbondy, J. E. (1998) 'Andean irrigation in history', in R. Boelens and G. Dávila (eds) *Searching for Equity: Conceptions of Justice and Equity in Peasant Irrigation*, Van Gorcum, Assen, The Netherlands, pp210–215

Starn, O. (1991) 'Missing the revolution: Anthropologists and the war in Perú', *Cultural Anthropology*, vol 6, no 1, pp13–38

Tambiah, S. J. (1985) *Culture, Thought, and Social Action: An Anthropological Perspective*, Harvard University Press, Cambridge, MA

Trawick, P. (2003) *The Struggle for Water in Peru: Comedy and Tragedy in the Andean Commons*, Stanford University Press, Stanford, CA

Treacy, J. M. (1994) *Las Chacras de Coporaque: Andenería y riego en el Valle del Colca*, Instituto de Estudios Peruanos, Lima, Peru

Turino, T. (1991) 'The state and Andean musical production in Peru', in G. Urban and J. Sherzer (eds) *Nation-States and Indians in Latin America*, University of Texas Press, Austin, TX, pp259–285

Urban, G. and Sherzer, J. (1991) 'Introduction: Indians, nation-state, and culture', in G. Urban and J. Sherzer (eds) *Nation-States and Indians in Latin America*, University of Texas Press, Austin, TX, pp1–18

Valderrama, R. and C. Escalante (1988) *Del Tata Mallku a la Mama Pacha: Riego, sociedad y ritos en los Andes peruanos*, DESCO, Lima, Peru

Valderrama, R., C. Escalante, P. H. Gelles and G. Martínez (eds) (1996) *Andean Lives: Gregorio Condori Mamani and Asunta Quispe Huamán*, University of Texas Press, Austin, TX

Van Cott, D. L. (1994) 'Indigenous peoples and democracy: Issues for policymakers', in D. L. Van Cott (ed) *Indigenous Peoples and Democracy in Latin America*, St Martin's Press, New York, NY, pp1–28

Weismantel, M. J. (1989) *Food, Gender, and Poverty in the Ecuadorian Andes*, University of Pennsylvania Press, Philadelphia, PA

Wittfogel, K. A. (1957) *Oriental Despotism: A Comparative Study of Total Power*, Yale University Press, New Haven, CT

Zuidema, T. (1986) 'Inca dynasty and irrigation: Another look at Andean concepts of history', in J. V. Murra, N. Wachtel and J. Revel (eds) *Anthropological Histories of Andean Polities*, Cambridge University Press, Cambridge, MA, pp177–200

7

Land, Water and the Search for Sustainable Livelihood in the Andes

Annelies Zoomers

Introduction

Recent debates in Bolivia and in many other developing countries have emphasized the need to improve the natural resource base of the poor as an instrument to combat poverty. Improving the access to and control of water and land is considered a crucial factor in improving agricultural production (and, thus, incomes), improving health and realizing the agro-ecological goals of sustainable land use (World Bank, 1996; UN, 2003). In a previous government's *Poverty Reduction Strategy Paper* (PRSP), which is aimed at combating extreme poverty in rural areas, it is argued that there is a close connection between land and water scarcity, on the one hand, and poverty, on the other, and that actions to address the former are the key to fighting poverty (PRSP Bolivia, 2001). The paper also emphasizes the need to 'regulate rural landownership and the modernization of the system for the registry of property rights and land registry' and to 'create irrigation systems' (PRSP Bolivia, 2001, pp100–104). 'Insufficiently defined ownership rights to land and natural resources have been a source of uncertainty and inefficient land use. In addition, water shortage has restricted productivity gains and expansion in the scale of production, and has made producers more vulnerable' (PRSP Bolivia, 2001, p4). Sound water and land management and clearly defined rights are presented as key elements in helping poor people to improve their lives. Such ideas, now converted into a more community-centred focus respecting existing water rights and the '*usos y costumbres*' approach (see also Chapters 2, 3 and 15 in this volume), also continue to feed current policy proposals.

In the Andes region, most rural poor are *minifundistas* who have access to small parcels of eroded land but no access to sufficient water resources. Although much of the land is communally owned, people work most fields individually. In recent decades, the increasing population pressure has contributed to ongoing processes of land fragmentation and environmental degradation. At the same time, competition for water has intensified, 'caused by endogenous demands linked to irrigated agriculture and urban population living in the Andean catchments and basins, and exogenous demands, especially from large urban centres and industrial, mining and energy-sector business' (Concertación, 2005, pp3–4). Furthermore, apparently as the result of climate change, there has been a reduction in the quantity of available water. The retention and storage of water have been reduced, while peak flow rates have increased during the rainy seasons. 'Competition, moreover, is not limited to the availability of water – the quantity, place and timing of delivery – but is also strongly related to the issue of worsening quality. Increasingly, activities affecting water pollute it, rendering the resource useless for many or most purposes' (Concertación, 2005, p5).

At the extremes of a continuum of approaches (see also Boelens and Zwarteveen, 2003), two distinct views are expressed in the debate about how to improve the land and water situation. At one extreme, until recently very influential in pre-Movimiento al Socialismo (MAS) Bolivia, there are the adherents of market liberalization, who are in favour of imposing market rules on the communities and externalizing water and land control. Modern 'transparent' policies – namely, those aimed at creating an active free market for water and land rights – are assumed to be powerful instruments for making resource use more efficient. 'Regulatory regimes must allow clear and transparent transactions between stakeholders in a climate of trust, together with shared responsibility for safeguarding water resources and making a sustainable use of land' (World Bank, 1996; UN, 2003, p11). The allocation of free, clearly defined and transparent individual rights is believed to encourage people to make optimal use of land and water, which contributes to poverty alleviation. 'Water and land rights should be tradable and, to the extent possible, there should be direct compensation of individuals willing to transfer their rights to others which is commensurate with the rights transferred; water and land transfers should be transparent and verifiable' (UN, 2003, p11).

At the other extreme, there are those who believe in essentialized and reified versions of community survival and collective action. They stress the advantages of Andean technology and the strengths of indigenous systems, which they see as the optimal model for resource management in the Andes because it has lower transaction costs and makes better use of local knowledge. Customary collective rights provide people with optimal possibilities for maintaining vertical land control, collective work, fallowing systems, etc. Although customary rights in the Andes are less transparent, they are better adapted to the local situation. According to the adherents of this view, indigenous people perceive their environment holistically – namely, they see a harmonious interplay among people (and animals and plants) and the mountains and other elements. Natural resources are interlinked and form part of one system. Land and water cannot be separated.

In Bolivia during the mid 1990s, there was a shift away from 'pure' neoliberalism (i.e. market liberalization) towards 'indigenous systems'. The previous government's neoliberal reforms were accompanied by legislative measures recognizing indigenous identities, languages and organizations and ascribing to them new categories for participation within formal institutions of state governance.[1] This was preceded by periods of protests and the *Marcha por Territorio y Dignidad* (March for Land and Dignity) in 1990. These developments have allowed room for the particularities of the Andes region. Now, priority is given to strengthening Bolivia's position as a multi-ethnic and multicultural society (Strobele-Gregor et al, 1994; Gustafson, 2002; see also Assies, 2000, p101), culminating in the election of the current administration and Evo Morales as the first indigenous president. The indigenous population is increasingly successful in defending its rights and modes of organization (which resulted in adaptation of the irrigation and land reform legislation). This is usually regarded as a step forward in the struggle to reduce poverty. It is less clear, however, to what extent the recognition of indigenous rights will directly help people to improve their situation.

This chapter analyses how regulating land tenure and expanding irrigation could help indigenous groups to develop sustainable livelihoods. In current research, land and water are usually described as 'natural capital'; but important questions are: what is the importance of land and water in comparison to other assets (e.g. human, social, cultural and produced capital) and how will it contribute to sustainable livelihoods and poverty alleviation? In addition, given the characteristics of land and water tenure today, to what extent should natural resources in the Andes be considered as natural capital or, to the contrary, as a socio-economic constraint?

Empirical research was carried out in 41 *comunidades* in the departments of Chuquisaca and Potosí (southern Andes in Bolivia; see Figure 7.1). Until the land reforms of 1953, most people were *peons*, semi-serfs living on the haciendas who, in return for usufruct rights to a small parcel of land, laboured for the *patrón* and performed other services (McEwen, 1975, p448).[2] The land reforms brought these feudal relations to an end. The *indios* living in communities became the collective owners of the land, most of which was cultivated individually with family labour, while only part of the land was used for collective purposes. Between the 1960s and 1980s, many attempts were made to help these groups through rural development programmes implemented by the state and non-governmental organizations (NGOs); but these were not successful in solving the structural limitations. Many people decided to leave and move to the cities (Sucre, La Paz, Santa Cruz), which resulted in a rapid urbanization; but many were not capable of improving their situation. People who stayed in the rural communities are still among the poorest in the Andes.

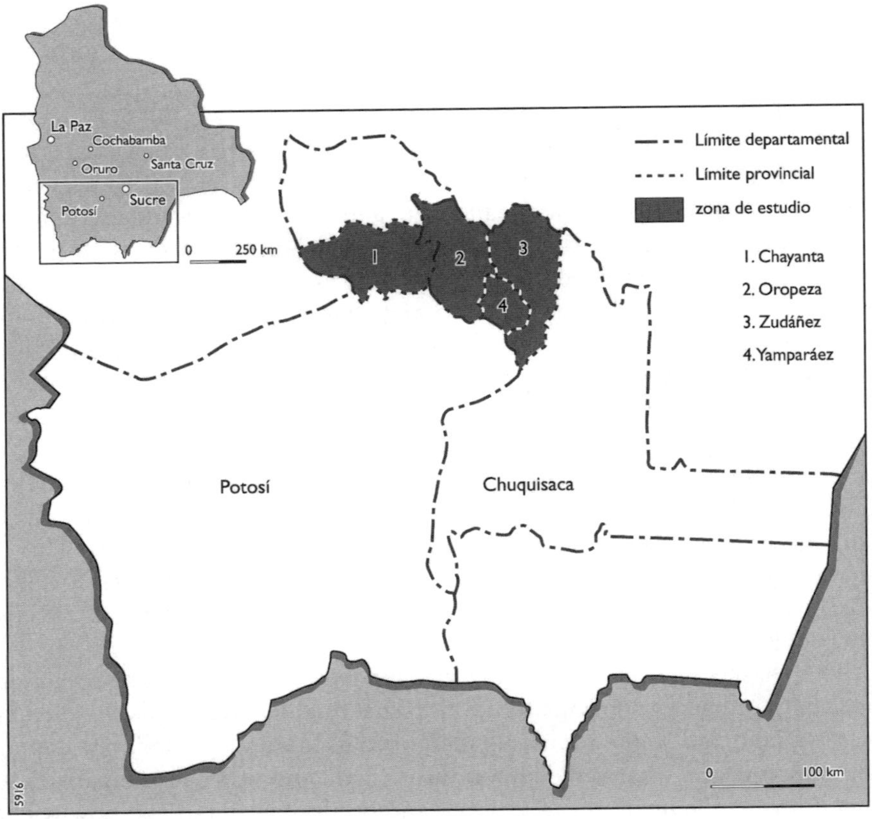

Source: de Morrée (2002, pp22–23)

Figure 7.1 *Southern Bolivia*

During 1995 to 1997, we carried out in-depth research among 136 families in 17 *comunidades* in order to analyse the resource base of the rural population in relation to their livelihood strategies.[3] Then, in 2002 to 2004, we collected additional information about the families' upward and downward social mobility by interviews in the *comunidades*. This second round of research focused on establishing the extent to which access to land and water and other factors are crucial for the capability of people to develop sustainable livelihoods or to escape from poverty.

After describing the 'Andean attitude' to natural resource management, this chapter discusses current land and water tenure patterns and analyses how these are reflected in the predominant livelihood systems. It then explores the importance of land and water in determining class relations and how they help (or fail to help) people to escape from poverty. Finally, some conclusions are offered concerning the kind of policy that could support people in their efforts to improve their livelihoods.

The 'Andean attitude' to natural resources: The holistic picture

Much has been written about natural resource management in the Andes. In the literature, the Andes region is often cited as an example of an ancient irrigation civilization (Gelles, 2000; Morris, 2004); highland irrigation is usually described as both ancient and widespread, and it is often stressed how in the Andes efficient use is made of a limited amount of water.

The Andean population is well known for its adaptive ways to overcome the constraints imposed on them by adverse agro-ecological circumstances. 'Because of the vertical series of environmental zones, each with different conditions, water goes at successive times to different places to be used for different purposes, and each region benefits according to its requirements in a most economical dovetailing of functions' (Mitchell, 1976, p39).[4] In the Andes, people seek complementarities between different production zones – that is, the high level (*puna*), where land uses include the cultivation of bitter potatoes and the breeding of sheep and llamas; the middle section, or valley heads (2200m to 3400m), where most of the cultivation is carried out (commonly potatoes, and the standard subsistence crops of maize, peas and beans); and the lower section (i.e. from the 2200m line down to the river), where hard maize for cattle feed, corn on the cob, squash and other crops are grown (e.g. *Capsicum pubescens*, *Arracacia xanthorrhiza* and *Xamthosoma sagittifolium*), and some beef and dairy cattle are bred. Often, the vertical access to different lands is important because of variability of the weather and hydrology (Valladolid-Rivera, 1998; Morris, 2004; see also Table 7.2).

In order to secure a harvest in the diverse, variable Andean environment, people have a broad repertoire of strategies, such as planting each crop in a companion-crop mixture in different places or at different altitudes and at different times. The wide variety of crops that are produced according to complex cultivation schemes reflects a great diversity of ecological altitudes and climates. For example, there is a broad range of potatoes, maize, apples, broad beans and wheat depending on the altitude and the microclimate. Among the mixture of varieties and ecotypes are some that are resistant to drought conditions or to excess humidity. Thus, when the mixture is cultivated in a year with little rain and with droughts and frosts, the varieties resistant to these conditions produce more than those that are not resistant; similarly, in a rainy year and with the concomitant diseases, the varieties resistant to these conditions produce more. In other words, in both cases there is an assured stability in food production despite the variability of the Andean climate (Valladolid-Rivera, 1998, p72).

> *People use a well-established fallowing system under which the community agrees to maintain their cultivated land, even when individually owned, in several* aynokas *(sectors), the use of each being agreed in advance of the growing season and for many seasons to come. These aynokas are rotated in what is termed a*

> *'sectoral fallow system' (Pestalozzi, 2000; see also Morlon, 1992).* (Morris, 2004, p108)

According to influential 'Andeanist' activists and scholars, Andean people have a holistic view with respect to how they regard natural resources and there is a harmonious interplay between people and their natural surroundings. This perspective, which is critically examined later in this chapter, is strongly promoted by, among others, Proyecto Andino de Tecnología Campesina (PRATEC), a group of Peruvian 'Andeanist' academics. These scholars maintain that:

> *... all that exists in the Andean world is alive. Not only man, animals and plants, but rocks, rivers, the mountains and everything else. One often ... hears the peasants, the elder ones especially, refer to the* chacras, *the plots of land, as personified beings, which show their own individuality through manifestations like the fact of needing affection, esteem, consideration... For that reason, commenting on the care and the arrangement of the fields, the peasants say that 'the* chacras *are also nurtured', thus placing the field on the same level of perception as the animals.* (Grillo-Fernández, 1998, p222)

Valladolid-Rivera (1998, p51) expresses it thus:

> *One must live the life of the Andean countryside in continuous conversation with the stars, rocks, lakes, rivers, plants and animals both wild and cultivated, with the clouds, the frosts. One must relish the taste of the rains, listen to the corn growing, observe the colour of the winds – feeling oneself at all times accompanied by our deceased ancestors. The nurturing of the chacra is the heart of Andean culture which, if not the only activity carried out by the peasants, is the one around which all aspects of life revolve.*

According to this view, land and water cannot be separated. Water plays a central role in the 'nurturing of the chacra'. Water is nurtured by digging wells and filtering galleries, and by protecting the *puquios* (springs). The nurturing of the microclimate in the chacra is done from the very moment of tilling the soil:

> *Tilling assists the retention of rainwater or irrigation water since the tilled soil acts like a sponge, preventing rapid drainage and thus modifying the original microclimate. Microclimate in the chacra is also modified by building walls of stones or trees or mud, which protect it from wind and cold. In the same way, levelling the field, fertilizing with sufficient guano, and the*

> *companion-cropping and rotation of crops all contribute to nurturing an appropriate microclimate in the chacra.* (Grillo-Fernández, 1998, p222)

The climate strongly influences life in the Andes. Not only is the climate diverse, varying from place to place, but each of the climates is quite variable from month to month and over the years. The population recognizes two distinct climatic periods in each year: a cold, dry period, during which the nightly minimum temperature can reach 0°C and precipitation is minimal; and a hot, rainy period, during which the minimum temperature and the pluvial precipitation rise, making possible the growth and development of plants and animals. The timing of each varies from area to area as well as within areas depending on the climatic characteristics of each year (Valladolid-Rivera, 1998, pp53–54). In this 'Andeanist' ideological view, for the Andean peasant there is no good year or bad year:

> *In both he receives a harvest for himself and his* ayllu, *either by production, or through using reserves (from years with levels of production which allowed them to conserve food for several years) or reciprocal relationships between neighbours and/or communities between different agro-ecological zones. Because the Andean peasant understood that the soil was limited, he 'nurtured' it by constructing* andenes *(terraces), and slowly forming terraces,* qochas, *and basins,* waru warus, *and as he nurtured them and still nurtures them, he knows full well their diversity and variability.* (Valladolid-Rivera, 1998, pp53–54)

Thus, in the Andes it is not a matter of resigning oneself to alternations between good (rainy) years and bad (dry) years. The Andeans accept the year as it is: 'They converse with it through a large repertoire of knowledge in such a way that they always obtain a harvest. If they were not so, Andean peasants would have disappeared from the Andes many years ago and with them the Andean culture' (Valladolid-Rivera, 1998, p54).

Current patterns of land and water tenure: A basis for sustainable livelihoods?

This section focuses on an analysis of current patterns of land and water tenure in relation to people's efforts to overcome the restrictions of the Andean environment. What is the quality of 'natural capital' and what is its importance for the capacity of people to develop a sustainable livelihood?

An analysis of the current resource base of the research population reveals that in contrast to what was described above, there are not many examples of overall 'adaptive engineering structures'. There is little evidence of large-scale irrigation systems, raised fields, *waru warus*, etc., and hardly any new terraces

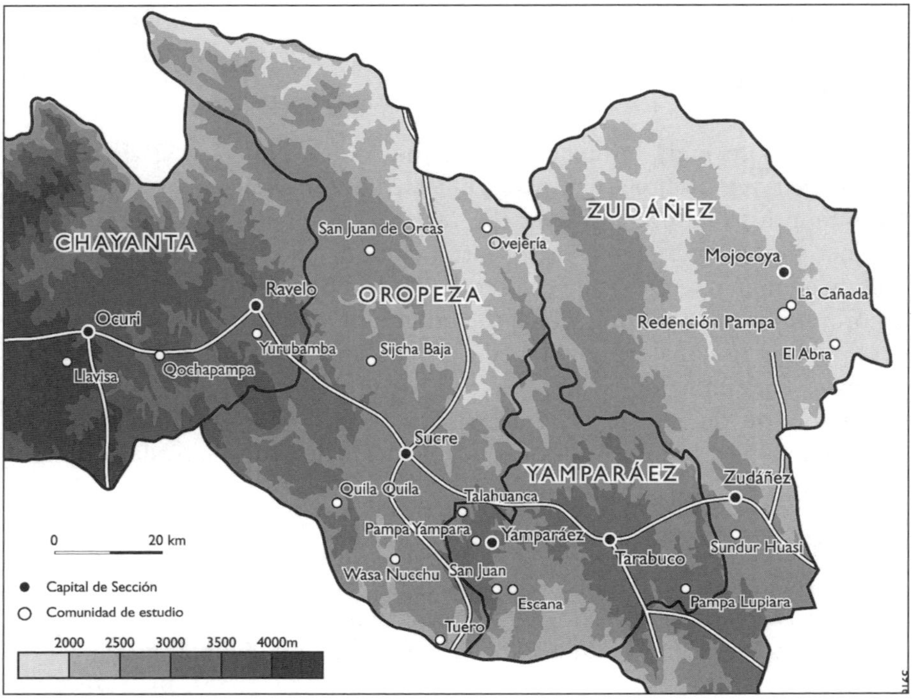

Source: de Morrée (2002)

Figure 7.2 *Research area*

are being constructed. People are confronted with many climate risks, including storms, hail, droughts and frosts. There is little or no vegetation, the soils are extremely rocky and eroded, and in the communities which were studied, there are hardly any possibilities for irrigation.

In the departments of Oropeza, Yamparáez, Zudáñez and Chayanta, within the *municipios* that form part of our research area we found 310 irrigation schemes, covering a total area of 6157.5ha and belonging to 717 communities. This means that the average area amounted to 19.8ha per scheme and 8.6ha per community (see Table 7.1). When focusing more closely on the 41 communities, we found, however, that only a few villages have access to irrigation systems, which means that most of the population has to deal with drought and adverse production circumstances.

In the research area, irrigation is mainly found in the lower-located communities (the *pampa templada* and below; see Table 7.2). Most irrigation systems tend to be rather small scale (0.1ha to 2.7ha) and simple: a restricted number of canals bring the water to the *comunidad*. Most irrigation canals and reservoirs are made of earth (cement is used only occasionally); usually they are dry and are put into use only as a particular field is irrigated. Most irrigation systems are communally owned and the number of beneficiaries is rather limited. Whether or not people have access to irrigation water depends on the location of their land and particular local rules and obligations for each irrigation system.

Table 7.1 *Inventory of irrigation systems in the research area*

Province	Municipio	*Number of communities**	*Total irrigated area (ha)*	*Average area per scheme (ha)*	*Total number of irrigation schemes*
Oropeza	Poroma	71 (5)	0	0	0
	Sucre	94 (4)	1262.5	19.7	64
	Yotala	26 (2)	294	12.8	23
	Subtotal	191	1556.5	17.8	87
	(Total Oropeza)				(121)
Yamparáez	Yamparáez	35 (5)	461.5	30.7	15
	Tarabuco	75 (1)	284	13.5	21
	Subtotal	110	745.5	20.7	36
	(Total Yamparáez)				(63)
Zudáñez	Zudáñez	26 (5)	1673	16.9	99
	Mojocoya	29 (6)	460	38.3	12
	Presto	31 (1)	560	15.1	37
	Subtotal	86	2693	18.1	148
	(Total Zudáñez)				(238)
Chayanta	Ocori	82 (3)	0	0	0
	Ravelo	93 (4)	978.5	31.5	31
	Colquechaca	155 (5)	184	23.0	8
	Subtotal	330	1162.5	29.8	39
	(Total Chayanta)				(55)
Total*		717	6157.5	19.8	310 (477)

Note: * These are the municipalities related to the 41 comunidades in the research area.
Source: www.aguabolivia.org

In most cases, irrigation tends to be rather informal and there are usually no special authorities (such as the *Varayoc*) to manage the system. The beneficiaries of the water are collectively responsible for the maintenance of the principal irrigation canals. A person who has not helped to clean the irrigation system is not given water unless he pays a fine. Even during the formal water distribution, it is the individual's responsibility to get the water to his fields, and it is the individual's responsibility to prevent the theft of water as there are no formal sanctions against stealing it. In the cases that were investigated, it was only during the most critical period of water distribution that the *comunidad* played an active role in water distribution; during the rest of the year, water was distributed informally by those who needed it.

In many of the villages, irrigation is relatively new or was recently improved. Between 1983 and 1995, irrigation projects were implemented in El Abra and Escana (six projects each); Quila Quila (three projects); Tuero Chico and Wasa Njucchu (two projects); and on a smaller scale in San Juan, Sundur Wasi, Yurubamba, Ovejería and La Cañada. Many of these projects generated a number of benefits, but often went hand in hand with conflicts (see Box 7.1). In general terms, the implementation of irrigation systems has contributed to extending the length of the growing season and to expanding the cultivated areas (in the upper parts, by planting an early dry-season crop); it has also

Table 7.2 *Access to land and water in 17 communities of the research area*

Agro-ecological zone	*Name of community*	*Average area of land per family/vertical orientation*	*Access to irrigation*	*Dominant crops/ livelihood system*
Puna alta	Llavisa	7.4ha 6 parcels/2 zones	Not irrigated	Not significant, mainly livestock-raising (*llameros*)
*Puna baja**	Qochapampa*	5.1ha 23 parcels/2 zones	Not irrigated	Commercial potato production, barley, wheat
*Pampa de puna**	Yurubamba* Pampa Lupiara* Pampa Yampara*	4.8ha 9 parcels/3 zones	Not irrigated	Commercial potato production, wheat, barley, maize
Pampa templada	La Cañada	6.9ha 8 parcels/4 zones	0.7ha, 15% of parcels/12% of cultivated area	Not significant, mainly wheat, potato, fruits (subsistence farming)
Valle alto	San Juan San Juan de Orcas Talahuanca	3.0ha 9 parcels/4 zones	0.1ha, 2% of parcels/0.8% of cultivated area	Not significant, mainly potato, wheat, barley, maize (subsistence farming), migration
Valle templado	Sundur Huasi; Tuero Chico; Wasa Njucchu;* Escana;* Sijcha Baja; Ovejería; Quila Quila*	4.2ha 9 parcels/4 zones	0.7ha, 20% of parcels/15% of cultivated area	Not significant, mainly fruits and horticulture; maize, potato, wheat, barley (part commercial), processing (wood, bricks), cattle
*Valle bajo**	El Abra*	2.7ha 5 parcels/1 zone	2.7ha, 100% of parcels/100% of cultivated area	Commercial fruit and horticulture production; potato, maize

Note: * Zones/*comunidades* with high production potential, selling to the market.
Source: Pied Andino (1998), cited in Zoomers (1998)

helped people to overcome periods of droughts by making production more secure in the lower parts during the growing season itself. Irrigation water is also used for domestic purposes (drinking water and sanitation), and this has important effects on labour availability and health.

Most villages that were part of the research – especially those in the higher parts – do not have access to irrigation and the production capacity is rather restricted as a result of the hostile (risky and eroded) environment. In most *comunidades*, the hydrology, topography, soils and microclimate are not favourable for farming. These ecological disadvantages are directly visible in the production orientation of the different villages (see Table 7.2). Commercial agriculture is found only in a restricted number of niche areas (see Figure 7.2): commercial potato production is concentrated in *pampa de puna* and *puna baja* (Yurubamba, Pampa Lupiara, Pampa Yampara and Qochapampa) and specialized fruits and horticulture are found in *valle bajo* and *valle templada* (such places as El Abra, Wasa Njucchu, Escana, Quila Quila, etc.), which have sufficient possibilities for irrigation. In most of the other villages, however, the agricultural potential is low and people are mainly dependent on a combination

Box 7.1 Some examples of irrigation projects and their impacts

In Tuero Chico and Wasa Njucchu, irrigation management is relatively simple. In both villages, community members purchased irrigated land from the *patrón* and expanded the irrigation with support from an institution. In both *comunidades*, water is plentiful year round and there is no need to schedule shifts. In this situation, the community organization's role is to ensure system maintenance. To clean ditches, the union secretaries-general organize community work days and monitor everyone's participation. They also inspect walls and deal with damage caused by animals. It is their job to get institutional support to expand or improve the system through the community organization.

Escana has an irrigation tradition: some of the ditches have been there for 40 to 50 years. This water comes from several rivers and small creeks and used to supply just a few families. During the 1980s and 1990s, irrigation was expanded by lining ditches and building water catchment facilities. On small ditches for a few families, they arrange with each other; but for medium ditches with a larger number of beneficiaries, there are problems in distributing water. In order to decide and monitor water shifts, there is a water judge for the three largest ditches who is elected by users at the union meeting for a year's term. The judge has to settle water disputes; he fines anyone who steals water or requires them to forfeit a turn; he also punishes those who do not help to clean the ditch. When the water judge emigrated to Argentina, there was no one to settle water theft problems, and many quarrels occurred. One source of problems is that older irrigators receive the same amount of water, although they contributed more workdays for building than newer members. On temporarily irrigated plots, another problem is crop damage by livestock. The community mayor is the authority responsible for solving this type of problem.

In Quila Quila there are several irrigation ditches and others are planned. Some years ago the main ditch of the Tajchi *ayllu* was lined. Rainfall waned, new irrigators joined and this higher number of participants complicated water management. Some of them took advantage of what little water there was; so, in 1995, irrigators decided to introduce a shift system. Irrigators monitor each other and anyone caught stealing water forfeits a shift. People who do not take part in the union meeting or cleaning ditches also lose their turns. In the ditches of Picachulla and Leopaya *ayllus* there are similar water management problems; the irrigators here also discussed finding a solution to distributing irrigation water in a balanced manner. In order to apply for the irrigation projects, villagers elected several irrigation committees, one per ditch. After getting the project approved, the committees vanished because the villagers felt that this would mean having too many meetings.

Source: de Morrée (2002, p75)

of subsistence farming and non-agricultural sources of income. Migration is an important source of income in almost all the communities (Zoomers, 1998).

By focusing in more detail on the resource base in the 17 villages, one can see that even though in earlier times the situation might have been more favourable, most of the population has to cope with small areas of steep, stony and eroded land. Access to land is extremely fragmented: people have access to between 5 and 23 parcels (comprising 2.7ha to 7.4ha of land). Even though in the Andes literature much emphasis is put on the importance of a vertical orientation for risk reduction (keeping land in the different ecological zones, from the *puna* down into the lower valleys), my results indicate that most families have access to only two zones – namely, the lower area (for farming) and the higher zones (for pasturing). It is only in the lower zones that people still show some degree of vertical orientation (i.e. having access to the various agro-ecological zones). Most people obtain their land through inheritance, and the possibilities of acquiring additional land are extremely limited.

An analysis of the way in which people adapt by applying different types of companion cropping and rotation schemes shows that these techniques should not be overestimated. The most common crop association is potatoes with legumes and gourds, and sometimes with maize, beans or *quinua* (Zoomers, 1998, p587). In most villages, there are three crop rotations – for example, potatoes–maize–wheat, then wheat–barley, and then potatoes–maize. Fallow periods were found in only 55 per cent of the villages. In most *comunidades*, the number of crops is rather restricted and the number of indigenous crops or varieties grown has dropped considerably in recent decades.

In most areas, the quality of the natural environment has deteriorated in recent decades. Soil degradation and erosion are apparent virtually everywhere, and pests and diseases are increasingly common. Particularly high-altitude areas are suffering from drought and, in many places, the damage caused by hailstorms has also increased. There is less farming in the higher areas as agriculture moves towards the low-lying zones. Only in the highest zones is there an 'upward shift' in farming activity because of an increase in temperature. Furthermore, fallow periods have been reduced by population pressure. Farming seems to have become riskier, partly because many domestic crops are no longer grown and seed quality has deteriorated. As agricultural land becomes scarce, more and more pasture is farmed. Land fragmentation and progressive erosion are also reducing the amount of pasture available. Lack of sufficient animal feed has reduced the quantity of livestock and upset the balance between animal husbandry and agriculture. In communities oriented towards potato production and in villages specializing in horticultural crops, purchasing fertilizers compensates for the shortage of manure. Near the Pilcomayo River, serious water pollution from mining threatens agricultural production and exacerbates health problems. Almost everywhere, deforestation makes firewood more difficult to obtain. Gathering wood requires more time and money as it now comes from further away, and many people have switched to bottled gas or purchase their firewood from transport agents.

Climate change means that the old climatological indicators are increasingly difficult to use and that droughts and strong winds affect people more adversely than before, especially in the higher zones. Predicting the weather seems more difficult than before. People in the lower zones complain about the incidence of pests, which nowadays seem to occur more frequently than in previous periods.

Land and water as criteria for well-being and social mobility

In order to gain more insight into the importance of land and water for well-being, we asked people in the 17 villages to classify the local population into four groups – namely, the poorest, the poor, the intermediate rich and the rich –

and we collected information about their resources base. It is interesting to see that access to land and water is a relative matter: whereas in some villages people with less than 2ha appear as 'rich', people owning the same amount at other locations are 'poor'. Rather than expressing well-being (or poverty) on the basis of the 'area of land' in general terms, people always take into account the quality of the land: reference is made to the area of cultivatable land, which very much depends on the geographic location. Access to irrigation water was not mentioned as an indicator for well-being simply because irrigation did not exist in the majority of cases. In the communities that do have irrigation, water use is usually a reflection of landownership.

The 'rich' and 'intermediate rich' do have larger areas of cultivated land at better locations; but differences in welfare are expressed mainly on the basis of other criteria. For example, the 'rich' are characterized by having more animals (access to pastures); the production of dung and the availability of animal traction help them to realize higher production levels. Richer people are also characterized as having the capacity to work hard, being well educated, performing communal tasks (*cargos*) or being a responsible community member. Larger families often have a higher income level because they have more members and each of them contribute. The poorer groups usually have worse land (erosion) and less of it; but, importantly, they have a small number of animals (no pastures), have a lower production level (lack of dung and dependency on renting labour or tractor services), are considered to be incapable of working hard ('uneducated', 'lazy', 'drunk') or are excluded from social relations.

In other words, within the *comunidades* the differences in well-being are not so much a reflection of access to land or irrigation. Instead of referring to access to natural capital (land and water), differences in well-being are mainly reflected in human capital (people's capacity to work hard and level of education), produced capital (ownership of cattle, a tractor, etc.) and financial capital (mainly originating from migration). The importance of the different forms of capital cannot be understood without taking into account locational aspects and, more specifically, the agro-ecological situation. There is a clear distinction between villages with agricultural potential (commercial production of potatoes, fruits and horticultural crops) and the other villages, where the population is devoted to subsistence farming, migration or non-agricultural income sources.

At the micro level, analysing in detail patterns of social mobility in relation to the kind of activities pursued between 1998 and 2004,[5] factors other than access to water and land also play a more crucial role in determining whether people can or cannot escape from poverty. An analysis of the characteristics of the more successful group (those experiencing upward mobility) shows that for most people the first step in success is inheriting sufficient land at the right location (especially in zones with agricultural potential). In the more marginal areas, the first accumulation is usually based on incomes originating from migration. Whether or not people will experience upward mobility after this first accumulation depends on subsequent patterns of investment.

Table 7.3 *Triggers and causes of upward and downward social mobility*

Examples of upward social mobility, ways out of poverty	*Examples of downward social mobility, ways into poverty*
Inheriting land	Death of family members/loss of labour force
Renting land to others (and/or collaborating with neighbours (sharecropping)	Illness and accidents (medical costs)
Investments in irrigation systems and/or erosion control, and intensification of production	Bad weather conditions/droughts/mis-harvest Erosion and declining production/lack of fertilizers, etc.
Expanding the cultivated area by purchase of land and/or sharecropping	Giving away land (anticipating inheritance)
Introduction of new crops and/or successful commercialization of crops and/or other goods	
Expanding animal stock	Selling animals/death of animals
Purchase of tractor (and servicing the rural population)	
Purchase of mill/milling cereals for the community	
Purchase of *motosierra* and processing eucalyptus wood	
Migration to Argentina and/or Santa Cruz and/or other destinations	Migration – bad luck

Source: author's fieldwork, Zoomers (1998)

Looking more specifically at the investment behaviour of the successful group after the first accumulation (see Table 7.3), we see some people focus on buying additional land or a *motobomba* (pump) for irrigation. The majority, however, are not capable of doing so because they have no access to water or no land is available for further expansion. They prefer an exit strategy when there are no opportunities for upward mobility without leaving. A very common strategy among these people is to use their money to purchase an urban plot in Sucre and to educate their children. Upward social mobility seems to be linked to outward-oriented strategies in which parents give priority to facilitating the departure of their children; rural life is seen as something inferior – as a non-future – and only a small group still has the desire to become full-time farmers.

When focusing on the other group – namely, those experiencing downward mobility – it becomes clear that the capacity of people to recover from crisis (e.g. death and accidents) and to overcome misfortune depends on their animal stock (which they can sell in an emergency), their flexibility in reallocating their labour (i.e. migration), their educational level (which very much determines a person's ability to find a job) and their social network. Most of those who are trapped in a vicious downward circle are older people and single-head households (small and 'incomplete' families) without reserves (animals) and insufficient labour to compensate for the loss (see Table 7.3).

In the longer run, the evolution of wealth (and of poverty) will depend on inheritance systems – namely, the passing on of land, water, animals, etc. to the next generation. Each community seems to have its own institutional regime: in some cases, all of the children (i.e. both sons and daughters) receive an equal share of each parcel and the accompanying irrigation rights; in other communities the sons receive the best part of the land; while in yet others the best land goes to the oldest children or remains with the youngest child. There is no fixed, local or indigenous 'rule': the inheritance system is largely dependent on location and varies in the extent of land inherited.

In conclusion, people with access to better land in an area with agricultural potential have a better starting position for accumulating capital. People without such a favourable position depend on migration for their first accumulation. In the subsequent phase, however, the majority do not buy additional agricultural land or invest in irrigation (opportunities for expansion not available), but buy an urban plot in Sucre and invest in the education of their children (exit strategies). Most of the 'rich' show patterns of dual residence: they keep their land in the community while investing in their Sucre life. Leaving the village is the ideal sought by most people who are experiencing upward mobility. After arriving in the city, however, many still see themselves confronted with a difficult situation and keep their land as a safety valve.

Conclusions and policy implications

In the current debate about how to reduce poverty, much attention focuses on the need to optimize the use of natural resources. Government proposals for regulating landownership through modernization of the registry of property rights and the creation of irrigation systems promote rural development. The 2009 constitution land reform provisions also include downsizing large land tenure rights. At the same time, in Andean policy and development circles, there is continuous debate about what should be done: impose market rules on the communities and externalize water and land control, as increasingly happens in neoliberal land and water policies of neighbouring Peru, or believe in community survival and collective action (as promoted in 'Andeanist' currents).

This chapter shows that 'indigenous' patterns of land and water tenure are restrictive by not providing opportunities for upward mobility. Access to land is often fragmented and the majority do not have access to water resources. People are not or are no longer able to adapt to, or compensate for, the ecological disadvantages, as is often suggested in the literature. Natural resources have become extremely fragmented and people are faced with both scarcity and degradation.

In the current situation, it does not appear that neoliberal land policies and improving individual land titles' registration will alleviate poverty for most families. In most communities investigated in the case study, people were not concerned about the insecurity of tenure. Even with a property title, many are not in a position to get credit for agricultural production because their property is a marginal plot of eroded land. In addition, many do not aspire to become commercial farmers. People who decide to leave frequently cannot afford to sell their lands. They leave it to family members or neighbours as a safety valve. The people left behind are often not in a position to pay for the land. Many people in the Andes do not claim formal, individual (private) property titles to land and water because they fear outside interference (land taxes, water tariffs, etc.). The community has its own ways of recognizing the validity of individual entitlements in terms of rights that establish access and control of resources within the community. In the current debate, much emphasis is given to the importance of indigenous rights and territorial autonomy; but little attention is paid to how to improve land management practices within a community (e.g. re-establishing the vertical orientation, implementing programmes to combat environmental degradation, etc.) and how to deal with people whose aspiration is to leave rural life and improve their welfare.

In water policies, there is a great deal of emphasis on the need to expand irrigation schemes; but the possibilities for irrigation are restricted and only a small number of places have agricultural potential. In addition, many villages do not have a history of irrigation. They are more experienced in dealing with droughts than with water. In many cases, irrigation institutions have weakened or disappeared because of drought, fragmentation of water rights, migration, etc. Discussing the possibilities for irrigation projects is usually wrongly presented as a technical solution; but it can be a highly sensitive cultural matter. Thus, interventions will often be accompanied by conflicts and contrasting interests (see Box 7.1). It is only in a few places where projects to build new irrigation systems are feasible from the economic point of view (i.e. in the areas with agricultural potential). At the same time, however, it is important to acknowledge that there are many low-cost ways to improve people's access to water (e.g. rain harvesting, multiple-use systems, water rights redistribution, shortening irrigation intervals, etc.). Small investments might help rural people to improve their production, cope better with droughts (risk reduction) and develop drinking water (sanitation). Illness and death are among the major causes of crisis and downward mobility, and so investments in drinking water and sanitation will help people to become more productive, earn higher incomes and reduce their costs by avoiding the need for medicines.

In the Andes, land and water are important elements for 'survival'; but in most places improving access will not help people in their attempt to escape from poverty. Although it has been suggested that understanding irrigation is necessary to understand Andean communities (Gelles and Boelens, 2003), Andean life is often shaped in response to drought. Whether people are capable of realizing upward mobility depends more on migration opportunities and access to education, which are important for local identity, production and natural resources management.

In recent decades, land fragmentation and the fragmentation of water rights have eroded Andean technology, which is not going to magically reappear in response to new land and water policies. Given the current patterns of land and water management in the research area, it is clear that regulating land tenure and improving irrigation might be attractive to people who want to specialize in the production of potatoes or horticultural crops; but it is important to acknowledge that opportunities for such strategies are restricted to a small number of locations. In most of the villages included in the above research in Chuquisaca and Potosí, the opportunities for agricultural intensification are extremely limited, and rather than investing in land or water, people prefer to invest in a future elsewhere by giving their children a good education or buying an urban plot in Sucre.

Notes

1 Examples of new laws that 'blend liberal and pro-indigenous tenets' launched during this period include the *Ley de Reforma Educativa* (education reform) in 1994, the *Ley de Participación Popular* (municipal decentralization) in 1994 and the *Ley INRA* (land reform) in 1996 (Gustafson, 2002, p268).

2 Part of the population lived in the *comunidades libres* and did not have to work for the *hacendado*. But this group also had to deal with feudal relations and was largely excluded from civil society.

3 This research, called the PIED Andino Project, was carried out by a team of Bolivian and Dutch researchers (see Zoomers, 1998). Participating researchers were Antonio Aramayo, Édgar Guerrero, Miriam Vargas, Dicky de Morrée and Jan Willem le Grand. Miguel Morales and the author were the project coordinators.

4 In addition to irrigation, there are other examples of an adaptive engineering structure, such as raised and sunken fields, including *qochas* (systems of ponds in areas of limited water supply) and *bofedales* (small hollows in highlands with deep black soils, which provide especially rich pastures that do not dry out in the dry seasons) (Morris, 2004).

5 During 2002 to 2004, I returned to the 17 villages in which I had conducted my research and tried to identify the people who had shown upward or downward mobility.

References

Assies, W. (2000) 'Land, territorities, and indigenous peoples' rights', in A. Zoomers and G. van der Haar (eds) *Current Land Policy in Latin America: Regulating Land Tenure under Neoliberalism*, KIT Publishers and Iberoamericana/Vervuert Verlag, Amsterdam, Frankfurt and Madrid, pp93–110

Boelens, R. and M. Zwarteveen (2003) 'Water, gender and 'Andeanity': Conflict or harmony? Gender dimensions of water right in diverging regimes of representation', in T. Salman and A. Zoomers (eds) *Imaging the Andes: Shifting Margins of a Marginal World*, CEDLA Latin America Studies no 91, Aksant, Amsterdam, The Netherlands, pp145–166

Concertación (2005) *Interdisciplinary Research and Capacity Building Program on Peasant and Indigenous Water Management in the Andes*, Project proposal, Sub-Department Irrigation and Water Engineering, Wageningen University, Wageningen, The Netherlands

de Morrée, D. (2002) *Cooperación campesina en los Andes: Un estudio sobre estrategias de organización para el desarrollo rural en Bolivia*, Netherlands Geographical Studies 298, KNAG, Utrecht, The Netherlands

Gelles, P. (2000) *Water and Power in Highland Peru: The Cultural Politics of Irrigation and Development*, Rutgers University Press, New Brunswick, NJ, and London

Gelles, P. and R. Boelens (2003) 'Water, community, and identity: The politics of cultural and agricultural production in the Andes', in T. Salman and A. Zoomers (eds) *Imaging the Andes: Shifting Margins of a Marginal World*, CEDLA Latin America Studies no 91, Aksant, Amsterdam, The Netherlands, pp123–144

Grillo-Fernández, E. (1998) 'Development or decolonisation in the Andes?', in F. Apffel-Marglin (ed) *The Spirit of Regeneration: Andean Culture Confronting Western Notions of Development*, Zed Books, London and New York, pp193–243

Gustafson, B. (2002) 'Paradoxes of liberal indigenism: Indigenous movements, state process and intercultural reform in Bolivia', in D. Maybury-Lewis (ed) *The Politics of Ethnicity: Indigenous Peoples in Latin America*, The David Rockefeller Center Series on Latin American Studies, Harvard University, Harvard University Press, Cambridge, MA, and London, pp267–305

McEwen, W. (1975) *Changing Rural Society: A Study of Communities in Bolivia*, Oxford University Press, New York, NY

Mitchell, W. P. (1976) 'Irrigation and community in the central Peruvian highlands', *American Anthropologist*, vol 78, pp25–44

Morris, A. (2004) *Raised Field Technology: The Raised Fields Projects around Lake Titicaca*, King's SOAS Studies in Development Geography, University of Glasgow, Ashgate, Aldershot, UK

PRSP Bolivia (2001) *Poverty Reduction Strategy Paper*, Republic of Bolivia, La Paz, http://siteresources.worldbank.org/INTPRS1/Resources/Country-Papers-and-JSAs/bolivaprsp.pdf

Strobele-Gregor, J., B. Hoffman and A. Holmes (1994) 'From *Indio* to Mestizo … to *Indio*: New indianist movements in Bolivia – social movements and political change in Latin America', *Latin American Perspectives*, vol 21, no 2, pp106–123

UN (United Nations) (2003) *Water for People, Water for Life*, World Water Development Report, www.unesco.org/water/wwap

Valladolid-Rivera, J. (1998) 'Andean peasant agriculture: Nurturing a diversity of life in the chacra', in F. Apffel-Marglin (ed) *The Spirit of Regeneration: Andean Culture Confronting Western Notions of Development*, Zed Books, London and New York, pp51–88

World Bank (1996) *Bolivia: Poverty, Equity and Income: Expanding Earning Opportunities for the Poor*, Country Operations Division I, Country Department II, Latin America and the Caribbean Region, Washington, DC

Zoomers A. (ed) (1998) *Estrategias Campesinas en el Surandino de Bolivia: Intervenciones y desarrollo rural en el Norte de Chuquisaca y Potosí*, KIT–CEDLA–Centro de Información para el Desarrollo (CID), PLURAL, La Paz, Bolivia

Part III

Tensions and Mergers among Local Water Rights and National Policies

8

Water Laws, Collective Rights and System Diversity in the Andean Countries

Jan Hendriks

Introduction

In the Andean region, increased attention is being paid to the local management of water resources.[1] Indeed, community control of rural and indigenous territories and of their historic irrigation systems is impressive. Academic research and training activities, such as those of the Water Law and Indigenous Rights (WALIR) programme, have contributed to the understanding of local water rights. These activities have revealed the divide between national legal frameworks and local and customary-law norms. Many studies evinced the need for the Andean countries to integrate legal pluralism within their regulatory frameworks. Research promoted by the WALIR programme provides a vast body of documentation on indigenous legal concepts, peasant community water management and case studies from several countries.[2] This chapter discusses the relative legal security and the complex relationship between individual rights and collective system in four Andean countries, namely Bolivia, Chile, Ecuador and Peru. Two core issues are explored: recognition of management system diversity, and the advisability of granting water use rights to the system as a legal entity rather than to individuals in order to recognize local law collective rights consistent with legal pluralism in the Andean region.

This chapter does not deny the importance of a national-level legal framework for regulating water affairs. But to the extent that the national framework makes room for local notions of law and organizational arrangements, it will strengthen the identity of the people within the user system.

Such recognition of local norms and practices could create a national water culture that would contribute better to people's well-being and a sense of equity while promoting stewardship of water resources and environmental sustainability.

Legal security of water rights

Water legislation in the four countries generally pursues the following aims:

- to provide a framework of legal security for water use by individuals and corporate bodies who need water for their personal use and economic activities, while protecting the rights of third parties (not necessarily users) who might be harmed by such use;
- to promote water use that is technically and economically efficient in allocation, distribution and application; and
- to encourage development, protection and conservation of water resources, user systems and other related resources.

These aims coincide with water's three basic functions: the social function, the economic function and the environmental function. Almost all water laws in the Andean countries allocate water to individuals and corporate bodies. This implicitly discriminates between users in terms of opportunities for economic development. Water laws provide only simple declarations regarding the environmental function. They lack enforceable provisions regarding watershed management.

The relative legal security provided by a law for rights of water users depends on the type of right granted. From greater to lesser legal security, the different types of rights in the Andean countries can be ranked as follows:

1 *Property right.* This is the right to use water as a real property entitlement, making the water use 'asset' the property of the holder for an indefinite period, with the full range of powers to engage in transactions recognized by private civil law.
2 *Concession.* This type of right is officially recognized in public registers, giving the holder a fairly good 'horizon of security' in terms of time and power to use the water.
3 *Licence.* This is lower-ranked in terms of legal security because the government entity or official granting the licence retains power to change, revoke, expire or terminate it.
4 *Permit.* A permit has less legal security and provides less water availability ('surplus water') than the above.
5 *Authorization.* In principle, all of the above categories involve an authorization; but, as in the General Water Law of Peru, an authorization refers to the right to specific *ad hoc* use of water for a temporary activity (e.g. to build a project).

6 *Registration.* This concept recently has been introduced in Bolivian legislation, specifically for rural communities and indigenous peoples.
7 *Other documentary grounds.* A user must possess some type of documentation of a right (e.g. by custom), such as historical deeds, inheritance, ordinances etc.

This typology refers to the security that the holders may have under national laws according to notions of formal or 'positive law'. These rights may not necessarily coincide with local socio-legal realities of management, authority and security.

The degrees of legal security for the different categories may vary according to the person who holds the right. Inequalities among users can affect security as the following examples illustrate. One factor that affects security is economic inequality. The transactional costs, time and advisory assistance required for a holder to register these user rights provides an advantage for people with greater economic capacity and a disadvantage – and less security – for people who cannot afford or who lack the skills to pursue the legal paperwork. This is the case under Chile's water law, where obtaining water use rights involves meeting a series of technical and legal requirements that are costly to the applicant (see Military Government of the Republic of Chile, 1981). In some cases users must pay legal costs to protect their rights because, although they hold water use rights, third parties can submit competing applications. For example, several user organizations in northern Chile have often had to use most of their annual budgets to fund lawsuits to defend their water rights. In the last decade, the National Corporation of Indigenous Development (CONADI), the National Irrigation Commission and other public institutions in Chile have subsidized the defence of water rights of low-resource people. Some user groups, not necessarily indigenous, are still legally vulnerable in their water use rights based on use, customs or on historical claims and territorial legitimization.

A second element reducing legal security of water rights is the degree of knowledge required to understand complex legislation. The Water Code of Chile is precise and internally consistent, but very legalistic and often in conflict with both other laws (e.g. the politically supported Mining Code) and the 'weaker' laws that are difficult to enforce in practice (e.g. the Indigenous Law). The situation in Bolivia, Ecuador and Peru is worse, with a huge body of complementary regulations and clashes with sectoral laws (see, for example, Military Government of the Republic of Peru, 1969; Military Government of the Republic of Ecuador, 1972; Congress of the Republic of Peru, 2002; Boelens et al, 2005; Bustamante, 2006; Palacios, 2006). Because socially and economically vulnerable groups generally have less formal education and literacy, they are disadvantaged by not knowing or being able to determine the scope of regulation. Especially in Bolivia, Ecuador and Peru, many indigenous and other user groups do not know that their longstanding rights must be registered or declared before some agency to become legally enforceable. In the event of a lawsuit regarding scarce water resources, these vulnerable groups are powerless against those who have greater legal knowledge, technical and economic capacity. This was demonstrated when the neoliberal Chilean Water Code was introduced in 1981, and now is an increasingly controversial issue in Peru with its introduction of many neoliberal decrees.

A third factor seriously affecting the security of water rights is the consistency and capacity of administrative and judicial institutions. With recent neoliberal policy implementation, the capacity of institutions responsible for granting and administering water use rights, as well as staffing, professional expertise and operational resources have been sharply cut back. Applications and complaints are backlogged and there are no resources for measurements or other technical inspections. This combines with increasing complexity and increased problems regarding availability, quality and uses of water, intensified economic and industrial activities, and more complex legal norms. These factors all negatively affect security and justice in water rights and transparent administration of water uses (Hendriks et al, 2003).

After having exhausted administrative appeals and attempts at reconciliation, the parties affected must turn to the courts. However, the fragile nature of the judicial system in most Andean countries is well known. Accordingly, the types of rights based on lesser legal security often call for more judgement and intervention by cognizant authorities, through their officials. For this reason, these cases lend themselves to arbitrary decisions. For example, in Peru, during the last decade a large number of licences were improperly granted. Since 2004, Peru's Ministry of Agriculture has been implementing a Programme to Formalize Water Use Rights (PROFODUA) in order to partly overcome the prevailing informality in granting water rights. However, such formalization also entails homogenization and denial of the variety of local contexts and rights (see Guevara-Gil, 2006; Boelens, 2009), and will certainly trigger judicial complaints.

A fourth element meriting attention is the fact that much of the analysis and a considerable part of the conflicts concentrate on the issues of legal security with regard to granting of rights to use water, rather than verifying whether distribution and use of water are enforced according to the rights acquired. This requires an analysis of whether day-to-day practice follows the law. In administrative and judicial proceedings for granting rights, users and authorities each make great efforts; but little is done to check whether the rights thus obtained are actually applied in water distribution. It seems that legal attention tends to focus on legalizing the right more than on actually enforcing the use of the volume of water to which the user is entitled.

Water laws in almost all Andean countries call for water measurement devices in user systems. However, most user systems (for irrigation and for drinking water supply catchments, domestic connections, etc.) have no control, measurement or regulation infrastructure. Moreover, user organizations' management capacity to implement and operate monitoring systems for water in their systems is normally quite weak or issues are resolved by informal collective social control.

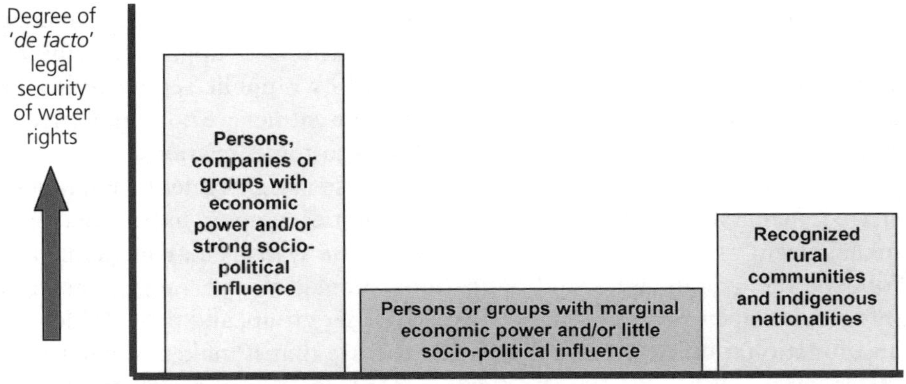

Source: author's elaboration

Figure 8.1 *Schematic presentation of three societal segments and their degree of* de facto *legal security*

Finally, economic models being implemented in the Andean countries legally and *de facto* favour the legal security of segments with greater economic power and socio-political influence because of their greater capacity to negotiate, argue with studies and to 'purchase' their rights. In many cases, Andean governments have attempted to give some degree of legal protection to rural communities and indigenous peoples through 'special' laws (see Boelens et al, 2005; Roth et al, 2005). International treaties such as the International Labour Organization (ILO) Convention No 169 on Indigenous and Tribal Peoples in Independent Countries have contributed to making progress. Nevertheless, the often essentialist content of these efforts has limited the benefits for most rural and indigenous families.

As a result of progress made in protecting indigenous groups and the already favoured position of economically powerful segments, it can be argued that other societal groups who are vulnerable but do not belong to rural communities or indigenous peoples are the great losers. They include, for example, the hundreds of thousands of smallholder and micro-holder farming families who attempt to subsist on the Peruvian coast where rural communities have faded. The hypothesis that there are three major diverging segments of society in terms of legal security for water rights merits discussion and is shown graphically in Figure 8.1.

Individual rights and system rights

Debates over legal frameworks for water resources in the Andean countries involve the fundamental issue of whether rights should be individually held or assigned to the entity managing the system for collective use. This debate requires defining the scope of a 'system'. A system could encompass an entire watershed as an integrated system or could refer simply to drip irrigation within a given plot ('high-tech irrigation system'). The degree of collectivity also varies.

Water use rights for a hydropower station are normally granted to a single holder who supplies electricity to many users. The same applies to drinking water supply systems where water is provided by a public service utility or company. Although they serve multiple users these entities are not a collective in organizational terms, but rather have a supplier–customer relationship.

In the context of indigenous communities, system rights refer to a notion of shared ownership and collective responsibilities among users regarding management of water. These collective water-use systems may be defined as follows: the overall rights and obligations, management arrangements and infrastructural elements are shared by a given user group, and provide identity and functional unity to the collective in the use that it makes of particular water sources. When this chapter mentions the term 'system rights', it refers to water use rights of a collective water-use system under the above definition.

During the last two decades, government proposals to change water laws in the region favoured individualized water rights. This fits the neoliberal spirit and is apparently driven by the notion that contemporary societies must grant sufficient legal security in order to ensure full opportunities to individuals. Such thought does not recognize that opportunities and security for individuals can prosper, in part because of community values and mechanisms for management of shared territories, natural resources and infrastructure in the region. These elements provide a frame of reference within which individuals can develop their own opportunities and local 'governance' that resolves conflicts. Moreover, the focus on individuality does not consider the state's limited ability to enforce individual rights given the fragile nature of public institutions in most Andean countries. Under such conditions, the existence and legitimacy of local systems is significant in preventing greater societal chaos regarding the use of water and other resources.

Chile is often cited as an example of the trend to individualize water rights (see Chapters 2 and 10 in this book). In fact, in Chile, private parties enjoy legal security for water rights, which is an important factor in economic development. However, in the country's history of progressive occupation of territory and resources by immigrants and assertion of private initiative, native ethnic groups have been marginalized. The reality of Chile is atypical because in the other Andean countries ethnic predominance and the existence of collective systems, uses and customs are so obvious. These systems often serve as a safety net amidst the frail institutions for local governance. Chile also has conflicts caused by individualizing water rights in Andean territories where there is water scarcity and a presence of indigenous peoples and disparities among economic sectors (see for example, Hendriks, 1998; Gentes, 2006). As in other countries, collective responses in Chile pursuant to the Indigenous Law have helped to establish collective rights to territories and natural resources, including water. The aim of these efforts has been to prevent deterioration of water systems.

In Peru, water use licenses and permits are individual in principle, but they are granted under a strongly system-based approach. A Crop and Irrigation

Plan (PCR) sets the allocation of water to each user for the season according to the total water demand in the system and forecasts of available water. The fee to be paid by water users is calculated volumetrically in proportion to demands in each irrigation sector. Allowing water use to be determined by local systems is important because there is insufficient state presence to enforce individual legal rights to water use rights (compare Del Castillo, 2001; Guevara-Gil, 2006; see also Chapter 9 in this book).

In Bolivia, state institutions are practically absent from water use regulation or control, except in those use sectors with political power such as mining, hydropower and, to a lesser degree, urban water utilities. In the sector consuming the most water – irrigated small farming, mainly in the highlands – user rights are based on a system rights approach to which individual rights are subordinated. This approach is accepted even by holders of private water rights for irrigation or other purposes in the same territory. Current legal and policy plans also favour collective system rights (compare Bustamante, 2006).

In Ecuador, the water law implicitly establishes water rights concessions for supply systems by stipulating that such rights will be granted to municipalities, provincial councils, public or private bodies and private individuals. For irrigated agriculture the law does not differentiate between individuals and corporate bodies. However, in practice water concessions are granted to private parties and legal entities representing collective systems. For irrigation projects, concessions are also granted to a 'project' (i.e. for a system that is built, and therefore to a municipality or organization of irrigators). This is also the case with state irrigation systems including those transferred to user organizations. Conflicts between the state and community systems with their notion of collective water rights (widespread, especially in the highlands) mainly arise when the state deals with individual users of established collective systems. They usually concern state attempts to impose requirements for applications, payments and other water rights procedures (Hendriks et al, 2003). The 2008 Constitution aims to provide greater autonomy and legitimization to both public and collective community water management, breaking away from the last decades' neoliberal policies. In 2009, a new water law should have been enacted, but by the end of the year this still was not accomplished. The new law concepts – present in the Constitution but to be elaborated in the new water law – consider water as a public good to be managed mainly by the state administration and by local community groups, and also grant water rights to nature. Allocation priorities will respond to social needs, as the new Constitution prescribes the need for redistribution of water rights within a two-years deadline. However, materializing this in one coherent legal body and in everyday practice is difficult because powerful economic interests continue to favour private water rights, and water accumulation in the hands of the few is a dramatic background reality.

Initiatives to adapt legal frameworks for water resources should take into account the reality of most Andean countries. They should establish preferences that reinforce the legal security of collective user systems without harming the recognition of individual rights. It is also important to grant these collective rights to existing organizations that represent the historic user systems rather than creating new organizations. This approach would have many advantages and make new legislative initiatives more viable for several reasons.

First, the system approach to formalizing agricultural water rights of users sharing water catchments is well recognized and socially accepted in the Andes. Water users identify with this arrangement and their culture has contributed historically to building such systems. Broad support for this concept will mean that legislative proposals adopting this approach will make such initiatives more socially and politically viable and they will work better than approaches focusing exclusively on individual rights.

Second, the system approach avoids the long, costly regularization and formalization procedures required if water rights are granted on an exclusively individual basis. Several reasons make it socially unacceptable and almost unworkable to grant rights individually. These include the existence of a large number of small farms in most Andean countries that each would have to pursue water rights certificates. The procedures are expensive, requiring technical and legal dossiers, an attorney, publications in the newspaper, procedural expenses, etc. Most small farmers are poor and they would be at a serious disadvantage if required to regularize their rights alongside those with greater economic power. If the state sought to ensure equal opportunity and justice it would have to create a costly subsidy system to enable the poor to participate in the legal and administrative system for individual rights to water. In a number of local areas, users belong to more than one collective system in different catchments and would have to register several water rights to gain legal security for irrigation water from an array of sources.

Third, legal recognition for system rights would reinforce the legitimacy of ownership and responsibility for the system's infrastructure. This is important where notions of ownership have been watered down and divided up among the state, local institutions and user groups, and where the ditches tend to belong to 'the territory', but the legal ownership is left in 'no-man's land'. Greater legitimacy for irrigation systems and the collective water rights and collective ownership of infrastructure would contribute greatly to ensuring the shared responsibility for maintaining and improving the system by the organization in charge. This should help in imposing water fees and encouraging users to contribute for the benefit of their water systems. Today, users tend to perceive fee systems established by legal norms as a tax that produces no direct benefits for the system. Greater legal legitimacy for collective water rights and legal ownership of the system's infrastructure would aid in identifying the fees as a source of benefits for the water use system.

Fourth, establishing individual water use rights with the possibility of transferring such rights outside the system undermines the system's wholeness and stability. It would enable reducing the volume of water in the system while another system could be overloaded to the point of exceeding its capacity.

Transfers also affect irrigation scheduling, flow rates at the 'tail' of the system, and distortion of the ratio of flow rate to infrastructure. This has occurred, for example, in Chile (Hendriks, 1998). If a system-rights approach is applied, the wholeness and operational logic of these parameters would be kept intact, even if the users vary or ownership changes through purchase or inheritance of cropland in the system.

Fifth, and directly related to the preceding point, a system rights approach would strengthen users' legal security, enabling indigenous and poor small-farmer families to keep and defend their water rights *vis-à-vis* outside stakeholders with greater economic power and influence.

Sixth, under collective water rights, no legal decisions are required regarding transfers of water among users within the system. The users' organization holding the collective rights decides whether a water transfer is allowed under its bylaws or regulations. This is more compatible with local realities.

The seventh advantage for the system rights approach is adaptability to the different local realities, thereby establishing conditions for genuine legal pluralism under legislative frameworks regulating water resources. The law must establish individual security in terms of justice, equity, protection and access to water for each user within the collective system rights. Respecting this framework of individual security, the legislation may establish criteria that recognize plurality in managing systems on the basis of collective rights.

It is not necessary to define a national legal criterion regarding the model for distributing water within a collective system. The law could indicate a range of criteria in which the preference for one or the other criterion could be defined within the cognizant decentralized agencies, according to the realities of the region. Next, such criteria might be defined by the organization in charge of the system or territory (e.g. the rural community or indigenous group). This would be consistent with ideals of legal pluralism by deferring to local preferences and ways of thinking.

Finally, granting rights to a system involves the structuring, administration and control of rights within a hydrologic unit such as a micro-, meso- or macro-watershed. In an individual water rights approach, it is difficult to coordinate all the legal participants in a watershed for purposes of administering rights, participating in consensus-building and resource development, conflict management, planning, distributing and controlling water supply and demand. The system rights approach makes it possible to structure governance much better by distinguishing between the internal self-management and problem-solving responsibilities within each system, while allowing for the system to participate in water management within the macro-territory.

Diversity of systems

As in the rest of the world, the Andean region has an enormous diversity of water use systems in the different usage sectors and within each sector. In generating hydroelectric energy, there is normally a single water rights holder (a private company or the public service utility) which runs high volumes of water through the turbines and returns them to the river. This is a non-consumptive use, but with an impact on the hydrological regime. Mining, fish-farming and tourism are also usually undertaken by a single rights holder and they often have negative effects. Water supply system rights are held by a single party (company or service); but the system serves a large number of users who have no legal right to the water beyond a contractual relationship with the provider and relevant government regulations. Because of its importance, the rest of this section discusses diverse agricultural irrigation systems.

The concept of an 'irrigation system'

Normally, an irrigation system is visualized physically as a coordinated interdependent infrastructure that catches, conducts, possibly stores, distributes and applies water to cropland. It can belong to a private party or a group of users. The system occupies a given agricultural area with a network of ditches.

The diversity of irrigation systems in the Andean region creates a conceptual problem because their organization may not match the engineering concept of agricultural areas concentrated and served by a congruent network of ditches. Instead, they often serve areas scattered through the territory, sometimes linked with other sources, waterways and networks. What is considered a 'system' may actually be a group of relatively independent units having no interconnection of canals. So, for example, the irrigation system of the Upper Piura (Peru) consists of over 60 intakes from a single river, each supplying specific small areas. Almost the only interdependence among these subsystems lies in the fact that there are agreements for scheduling to rotate the intake of water at each point. Aside from that, subsystems are independently operated, maintained and organizationally managed. Because this relative independence complicates management of the overall system in a territorially continuous coastal valley, it is easy to understand why the legal status and functions of the 'users' board' in the Peruvian highlands are hard to enforce. It is often a challenge to integrate independent irrigators' committees and groups of users scattered through the territory with regard to the water sources that they use. Accordingly, some users' boards in Peru's highlands seem to exist for other purposes, services or advocacy roles, without necessarily performing the functions of operating or maintaining the 'irrigation system'.

This shows why we need to analyse the dysfunctionalities that may arise between the physical system and the organization that must manage it. Some problems have arisen from the organizational models introduced by legislation or related bureaucratic demands, or as an outgrowth of local adjustment to them. For example, Peruvian law applies the same organizational arrangements and administrative requirements to a users' board with almost 25,000 users and 119,000ha (case: Chancay-Lambayeque on the northern coast), as for a board that has only 135 users and a total of 1500ha (case: Santa Rita de Siguas in southern Peru).

Diversity in irrigation systems goes beyond the territorial scope. In much of the Andean region, the concept of system is not understood so much as the physical location of farmland and ditch networks, but as a system of rights, obligations and culturally based management of one or more water sources shared among users. Many of these systems date back to pre-conquest times and each has evolved uniquely even as the territory has been filled with more contemporary systems. So, the infrastructural concept that a given agricultural area is served by a single irrigation system must be revisited, accepting that there may be many overlapping and complementary systems for catchment, conduction and distribution comprising a system to provide irrigation for a particular territory and to users who may belong to several systems at once.

Diversity and coexistence of systems

Many irrigation systems are distinguished because they carry different 'kinds' of water with regard to the source of water or the structure of the system. So, for example, in the communities of Ispacas and Yanaquihua (province of Caylloma, region of Arequipa, Peru), there is 'pond water', 'mountain water' and 'hacienda water', and the rights, obligations, management arrangements and universe of users for each type of water comprise a well-identified irrigation system. Similar situations have been reported, for example, in the Valley of Colca, Arequipa, in the region of Cusco, etc. (e.g. Valderrama and Escalante, 1988; Gelles, 2000; Guevara-Gil et al, 2002; Bustamante, 2002).

A good example of this 'coexistence of systems' is the situation in the locality of Tiquipaya (Cochabamba, Bolivia), where as many as eight irrigation systems have been identified in a single local territory, each managed according to different parameters:

- The *Machu Mit´a System*, taking the water discharged from the Khora River. This system has been used since pre-Incan times. During the 16th century, allotments were structured on the *ayllus* of the 'Royal Town of Indians' (Pueblo Real de Indios), so the water would reach most of the irrigators in Tiquipaya (about 830 users with nearly 1030ha, distributed among 55 allotments, or *asignaciones*, in 13 communities).
- The *Lagum Mayu System*, comprising the water dammed in the lake of the same name. Its origin is probably pre-Colombian, but its reservoir capacity was increased by hacienda owners in the late 19th and early 20th centuries (1924). Water rights from Lagum Mayu total 48 allotments in 9 communities, representing about 450 users on some 650ha of cropland. The reservoir operates by brief releases, called *largadas*, six or seven times a growing season, at intervals of 14 to 28 days, until the stored volume is depleted.
- The *Sayt´u Kocha System*, built in the late 19th century, serves allotments in the southern part of Tiquipaya. Water rights for this system depend on the contribution originally made by each hacienda owner or landowner.

- The *Chankas System* (11 lakes), established in 1967 by the communities of Montecillos and Sirpita (both within Tiquipaya), in which each user made equal contributions to the construction, so each has the same water right, regardless of the land area that they have.
- The *El Tajamar System*, where the water rights are also based on the contribution to construction. The system consists of an underground water catchment gallery on the Khora riverbed, managed by some 80 rural families, ex-colonists and *piquero* settlers from the sector of Montecillos.
- The *National Irrigation System No 1 (SNR-1) of the La Angostura Dam.* In 1946 National Irrigation System No 1 began operation, with water that came through the La Angostura Dam, with a reservoir capacity of 85 million cubic metres to provide irrigation water for large areas of cropland around Cochabamba. The SNR-1 System is administered by the government and serves some 530 users in 8 communities in the territory of Tiquipaya, with approximately three to six opportunities of irrigating per season.
- The *Springs System* exists in areas where the aquifer surfaces in the central part of Tiquipaya (about 30 to 40 local springs without any surface interconnection). They are utilized by roughly 850 users. Each spring, with its own pond, is tapped by an independent group of users having its own internal irrigation roster.
- *Private wells.* Some farmers have drilled wells in order to have totally independent access to water for their crops, livestock, etc.

According to the historical origin of these eight irrigation systems in Tiquipaya, at least four or five different types of water rights may pertain to irrigation on a single farm:

1 Irrigation water rights originating in the 15th and 16th century from the Inca configuration of *ayllus* before and during the Royal Town of Indians period and subsequently expanded to peripheral sectors. These rights were granted to the allotments within all these territorial units.
2 Rights originating more recently (19th and 20th centuries) through investment of labour or capital in building or expanding irrigation systems by hacienda owners. Water was distributed by a different territorial pattern from the allotments, reflecting hacienda configurations and correlations of power among them.
3 Local old rights related to the presence of springs for irrigation use. The nearby allotment (understood here as a territorial unit) and the group holding it may be in the area of influence of one or more springs.
4 Reallocation or incorporation of water rights by implementing and operating projects recently built to improve infrastructure for catchment, conduction and distribution of irrigation water.
5 In some cases, water access rights acquired by drilling a private well may be established.

Several overlapping irrigation systems, applying different concepts of rights and distribution, have generated a very complex situation as to when, in what order and how much water a given irrigator can use on which land from the different systems. Furthermore, a single user may belong to several irrigator organizations at once. It is curious that the different systems partly use the same network of ditches, so it is easy to mistake it for a 'single system'.

The example of Tiquipaya, as well as many situations found in Peru, Ecuador and some in Chile, shows how the origin of water rights and criteria for water distribution are often very different among local irrigation systems and may not fit into legislative moulds. There are customary law criteria for local allocation of water rights and for distribution within a system and among users. These criteria may vary during the year or, for example, between the period of plenty and the period of *mitas*. Normally, local criteria are more binding on users than official legal norms. The following water rights criteria are reflected in their respective distribution arrangements:[3]

- The criterion of allocation as a function of crop water demand and land areas planted or established. This criterion of 'volumetric distribution' is the official norm, at present, in Peru (Crop and Irrigation Plan), but is inconsistently applied. Officially, a similar criterion is applied in government systems in Ecuador.
- The criterion called 'proportional distribution', allocating water in proportion to the cropland as an aliquot among a system's users. This criterion is widely applied, especially in southern Peru.
- The criterion of allocating in proportion to the contribution that each farming unit (family or community member) makes to build, improve or maintain the irrigation system (often applied in highlands communities).
- The criterion of allocation by equal volumes to all users sharing the system (often used with small community dams).
- The criterion of allocation according to timing arrangements ('shares' of time).
- The criterion of allocating first to certain ecological niches or crop types. In several localities this criterion is combined with allocation by time.
- The criterion of allocating when there is surplus water applicable to areas or users who do not have permanent rights (such as the 'water permit' in Peru).
- Finally, there are other local and regional criteria.

It is almost inconceivable amidst such a diversity of systems, organizational arrangements and variety of local criteria based on customary law rights and arrangements that official legal frameworks continue to promote a blanket nationwide norm or system for allocation and organization of water distribution.

The different modes of rights and the variety of local mechanisms for acquiring and managing rights produce heterogeneity within and among the multiple systems in a single watershed. This also extends to organizational forms and authorities enforcing these rights.

It is legitimate for the state to apply certain criteria and management models to those systems designed and funded by its own institutions, such as governmental irrigation systems in Ecuador or large water projects in Peru. But even for these 'government' systems, serious thought must be given to the advisability, feasibility and functionality of imposing provisions for management that may be unreasonable or difficult from the standpoint of user organizations or users. As shown by the Bolivian, Peruvian and Ecuadorian cases analysed in Boelens and Hoogendam (2002), many state systems overlap with local systems and the territories of rural and indigenous communities where the state is not the 'landlord'.

Above all, there is an urgent need to question the current degree of governmental intervention and normative imposition on and within individually and communally owned systems that were historically built by the users and managed on the basis of local culture, creativity and input. This does not contradict the need for legislation to establish frameworks of legal security to ensure individual users comprising a system that is sufficient in terms of equity, justice, democracy and operational efficiency.

Finally, the diversity of systems is also reflected in local codes that define scheduling and allocation of water volumes despite official laws prescribing in technical language how water rights are to be granted (flow rate in litres per second or cubic metres per second; volume in cubic metres).[4] The legal norms of the different countries usually grant usage rights by defining a measurement of flow rate or volume that does not reflect local realities of allocation of scheduled times or flexibility in practice. This insufficient distinction between the abstract concept of 'usage rights' and the practical concept of 'irrigation scheduling' has often caused excessive rigidity in the design of water distribution schedules,[5] one of the factors unnecessarily interfering with the desire for more frequent irrigation or for intensification and diversification of cropping.

Water, territoriality and community

In previous sections we have seen that a user may have water rights that cannot be analysed in isolation from the other users sharing a water source, the network of ditches, etc. In cases where two or more users share a water source, there must be a management structure and rules for distributing water and maintaining the shared infrastructure. But we have also seen that current laws fail to take into account the diversity of local systems existing in the Andean region.

In the case of rural communities and indigenous peoples, the notion of a system often goes far beyond the hydraulic facilities. Water management is an integral part of the geographical space where an indigenous people or community lives, reproduces and sustains its culture. This territorial/cultural system recognizes a local structure of community bodies, local rules of conduct, norms for enforcement, systems for penalties, and principles for resolving conflicts. It is the community or indigenous people who make decisions about natural resources, including water, and who decide how to allocate and use them. In principle, the sole holder *de facto*[6] of water rights is the community or indigenous people as a whole.

Understood in this way, collectively managing water use systems in a community or indigenous territory is part of community management. Local water management bodies (e.g. irrigators' committee) are part of the community governance structure. An irrigators' committee may be in charge of a single irrigation system, several systems or all the sources and irrigation systems in communal territory. Such concepts prevail in many Andean settings and are generally quite different from the management schemes envisioned in water laws. For example, water legislation in Peru ignores the irrigators' committee as a formal management structure (although it is well known and accepted, especially in Peru's highlands). It then imposes a blanket management model that is completely different, with an organizational arrangement of user boards and irrigator commissions having no relationship to other local management structures. In general, water law in the Andean countries is becoming obsolete as legal recognition of the communal and indigenous rights advances (among other reasons, pursuant to ILO Convention No 169).

This analysis leads to questions about whether the state water authority should grant water rights to third parties within community or indigenous territory, especially if the third parties are not community members. This issue relates to the understanding that 'the water belongs to the territory and the territory belongs to the community'. However, the counter-argument is that barring grants of rights to third parties could allow communities to monopolize all the water, even if there is clearly a surplus or the community is irrationally wasting water when third parties such as neighbouring communities need it. To deal with the possibilities of monopolizing water within a territory, the Indigenous Law in Chile grants relative legal security to collective indigenous rights to water, but does not 'totally close the faucet'. Article 64 (Congress of the Republic of Chile, 1993) states:

> *The water located in community lands, such as rivers, canals, ditches and springs, shall be considered property to be owned and used by the indigenous community established by this law, without precluding any rights that third parties have registered pursuant to the general Water Code. No new water rights shall be granted to lakes, ponds, springs, rivers and other aquifers providing water belonging to the various indigenous communities established by this law without guaranteeing beforehand normal water supply to the communities affected.*

Clearly, this law establishes that the community holds the water use rights throughout a territory, and not just collective rights to a given usage system (i.e. 'network of ditches').[7] However, it also provides for granting usage rights to third parties so that other communities are ensured a normal water supply.

Conclusions

Although quite a number of case studies have been conducted, the existence of multiple ways of managing collective water-use systems in the Andean countries has not been sufficiently appreciated in the design of official legal frameworks. Further research is called for in order to make these systems more visible and to examine their criteria and management approaches. Even from the knowledge gained about local water-use systems from existing case studies, however, it is clear that there are many better ways to define local uses, customs and rights in official legal systems. This plurality is not yet reflected in most Andean countries' water legislation.

The dysfunctionalities generated by the impacts of current legislative frameworks for the management of collective systems should be further analysed. Many such legal provisions are effectively unenforceable within user organizations. This undermines the systems' legal security and water rights because the law is not enforced locally. Future legal frameworks for water must incorporate collective rights for water use systems, not just individual rights. They must also provide for greater recognition of legal pluralism regarding organizations to operate these systems. At the same time, legal security must be ensured for individual users in terms of equity, justice, democracy and operational efficiency.

On the one hand, proposals to modify legal frameworks for water resources in Andean countries have focused on individual tradable rights, which, in practice, is in conflict with the collective rights of many local water-use systems. On the other hand, the right to transfer water temporarily among users belonging to a single system has been the tradition and practice in many Andean settings. This must not be confused with the discussion about the advisability and workability of structural transfer of rights or volumes of water among systems or between different usage sectors.

The advantages of a legal and policy approach promoting and granting preferences to collective water-use systems should be analysed in greater depth. The goal should be to achieve better governance at local watershed and sub-watershed levels. This approach will enable national water legislation to embody greater legal pluralism, more in accordance with the diversity of local realities in Andean countries, and would enhance the democratic content and local representation in the decentralization currently under way in these countries.

Notes

1 Elements of this chapter have been presented at the Water Law and Indigenous Rights (WALIR) International Congresses on Collective Rights and Water Policies in the Andean Region (Quito, October 2004) and Water Policies, Legal Pluralism and Identity Politics (Cusco, November 2006), and published in Urteaga and Boelens (2006).

2 The Water Law and Indigenous Rights (WALIR) programme (2001 to 2007) has developed as an international research and training alliance, with comparative research in the Andean countries, Mexico and the US. The WALIR network was coordinated by Wageningen University (WUR/IWE) and the United Nations

Economic Commission for Latin America and the Caribbean (ECLAC). For its studies, see, for instance, the WALIR book series published with Instituto de Estudios Peruanos (IEP) (Lima); the WALIR books with Fondo Editorial PUCP (Lima); Abya-Yala (Quito); Plural (La Paz); United Nations Educational, Scientific and Cultural Organization (UNESCO) (Paris); Rutgers University Press (New Jersey, US); and the WALIR study volumes (Wageningen University/IWE).

3 Regarding these principles or criteria about rights, see also Boelens and Dávila (1998, Chapter 8, p92).

4 At the local level, other units of measurement are used, such as 'a turn' (on Peru's northern coast, 160 litres per second for 1 hour = 576 cubic metres), 'a release' (defined *ad hoc* by the managers or water judge), an 'allotment' (24 hours of flow or 6-hour fractions), a 'share' (a fraction of total rights to the system), etc., and the 'irrigation module' (= litres per second per hectare according to international technical terminology, and cubic metres per hectare per season on Peru's coast).

5 For example, if the right to use were equivalent to two hours' watering per month (a 'share' type), this would not prevent the user from receiving twice a month (every two weeks) one shift of one hour's watering, or four times a month (weekly) a half-hour shift. Examples of rigidity in this regard have been found in all Andean countries.

6 This *de facto* right may have more legal backing: see ILO Convention No 169 on Indigenous and Tribal Peoples in Independent Countries, and the constitution of several Andean countries (including Bolivia, Ecuador and Peru).

7 It is interesting to compare the *Andean* region to US legal provisions (as done in the WALIR programme), where 'reserved rights' may be an instrument to defend not only contemporary water rights of indigenous peoples, but also to 'keep' extra rights (volumes) for future expansion of the universe of rights holders in the communities (see Getches, 2005).

References

Boelens, R. (2009) 'The politics of disciplining water rights', *Development and Change*, vol 40, no 2, pp307–331

Boelens, R. and G. Dávila (eds) (1998) *Searching for Equity: Conceptions of Justice and Equity in Peasant Irrigation*, Van Gorcum, Assen, The Netherlands

Boelens, R. and P. Hoogendam (eds) (2002) *Water Rights and Empowerment*, Van Gorcum, Assen, The Netherlands

Boelens, R., I. Gentes, A. Guevara-Gil and P. Urteaga (2005) 'Special law: Recognition and denial of diversity in Andean water control', in D. Roth, R. Boelens and M. Zwarteveen (eds) *Liquid Relations: Contested Water Rights and Legal Complexity*, Rutgers University Press, New Brunswick, NJ, and London, pp144–171

Bustamante, R. (2002) 'Water legislation in Bolivia', in *Indigenous Water Rights, Local Water Management, and National Legislation*, WALIR Studies, vol 2, Wageningen University and ECLAC, Wageningen, The Netherlands

Bustamante, R. (2006) 'Pluri-, multi-issues in the reform process: Towards a new water law in Bolivia', in R. Boelens, M. Chiba and D. Nakashima (eds) *Water and Indigenous Peoples*, WALIR–UNESCO, Paris, pp118–129

Congress of the Republic of Chile (1993) *Law No 19,253: Indigenous Law*, National Corporation of Indigenous Development (CONADI), Temuco, Chile

Congress of the Republic of Peru (2002) *Proposed Law to Amend the Constitution: Commission for the Constitution, Regulations and Constitutional Accusations*, Congress of the Republic, Lima, Peru

Del Castillo, L. (2001) 'El largo camino hacia una ley de aguas', *Agua y Riego: Revista del Instituto de Promoción de la Gestión del Agua*, no 28, supplement

Gelles, P. (2000) *Water and Power in Highland Peru: The Cultural Politics of Irrigation and Development*, Rutgers University Press, New Brunswick, NJ

Gentes, I. (2006) 'La interacción de los derechos locales e indígenas de agua con la legislación nacional chilena', in R. Boelens, D. Getches and A. Guevara-Gil (eds) *Agua y Derecho: Políticas Hídricas, Derechos Consuetudinarios e Identidades Locales*, IEP, Lima, Peru, pp255–283

Getches, D. H. (2005) 'Defending indigenous water rights with the laws of a dominant culture: The case of the United States', in D. Roth, R. Boelens and M. Zwarteveen (eds) *Liquid Relations: Contested Water Rights and Legal Complexity*, Rutgers University Press, New Brunswick, NJ, and London, pp44–65

Guevara-Gil, A. (2006) 'Official water law versus indigenous and peasant rights in Peru', in R. Boelens, M. Chiba and D. Nakashima (eds) *Water and Indigenous Peoples*, WALIR, UNESCO, Paris, pp126-143

Guevara-Gil, A., I. Vera, C. P. Urteaga and G. Zambrano (2002) 'Estudio de la legislación oficial peruana sobre la gestión indígena de los recursos hídricos', in R. Boelens (ed) *Indigenous Water Rights, Local Water Management, and National Legislation*, WALIR Studies, vol 2, Wageningen University and UN CEPAL, Wageningen, The Netherlands, and Santiago de Chile, Chile, pp101–131

Hendriks, J. (1998) 'Water as private property: Notes on the case of Chile', in R. Boelens and G. Dávila (eds) *Searching for Equity: Conceptions of Justice and Equity in Peasant Irrigation*, Van Gorcum, Assen, The Netherlands, pp297–310

Hendriks, J., R. Mejía, H. Olázaval, L. Cremers, M. Ooijevaar and P. Palacios (2003) *Análisis de la situación del riego en la República del Ecuador*, Mission Report, CONAM/Agreement ATN/SF-7743-EC, Quito

Military Government of the Republic of Chile (1981) *Decree with Legal Force No 1,122: Código de Aguas de Chile*, Santiago, Chile, reproduced on the website of the General Directorate of Waters, www.dga.cl

Military Government of the Republic of Ecuador (1972) *Supreme Decree No 369, RO/69: Ley de Aguas*, Quito

Military Government of the Republic of Peru (1969) *Decree Law No 17,752: Ley General de Aguas*, Lima, Peru

Palacios, P. (2006) 'Between customs and the establishment of juridical pluralism', in R. Boelens, M. Chiba and D. Nakashima (eds) *Water and Indigenous Peoples*, WALIR-UNESCO, Paris, pp118–129

Roth, D., R. Boelens and M. Zwarteveen (eds) (2005) *Liquid Relations: Contested Water Rights and Legal Complexity*, Rutgers University Press, New Brunswick, NJ, and London

Urteaga, P. and R. Boelens (eds) (2006) *Derechos Colectivos y Políticas Hídricas en la Región Andina*, Water and Society Series, Instituto de Estudios Peruanos (IEP), Lima, Peru

Valderrama, R. and C. Escalante (1988) *De Tata Mallku a la Mama Pacha: Riego, sociedad y ritos en los Andes peruanos*, Desco, Lima, Peru

9

Water Rights and Conflicts in an Inter-Andean Watershed: The Achamayo River Valley, Junín, Peru

Armando Guevara-Gil[1]

Introduction

What happens in a valley with clear blue skies and an idyllic landscape when new crops are introduced, a fish-farming operation expands and the population swells? The Achamayo River watershed's famous blue valley in Peru's central highlands is renowned for the beauty of its sky contrasting with the golden heights, the green eucalyptus foliage and the multicoloured patchwork fields of potatoes, artichokes and kitchen gardens. It is also the setting for an intense social conflict over water involving farmers, fish-farming operations, populated areas that are urbanizing, and even a hydropower plant. In theory, all of these disputes are covered and regulated by official water legislation that is binding on the authorities and all water users in the country. Therefore, conflicts ought to be processed by the bureaucratic system for water management, and users ought to abide by the dictates of official norms. However, an ethnographic approach to water conflicts reveals severe limitations and distortions affecting both the bureaucracy in charge of water management and the official regulations.

Accordingly, some observers have hastened to offer a gloomy assessment, depicting water landscapes characterized by lawlessness, ungovernability and irrationality. From this perspective, indigenous and other rural people are often viewed as among the main impediments to the creation of a unified 'water governance' regime because they represent backwardness that needs to be transformed. This is undoubtedly a slanted, prejudiced proposition, but widely espoused by officials and international technocrats striving to impose a modern, comprehensive and homogeneous management system that ignores the broad diversity of water management systems that are alive and well in Andean watersheds (see Chapter 8 in this volume).

However, beyond the prejudice and ideological slants, the ethnographic record shows that such constraints and distortions in the official water management system are the outgrowth of a number of factors. These include the weakness of the governmental system itself, the strength of indigenous/rural water management systems and the inadequate norms, institutions and policies characterizing state–society relations in Peru. This situation creates fertile ground for legal and institutional experimentation in which indigenous and rural societies define and administer water according to their own way of thinking, in a tense dialogue with the state bureaucracy, authorities and norms (e.g. Mitchell and Guillet, 1994; Boelens and Dávila, 1998; Gelles, 2000; Trawick, 2003; Oré, 2005; Verzijl, 2007; Boelens, 2008).

To illustrate the interesting inter-legal dynamics unleashed by water conflicts (see Chapter 13 in this volume), this chapter will describe the gaping distances between governmental design and the local water landscape. Second, it will summarize water organization and management in the Achamayo River Basin. Next, it will offer an overview of the main conflicts portraying the growing inter-sectoral competition for water resources. It will then conclude with a brief reflection on the relevance of national water norms and the need to process social conflicts over water from a redistributive intercultural perspective.

Official design versus local reality

In Andean watersheds, the water laws enacted by the state have a limited and spasmodic presence and are subject to the conditions imposed by irrigators, other water users and the longstanding shortcomings of the Peruvian state. This situation illustrates the difference between the official design and the social life of water management in Peru (Hendriks and Saco, 2008; see also Chapter 8 in this volume). Theoretically, there are two main foundations: the government and user organizations.[2] The ideal complementarity and synergies that should emerge between public administration and user organizations, however, are nowhere to be seen (Verzijl, 2005). On the contrary, recurrent dysfunctionalities and poor coordination create multiple opportunities for semi-autonomous water management systems to reproduce or emerge (e.g. Gelles, 2000; Trawick, 2003).

Interestingly, the gaps in the government system help to create this plural scenario. Supposedly, the country's 106 main watersheds are administered by the National Water Authority (ANA), which is under the Ministry of Agriculture. To manage these watersheds, the country has been divided into 68 water districts, under the same number of local water administrations (ALAs).[3] The Mantaro Water District, for example, is handled by a local water administrator (ALAM), who is the water authority in that jurisdiction. The problem is that this water district is huge, covering four regions (Lima, Junín, Huancavelica and Ayacucho), 20,000 square kilometres (twice the size of Lebanon and equal in area to El Salvador), and ranging from 5800m altitude down to 500m. To do their work, the ALA for the Mantaro Water District (DRM) has a staff of only 17 people, including administrative personnel. Moreover, the facilities, equipment and infrastructure for water management in one of the country's most important watersheds are grossly inadequate. This situation is not exceptional. In fact, most local water administrations are ill equipped to perform their functions.[4]

As for water user organizations, the law provides for a pyramid structure comprising a National Board of Users in Irrigation Districts (JUDRP), irrigation district user boards, user commissions and user committees. In the Mantaro watershed, although the irrigation district board (JUDRM) is officially recognized, as of 2007 only one of the 20 user commissions (previously irrigator commissions) had a water use licence and only one other had managed to file with the public register. The rest had not established their legal status or obtained a water use licence, prerequisites for defending their rights against other social sectors and stakeholders. Furthermore, in the Mantaro Valley, both the board and the commissions are weak, fragile organizations, with little public credibility, limited economic means and poor institutional development. In this basin, since the commissions are not duly operating, user committees (previously irrigation committees) for micro-watersheds and irrigation sub-sectors regulate and administer water on a day-to-day basis (see Verzijl, 2005, 2007).

This situation, clearly critical from the standpoint of the state and the official user organizations, can lead to different assessments: the water management crisis, social anomy, expanding informality and, in general, the failure of the state and official law to apply to local communities (e.g. Morales, 2002; Perú, 2004, p9; del Castillo, 2008a, p46). By using legal anthropology tools, this reality can be studied as an example of legal pluralism and inter-legality. This calls for abandoning the idea that the state is the only producer of law, and for recognizing the diversity of normative frameworks operating in a society. When we do this, we see how official law faces a series of competing normative systems (whether indigenous, rural, local or customary)[5] that are enforced as much or more than state law. The important thing, in any event, is to analyse the local socio-legal context, the role of state law in different social universes, and to determine how people reinterpret their norms and use them in local contexts.

Thus, studying law through anthropology reveals the great structural tension between the state and local societies, and the social, political and cultural pressures that produce such phenomena as inter-legality and legal plurality (Santos, 1995; Hoekema, 2002; see also Chapter 13 in this volume).

This is where official norms, policies and institutions for water acquire a different meaning from the role assigned by the state. The meaning is diverse and defined by the way in which local arrangements metabolize the influence of official law and, depending on their inner strength, affirm a set of standards based on their own norms, penalties and procedures.

To illustrate how legal plurality and inter-legality arise, this chapter will offer some ethnographic references on water organization, management and conflicts in an inter-Andean watershed of Peru's central highlands. In this scenario of tension and conflict, farmers, a fish farm, urbanizing areas and a hydropower plant compete for water use. Each basin is unique socially, politically, geographically and hydrologically. However, the organization, processes and conflicts identified here are illustrative of the dynamics of legal pluralism, inter-legality and the social life of state water law.

Water organization and management in the Achamayo River Basin

The Achamayo River Basin is a typical inter-Andean valley of Peru's central highlands. It is located in the province of Concepción, region of Junín, and its waters flow east to west, from the snow caps and lakes of the Andes range (4500m altitude) down to its mouth, on the left banks of the Mantaro River, near the town of Matahuasi (at 3262m). Local folk divide the basin into two parts: upper and lower. The upper basin comprises gullies and dry steep land. Dryland agriculture there (potato, *oca*, fava beans, barley) is complemented by irrigation using water from springs and wetlands. The lower basin begins at Ingenio and flows through the districts of Quichuay, Santa Rosa de Ocopa (3376m), Santo Domingo and Concepción (3290m). The river's volume is highly variable, ranging from 120 cubic metres per second during the rainy season to 1.5 cubic metres per second from June through August. The basin's total area is estimated at 248 square kilometres and its waters are used for municipal water supply, agriculture, fish-farming and hydropower (Antúnez de Mayolo, 1990, pp2–15). Irrigation is intensive, controlled by a local irrigators' organization during the dry season, but free for all during the Andean winter.

According to state prescriptions, the Achamayo River Basin is an irrigation sub-district belonging to the larger water district of the Mantaro Valley. Its members are formally organized into a users' commission that represents them before the Users' Board of the Mantaro Irrigation District (JUDRM) and before government authorities such as the Local Water Authority of Mantaro (ALAM). Locally, irrigation water management is organized on the basis of the 11 canals fed from the Achamayo River. Each of these main canals has a users' committee, which is responsible for administering water use. Each committee has a president, treasurer, secretary and several intake officers who are elected every three years by the general assembly of the committee members. The canal system fed by the Achamayo River is estimated to benefit some 5000 irrigators, most of them small farmers (some farming just a few furrows).

Out of these 11 canals, this discussion will focus on the Quichuay–Santa Rosa de Ocopa–Huanchar–Huayhuasca Canal. This canal is 10km long and irrigates some 330ha (30ha in Quichuay, 150ha in Santa Rosa and another 150ha in Huanchar and Huayhuasca). The users' committees have established and coordinate a rigid sequence of daily turns to prevent water theft and violence among irrigators. Those responsible for enforcing the distribution system are the intake officers, or *tomeros*. These *tomeros* issue irrigation orders, set schedules on the basis of the land area to be watered and ration water when it is scarce.

Water use rights and access to irrigation infrastructure are generated by landownership and participation in collective work parties.[6] Everyone, man or woman, who contributes in these collective efforts is entitled to irrigate, regardless of the size of their farm. Additionally, to access their water allotment they must pay the local fee set in the general assembly by the committee members (different from the official fee set by the ALAM), apply for and respect the shifts assigned by the officer, and participate in committee assemblies.

Conflicts among irrigators are generally settled by the users' committee leadership, and problems among committees (e.g. Santa Rosa versus Quichuay) are settled by the water users' commission president for the entire Achamayo Basin. Decisions by these authorities are backed by fines and other penalties that may include suspending the irrigation turn or an obligation to purchase construction materials to maintain canals. Although no irrigator has been expelled for infractions, local water leaders request the national police and the governor's office to reprimand irrigators for waste and damage to others' property (e.g. failing to close gates and flooding a neighbouring field or home).

The most frequent problem among irrigators is 'water theft'. This often happens at night and, because such small plots are watered, sometimes the authorized irrigator does not realize that his or her flow has been reduced. Water theft makes it necessary to employ more people than technically required to monitor water use. One must control the main intake, while another must patrol the canal to detect leaks or theft. The irrigator himself must then actually distribute water on the land.

In addition to irrigators, the watershed has other important users. Demand for municipal water use has increased as the province of Concepción contains several growing towns, including the capital, with 15,000 inhabitants, and Matahuasi, with 8000. In general, the countryside is urbanizing, which generates a growing demand for water supply and sanitation services. Another major user is the privately owned Los Andes fish-farming operation. The third is the Ingenio Hydropower Station, which does not conflict with irrigation uses, because irrigation systems begin after the point where the power plant returns water to the river and it does not dramatically alter the river's flow regime. The hydropower station has, however, evidently altered upper basin water levels by transporting upstream water through pipes to gain head and releasing it downstream of the upper basin through its turbines. Finally, government agencies, non-governmental organizations (NGOs) such as Cáritas (a Catholic Church charity organization) and a couple of agro-export companies growing artichokes have a limited but influential presence in the watershed. The state water development agencies have no major activities in the zone; but irrigators have received some support from Cáritas to improve their canals and participate in the recent boom in artichoke exportation.

Conflicts over water and legal pluralism in an inter-Andean watershed

A quick review of the main conflicts among water users will illustrate how legal pluralism and inter-legality work in the Achamayo Basin.

Irrigators' issues

One of the most significant conflicts is among irrigators, especially on the Quichuay–Santa Rosa–Huanchar–Huayhuasca Canal, and the Los Andes fish farm. The company usually uses more water than assigned by the Mantaro ALA because trout exports are booming and Los Andes has expanded its facilities. Overuse decreases flow for downstream irrigators, who get less than their share. The Mantaro Local Water Administration has mediated the conflict with only minor success. During the dry season the company and the farmers wage an ongoing furtive battle to get more of the Achamayo's water by nightly manipulating the rustic water gate that they share. Farmers complain about this pilferage and constantly confront fish-farm employees. Here, the rustic nature of the gates and lack of volumetric gauges make the conflict worse during the Andean summer.

During these dry periods, even the amount officially allocated in the fish farm's water-use licence is challenged because farmers need more irrigation water. In theory, no one can dispute the water rights granted in a usage licence and the fish farm should not have this sort of conflict if it has a licence and has paid up the water use fee set by the Mantaro ALA (del Castillo, 2008a, p41; Hendriks, 2008b, pp57–60). However, during these critical months, irrigators invoke their 'ancestral rights' and assert the strength of their local organization to challenge and reduce the water rights officially granted to the company. This forces the company to dispute and renegotiate its water allotment constantly. In this scenario, the rights assigned by the water authority are not fully enforceable but are mediated by social forces and local rights. So, farmers on the Quichuay–Santa Rosa–Huanchar–Huayhuasca Canal challenge the overriding primacy of the rights granted by the state, obtain a redistribution of the available flow, and affirm the current value of their water rights, above and beyond governmental regulation and administration.

In 2003, the introduction of artichoke farming triggered another intense conflict over water. This issue lulled in 2008 when the agro-export companies decided to take their operations to the valleys and irrigation systems on the coast. The move, however, did not solve the water conflicts that they created.

At first the artichoke was introduced as a cash crop alternative to potatoes. It was very attractive because of its high commercial value, especially compared to the extremely low prices for potatoes. Since agrarian property is quite fragmented and companies could not consolidate large areas, they offered small farmers credit, seedlings and technical supervision to grow for them. The problem for small farmers was that the company did not adjust the purchase price as production costs rose (e.g. fertilizers and other chemicals; interest on loans), so the boom did not last long.

The problem is that artichokes need more water than other crops and this increases friction among irrigators on a single canal, among irrigators on different canals and between farmers and the Los Andes fish farm, as do other water-intensive crops with higher commercial value, such as alfalfa and maize/corn in other valleys (Guillet and Mitchell, 1994, pp7, 13). As an engineer from one company put it: 'an artichoke is a water pump'. Because agro-export companies do not grow the produce but only buy it from small farmers, they do not take part directly in the water conflict. But their actions do place extra pressure on water resources because farmers have to irrigate more frequently and with more water.

The pressure by agro-exports on water use has not been duly evaluated. It is clear that expanding artichoke farming has caused tensions for local water organizations, affecting the distribution of turns and the proportion of volumes assigned to land converted to the new crop. Although the law states that water scarcity must be faced by allocating water according to criteria of efficiency and equity, these notions end up slanted by the new commercial thinking that favours export crops over local food crops. In this case, the gravitational force of the market economy has made substantial changes in traditional distribution of water rights. In view of the minimal influence or concern of ALAM for local water management, water appropriation policy has evolved in a different direction than established by the official norm. Furthermore, the supposed enforcement of the Crop and Irrigation Plan, as a tool for organizing water management, did not prevent artichoke growers from expanding farms or intensifying irrigation.

Growth in municipal demand

Another important conflict in the watershed is due to growth in urban population in the province of Concepción and the large demand for water by the city of Huancayo (about 450,000 inhabitants). Municipal demand is placing strong pressure on available water resources. One project to relieve water shortages for Huancayo, for example, plans to take water from the lakes in the upper basin and pipe it to the city. This would severely affect water supply throughout the watershed and current water right allocations. There have also been conflicts among localities in the basin because some have attempted to get water use licences for their water supply systems, but have faced tenacious opposition from traditional water users. Others have not only obtained such licences, but have begun water projects that were interrupted because of the unmoving opposition by communities who felt that these projects would negatively affect them.

The community of Quichuay, for example, has confronted the town of Matahuasi regarding the installation of filter galleries on the banks of the Achamayo that border on their land. The project, authorized by ALAM and considered a high priority because it would supply water to a city, came to a stop because the community refused to authorize them to take 'their water' to Matahuasi and to use their land to access the river. After long, difficult negotiations involving the water authority, the community agreed to allow the work to continue in exchange for a series of benefits. Their position was vulnerable to pressure from government authorities because they were facing a more powerful societal stakeholder (a town) and a use (water supply) considered a top priority. Even so, they received compensation despite no mention of such restitution in official norms, which say that only the state can decide on water allocation.

In most less-visible conflicts, the parties negotiate directly rather than pursuing official water-use licences. Although the law states that all water belongs to the state and only the state can grant water use, residents of the watershed assume that the water sources located on their land belong to them. Under this socio-territorial concept of water rights, any use or diversion of water must be authorized by the residents and calls for compensation in return. This notion frustrates the local bureaucracy, but is so widespread that when communities and towns request them to participate in negotiations, they attend and even formalize these agreements. In such cases, the water authority (ALAM) waives jurisdiction and acknowledges local practices, playing a role quite unlike the one assigned by law.

Measurement issues

An interesting dispute also exists between the state and irrigators over the unit of measurement applicable to water use. Where the local water administration and the Mantaro District Irrigation Users' Board want to install volumetric gauges pursuant to law and neoliberal economics, farmers on the Achamayo consistently oppose this measure. They have already destroyed some gauges, stating that it is 'fairer' to measure by irrigation time and not by volume (for example, three hours to water 1ha). They argue that water is a natural resource and that the irrigation infrastructure was made by 'our grandparents' and 'neither the users' board nor the Ministry of Agriculture has ever spent a single penny to build it' (Walter Maraví, pers comm, 17 August 2002).

Interestingly, farmers, especially those on the Santa Rosa de Ocopa Canal, are the first to demand volumetric control for the Los Andes fish farm in order to prevent it from using more water than it is assigned. They refuse, however, to let the water administration apply this measurement to them. This is a typical case in which logical coherence gives way to strategic and instrumental reasoning because what matters to them is to have access to more water. When it is strategically necessary, they resort to contradictions and inconsistency. In any event, since volumetric control is a keystone for official water management, the insistence on control by hourly scheduling highlights the way in which local management notions and approaches conflict with state policy but end up prevailing.

Resistance to fees

Another conflict between the state and farmers on the Achamayo results from differing outlooks regarding water fees. Theory says that fees are paid to the state, recognizing that water is a natural resource belonging to the nation and that the state administers it with exclusive sovereign authority (Guevara-Gil, 2006; del Castillo, 2008a). The fee, supposedly paid to the users' board, is the payment for the water that is granted to the applicant (e.g. for agrarian use). Irrigators have resisted and the users' board and the Mantaro Local Administration have been pressuring them to catch up on their fees.

Farmers have systematically refused to pay the fees, in some cases since 1992 or 1999, for a number of reasons. First, they do not recognize these prerogatives of the state because of their socio-territorial concepts and rationales. Second, the money charged for fees has not been used to improve infrastructure, as provided by law. Third, the users' committees themselves charge a 'local' or 'customary' fee for farmers who get their irrigation turns from their own intake officers in charge of distributing water from secondary canals. The amount collected from this local fee is used to finance maintenance and improvement of water infrastructure, to hire watchmen to prevent other committees from stealing water, and to pay *tomeros* to distribute water.

General opposition to state control

Disagreement with the notion of state ownership is a fundamental legal and cultural objection that underlies challenges to the legitimacy of the state's water fees, the imposition of a supra-local organization and the application of volumetric control over water. While the constitution, the General Water Law (Decree Law 17752) and all manner of norms (Legislative Decrees 1081 and 1083) state that all water belongs to the state, farmers on the Achamayo maintain that water is subject to local control and their own concepts about riparian and socio-territorial rights. They bolster their argument with the collective memory of the work that their ancestors did without government support on the infrastructure, and they assert the legitimacy of their ancestral rights.

As the following statement shows, their claims to historical rights and opposition to state interference are quite clearly defined:

> *How long have we defended this canal, because this canal was made before by the chieftains, the Sarapura and the Bendezú! How long ago were the chieftains! Back in 1800, in 1700. They built it with the whole community of Huanchar, beyond Huayhuasca, from all these places. When they began [setting up the users' board of] Mantaro, they wanted to register us; but we and our elders refused because this canal belongs to the people; it is ancestral! It is very old, and the state did nothing here – nothing, nothing. Now the state comes along, wanting us to sign up to pay the water fees and all that. But now the modern people are practically agreeing to that, although our ancestors refused, saying: 'You can kill me, but we refuse to register.' This is a very old canal, and is maintained to this day by collective work parties – the state doesn't contribute a single cent here, nothing, absolutely nothing, not even technical support; we do it all by our own commonsense judgement.*[7]

Any attempt to turn water into an economic commodity as promoted by neoliberal reforms backed by international agencies and much of the water bureaucracy (Perú, 2004; compare Hendriks, 2008b) will run up against this local concept of water rights. Even now, sporadic efforts by ALAM and JUDRM to levy fees do not even generate enough money to cover the costs of collection. Moreover, irrigators in the basin express their marked interest in maintaining this situation of 'lawlessness' and 'water ungovernability' – epithets used by neoliberal reformers – because it benefits both the 'rich' and 'poor'. Both groups of irrigators pay very little to use the water, the volume control system they use is flexible, and each uses its resources or social networks to maintain its water rights. Here, interests and concepts blur, strengthening the local organization's autonomy, and producing a normative clash that gives rise to legal plurality in water management.

Conclusions

As the examples outlined and the ethnographic record cited show, there are drastic gaps between official water laws and policies, and indigenous, small-farmer, local management systems. This is not a radical dichotomy where the two extremes have no contact. On the contrary, local system managers are keenly attuned to the pulse of national policies and laws and the demands of the market economy because these forces threaten their existence and reproduction as semi-autonomous social groups.

So, it is no accident for indigenous people, peasants and small farmers to be the first to oppose neoliberal prescriptions proposing privatized water rights and distribution. This opposition is not driven by ignorance or ideology. Rather, it is an affirmation of their own lifestyle and the threats that they perceive to their water rights and livelihood. Nor is it surprising that they employ strategic inter-legal reasoning for day-to-day defence of their rights (such as against the fish farm) and invoke ancestral socio-territorial rights denied by official law in conflicts with other water users or with the state itself (such as in Matahuasi or in their opposition to pay official fees). If anyone is practising the 'inter-cultural' approach, which is the new buzzword in Peru's official discourse, it is precisely the indigenous people and small farmers who have decided to defend their livelihoods and well-being.

The problem lies, rather, in the deafness and blindness institutionalized by the state when preparing and enforcing its water laws and policies. The result is foreseeable. These norms and policies lack any social roots, fail to reflect the country's extraordinary diversity of water landscapes, and end up configuring a labyrinth that not even the water bureaucracy can fully understand or enforce. In view of the state's stubborn failure to recognize or value local collective systems to manage resources, indigenous people and small farmers develop their own ways of organizing, managing and resolving their water conflicts. Such a great investment of creativity, resources and time is possible only through collective action. This, in turn, is grounded in local criteria of equity and redistribution that set up a moral economy acceptable to the social group's members. Again, if anyone is putting into practice water management with redistributive criteria – another catch phrase in official discourse – it is the indigenous people and peasants who have decided to defend their way of life and welfare.

Therefore, the state and Peru's ruling elite would do well to learn some lessons from the indigenous and peasant societies in their own country. If they did so, this would open up true possibilities in policies and laws to process water conflicts under new perspectives, renewed by criteria such as inter-cultural relations and redistribution.

Notes

1 Based on Guevara-Gil (2008). I would like to express my appreciation for the support of the Water Law and Indigenous Rights (WALIR) programme (ECLAC and Wageningen University, The Netherlands) and the Pontificia Universidad Católica del Perú for conducting my fieldwork in the Achamayo River Basin, and my studies of water law in the Andes, from 2002 to 2006. My thanks, as well, to Cirilo Bendezú, a great irrigation leader and friend, and my *compadres* Walter and Consuelo Maraví for their generous hospitality in their town, Santa Rosa de Ocopa, which is now my town, too. Data on water management and conflicts in the basin come from my own ethnographic observations.

2 Del Castillo (2008a, 2008b) offers a very instructive overview of the legal system for water in Peru. Regarding indigenous and *campesino* rights versus official Water Law, see Guevara-Gil (2006).

3 Legislative Decrees 1081 and 1083 in 2008 have once again changed the institutional design governing water. Now the ANA replaces the Water Resource Superintendant of the Natural Resource Institute (INRENA), which was also under the Ministry of Agriculture, and the ALAs replace technical administrators

of irrigation districts (ATDRs). The term Water Districts is used instead of Irrigation Districts, although this last one has not yet been explicitly replaced but it seems it will (del Castillo, 2008b, p3; Hendriks, 2008a, pp108–110). Finally, in March 2009, after years of debate, Peru's Congress passed the Water Resources Law that repeals the above two norms due to political and social pressure from the National Board of Irrigation Users of Peru. The law declares that access to water is a human right, that water pertains to the nation and that it cannot be bought, sold or used as private property. However, it opens the door for the private sector to invest in, modernize and manage irrigation systems.

4 Similarly, the Cuzco Water District, including the river that has formed the famous Sacred Valley of the Incas leading to Machu Picchu, has just one administrator and seven employees (Verzijl, 2005, p56).

5 Debates on the nature, consistency and historicity of customary law have become endless. Here I take the position summarized by Wiber (1993, p24): 'I reserve the term customary law to refer to the transformed normative orders which result in local communities when indigenous law and state law interact over time.' Regarding the contingency and porosity of these state and local/customary 'systems', see, for example, von Benda-Beckmann et al (1997); regarding the hybrid nature of local, indigenous or customary law, see, among others, Santos (1995); Gelles (2000, p117); Hoekema (2002); and Boelens et al (2005, p7). As von Benda-Beckmann et al (2000, p11) put it:

Increasingly it is not so much the historical origin that counts, but rather the fact that people perceive regulations as belonging to them and based on local authority structures, rather than on external legitimate authority. Their 'local laws' are often hybrid legal forms that combine elements of state law and customary legal rules and principles.

6 Wiber (1993, p59) distinguishes between two models to explain the water–landownership relationship. Whereas in the 'Syrian model, irrigated land and irrigation water are inseparably linked', in the 'Yemeni model, land and water are kept distinct and each can be sold separately'. In the Andes, the correlation between landownership and access to water and to irrigation infrastructure is very tight. Moreover, collective work to build and maintain that infrastructure is essential to create and revalidate water rights (Boelens and Doornbos, 2001, p344; Beccar et al, 2007; Verzijl, 2007; compare Sodemba and Pradhan, 2000, p101).

7 Told by Pedro Maraví Aguilar, a farmer headed for a work party to clean the intake for the Quichuay–Santa Rosa–Huanchar–Huayhuasca Canal (pers comm, 17 August 2002). However, other versions, such as told by Cirilo Bendezú, indicate that the Quichuay–Santa Rosa–Huanchar–Huayhuasca Canal was built by the Santa Rosa de Ocopa monastery to set up a hydropower station (pers comm, 15 September 2008; see Tord et al, 1969, p44).

References

Antúnez de Mayolo, S. (ed) (1990) *Perfil antropogeográfico del departamento de Junín*, Fondo Editorial de la Sociedad Geográfica de Lima, Lima, Peru

Beccar, L., R. Boelens and P. Hoogendam (2007) 'Derechos de agua y acción colectiva en el riego comunitario', in R. Boelens and P. Hoogendam (eds) *Derechos de agua y acción colectiva*, Instituto de Estudios Peruanos, Lima, Peru

Boelens, R. (2008) *The Rules of the Game and the Game of the Rules: Normalization and Resistance in Andean Water Control*, Wageningen University, Wageningen, The Netherlands

Boelens, R. and G. Dávila (eds) (1998) *Buscando la equidad: Concepciones sobre justicia y equidad en el riego campesino* [*Searching for Equity: Conceptions of Justice and Equity in Peasant Irrigation*], Van Gorcum, Assen, The Netherlands

Boelens, R. and B. Doornbos (2001) 'The battlefield of water rights: Rule making amidst conflicting normative frameworks in the Ecuadorian highlands', *Human Organization*, vol 60, no 4, pp343–355

Boelens, R., D. Roth and M. Zwarteveen (2005) 'Legal complexity in the analysis of water rights and water resource management', in D. Roth, R. Boelens and M. Zwarteveen (eds) *Liquid Relations: Contested Water Rights and Legal Complexity*, Rutgers University Press, New Brunswick, NJ, and London

del Castillo, L. (2008a) 'El régimen legal del agua', in A. Guevara-Gil (eds) *Derechos y conflictos de agua en el Perú*, Concertación, WALIR and Departamento Académico de Derecho de la Pontificia Universidad Católica del Perú, Lima, Peru

del Castillo, L. (2008b) *Antes y ahora: Decretos Legislativos que le cambian el rostro al sector rural*, CEPES, CONVEAGRO y Grupo Propuesta Ciudadana, Lima, Peru

Gelles, P. (2000) *Water and Power in Highland Peru: The Cultural Politics of Irrigation and Development*, Rutgers University Press, New Brunswick, NJ, and London

Guevara-Gil, A. (2006) 'Official water law versus indigenous and peasant rights in Peru', in R. Boelens, M. Chiba and D. Nakashima (eds) *Water and Indigenous Peoples*, UNESCO, Paris

Guevara-Gil, A. (2008) 'Derecho de Aguas, pluralismo legal y concreción social del Derecho', in A. Guevara-Gil (ed) *Derechos y conflictos de agua en el Perú, Concertación*, WALIR and Departamento Académico de Derecho de la Pontificia Universidad Católica del Perú, Lima, Peru

Guillet, D. and W. Mitchell (1994) 'Introduction: High altitude irrigation', in W. Mitchell and D. Guillet (eds) *Irrigation at High Altitudes: The Social Organization of Water Control Systems in the Andes*, Society for Latin American Anthropology and the American Anthropological Association, Washington, DC

Hendriks, J. (2008a) 'Estructuras político-institucionales para la gestión de recursos hídricos en países andinos', in P. Urteaga (ed) *Gobernabilidad de los recursos hídricos: Legislación, administración y políticas hídricas*, Wageningen University, IHE-UNESCO, IPROGA, Lima, Peru

Hendriks, J. (2008b) 'Legislación de aguas y gestión de sistemas hídricos en países de la región andina', in P. Urteaga (ed) *Gobernabilidad de los recursos hídricos: Legislación, administración y políticas hídricas*, Wageningen University, IHE-UNESCO, IPROGA, Lima, Peru

Hendriks, J. and V. Saco (2008) 'Gestión local de agua y legislación nacional en el Perú', in A. Guevara-Gil (ed) *Derechos y conflictos de agua en el Perú*, Concertación, WALIR and Departamento Académico de Derecho de la Pontificia Universidad Católica del Perú, Lima, Peru

Hoekema, A. (2002) 'A new beginning of law among indigenous peoples', in F. J. M. Feldbrugge (ed) *The Law's Beginning*, Martinus Nijhoff Publishers, Leiden, The Netherlands, and Boston, US, pp181–200

Mitchell, W. and D. Guillet (eds) (1994) *Irrigation at High Altitudes: The Social Organization of Water Control Systems in the Andes*, Society for Latin American Anthropology and the American Anthropological Association, Washington, DC

Morales, J. (2002) *En busca de una gobernabilidad eficaz del agua*, Documento base para el Diálogo Nacional, Global Water Partnership, Asociación Peruana del Agua, Lima, Peru

Oré, M. T. (2005) *Agua: Bien común y usos privados. Riego, Estado y conflictos en La Achirana del Inca*, Fondo Editorial de la Pontificia Universidad Católica del Perú, WALIR, ITDG, Wageningen UR, Lima, Peru

Perú (2004) ENGRH, Estrategia nacional para la gestión de los recursos hídricos continentales del Perú, Comisión Técnica Multisectorial, Ministerios de Agricultura; Defensa; Economía y Finanzas; Energía y Minas; Vivienda, Construcción y Saneamiento; Salud; Producción, Lima, Peru

Santos, B. (1995) *Toward a New Common Sense: Law, Science and Politics in the Paradigmatic Transition*, Routledge, New York and London

Sodemba, I. and R. Pradhan (2000) 'Land and water rights in Thulo Sangrumba, Ilam', R. Pradhan, F. von Benda-Beckmann and K. von Benda-Beckmann (eds) *Water, Land and Law: Changing Rights to Land and Water in Nepal*, FREEDEAL, WAU, EUR, Kathmandu, Wageningen and Rotterdam, The Netherlands

Tord, L., F. Murrugarra, N. Gómez and J. Núñez (1969) 'Tenencia de la tierra en Santa Rosa de Ocopa', in Instituto Indigenista Peruano (ed) *Dos Estudios en el valle del Mantaro*, Instituto Indigenista Peruano, Huancayo, Peru

Trawick, P. (2003) *The Struggle for Water in Peru: Comedy and Tragedy in the Andean Commons*, Stanford University Press, Stanford, CA

Verzijl, A. (2005) *Two Water Right Systems in the Peruvian Andes; Case 17752: The State versus the People of Picol. Recognition of Local Water Rights in the Quesermayo Watershed, Cusco, Peru*, MSc thesis, Wageningen University, Wageningen, The Netherlands

Verzijl, A. (2007) *Derechos de agua y autonomía local: Análisis comparativo de los Andes peruanos y los Alpes suizos*, Instituto de Estudios Peruanos, WALIR, Lima, Peru

von Benda-Beckmann, F., K. von Benda-Beckmann and A. Hoekema (eds) (1997) *Law and Anthropology: International Yearbook for Legal Anthropology. Issue on Natural Resources, Environment, and Legal Pluralism*, vol 9, Martinus Nijhoff Publishers, The Hague, The Netherlands

von Benda-Beckmann, F., K. von Benda-Beckmann and R. Pradhan (2000) 'Introduction', in F. von Benda-Beckmann, K. von Benda-Beckmann K. and R. Pradhan (eds) *Water, Land and Law: Changing Rights to Land and Water in Nepal*, FREEDEAL–WAU–EUR, Kathmandu, Wageningen and Rotterdam, The Netherlands

Wiber, M. (1993) *Politics, Property and Law in the Philippine Uplands*, Wilfrid Laurier University Press, Waterloo, Ontario, Canada

10

Water Rights, Mining and Indigenous Groups in Chile's Atacama

Jessica Budds

Introduction

One of the most radical neoliberal reforms devised and implemented in Chile under the military regime led by General Augusto Pinochet (1973 to 1990) was the major revision of the Water Code in 1981. This legal and policy process was introduced in Chapter 2 of this book. Designed according to free market principles embraced by the 'Chicago Boys' – a group of Chilean economists who studied at the University of Chicago – the result was an extremely neoliberal water law, which introduced a system of private water rights that could be freely traded with almost no government regulation. In Chile, rights to use all surface and groundwater resources are defined as property rights, and the Water Code applies to all uses, including agriculture, drinking water supply, industry, mining and hydroelectric power. Furthermore, the Chilean Water Code has also become significant internationally for its application of neoliberal principles to water management to an unprecedented extent (e.g. Simpson and Ringskog, 1998). These principles corresponded to the free market reforms that were aggressively promoted by multilateral financial institutions (especially the World Bank) to governments of indebted low- and middle-income countries under structural adjustment programmes since the 1980s. In line with the influential 1992 Dublin Principles, water-sector reforms urged during the 1990s included privatization of urban water utilities and the use of water markets to increase the efficient usage of water resources (e.g. Budds and McGranahan, 2003).

Free market economists assert that tradable property rights will reduce the inefficient allocation and use of scarce resources because the market will price the resource according to its true scarcity value. So, when shortages occur, its higher price will encourage users to switch to cheaper sources and those who value it most highly will pay the price. In order to achieve this outcome, the market must be left to its own devices, with minimal state intervention or regulation (Simpson and Ringskog, 1998; Tietenberg, 2000).

Although the relationship between efficiency and equity within market mechanisms is much debated in economics, some authors have argued that greater efficiency will produce a range of other benefits, including positive social outcomes. For example, efficiency has 'the potential to increase the productivity of water use; improve operations and maintenance; stimulate private investment and economic growth; reduce water conflicts; rationalize ongoing and future irrigation development; and free up government resources for activities that have a public good content or positive externalities. It even can benefit the poor and help to conserve natural resources' (Thobani, 1995, p4).

Four arguments have been put forward to support the idea that water rights markets contribute to social benefits among water users:

1. State water management has tended to favour the wealthy (Thobani, 1995).
2. Farmers can use secure water rights as collateral for credit (Thobani, 1995).
3. Through realistic and/or higher pricing, government revenue from water taxes can fund poverty eradication policies (Gazmuri, 1994).
4. Greater water efficiency will increase water availability for the poor (Grant, 2001).

Many of these supposed benefits of water markets have been attributed to Chile's model of markets in tradable water rights (e.g. Gazmuri, 1994; Briscoe, 1996; Gazmuri and Rosegrant, 1996; Briscoe et al, 1997; Donoso, 2003). Since the early 1990s, the World Bank, in particular, has heralded the Chilean system as a successful water markets model and has incorporated it within its recommendations for water resources management (e.g. World Bank, 1992). The World Bank has also published several documents on the Chilean model, which have made bold, yet unsubstantiated, claims about the model that have since been repeated elsewhere (e.g. Gazmuri, 1994; Ríos and Quiroz, 1995; Briscoe, 1996). Furthermore, on this basis, the World Bank has strongly recommended, and in some cases actively supported, the replication of the Chilean water law and water markets model to other Latin American countries, including Argentina, Bolivia, Ecuador, El Salvador, Mexico, Nicaragua and Peru (ECLAC, 1995; Muchnik et al, 1997; Trawick, 2003; Assies, 2003; Bauer, 2004a, 2004b).

The social implications of the Chilean model have not been studied despite general concerns over the potential lack of social equity and monopolization of water by wealthier groups (Bauer, 1998a, 2004a; Romano and Leporati, 2002; Galaz, 2004). Indeed, some authors and studies have dismissed serious social impacts without any thorough field research (e.g. Ríos and Quiroz, 1995).

This chapter follows up on the broader analysis of neoliberalism and water policies by Achterhuis, Boelens and Zwarteveen in Chapter 2 and focuses in more depth on empirical facts of 'the Chilean case'. It examines the social outcomes of Chile's Water Code among poorer water-user groups based on empirical research in the Atacama region of northern Chile. Following a description of the key market-based characteristics of the Water Code, the chapter will then present the various claims of economic efficiency and social benefits associated with this mode of water resources management. Empirical evidence of the social impacts of converting water rights into tradable private property under the Water Code in the arid Atacama region is then presented to show that the water market has produced challenges for the protection of indigenous water rights by enabling commercial interests, especially the mining industry, to purchase traditional water rights from Atacameño indigenous groups. The final section draws conclusions on how water markets configure access to and control over water resources, and briefly reflects on the extent to which these issues are likely to be addressed by the modifications made to the Water Code in 2005.

Chile's 1981 Water Code

In Chile, water rights, or concessions allowing the exclusive use of water, have existed in various forms since colonial times (Muchnik et al, 1997; Bauer, 1998a). The first Water Code of 1951 defined water rights as fixed-term concessions that could be registered as private property. Applicants for water rights had to specify the intended use and location of the water, which could not then be changed, and prove that the water was being used. Private water rights could be sold or rented, and could only be cancelled with compensation unless they were not used, in which case the state could cancel them after five years (Muchnik et al, 1997; Bauer, 1998a; Alegría, 2004).

The reformist government of Eduardo Frei Montalva (1964 to 1970) embarked upon agrarian reform, which also required the redistribution of water. In 1967, the Water Code was revised in line with the 1967 Agrarian Reform Law to reinstate government control over land and water. After amending the 1925 Constitution to declare water 'national property for public use', the government was able to cancel water rights without compensation (Bauer, 1997). Having lost their status as private property, water rights could no longer be traded and were no longer filed in property registers (Bauer, 1998a; Alegría, 2004).

Following the coup that overthrew Salvador Allende (1970 to 1973), the military government under Pinochet immediately set out to reverse the policies implemented by Frei and continued by Allende, especially suspending land reform and confirming private titles to expropriated land. However, water was a secondary priority and the 1967 Water Code was not repealed until 1979. By this time, the situation with water rights was chaotic, partly because no records of water allocation had been kept since 1967, but mainly because the statist water law was incompatible with the new neoliberal political-economic order that had been implemented from 1975 (Bauer, 1997).

The new Water Code was passed in 1981, at a time when neoliberalism was at its most ascendant, and just before the economic crisis of 1982 to 1983 (Angell and Pollack, 1990). Its economic and market features were designed by the 'Chicago Boys' in order to make water a commodity like any other. This entailed separating water from land in order to attribute a specific cost to it (Bauer, 1998a). The reforms had three objectives:

1 to privatize water rights and prevent state intervention in land and water (reinforced by protection of property rights in the military's 1980 Constitution);
2 to increase the efficiency and productive value of water by enabling the allocation of scarce water resources to higher-value uses (intended to improve the efficiency of water use for Chile's export-oriented natural resources industries, especially agriculture); and
3 to foster private investment in water development to replace state funding (Bauer, 1997, 1998a, 1998b).

The economists considered water taxes to be a crucial element of the new Water Code by acting as an incentive to increase efficiency and transfers to higher-value uses. However, water taxes were strongly opposed by water users, especially the agriculture sector. Commercial farmers supported secure private water rights, but resented the idea of starting to pay for water. The Water Code was ultimately a compromise negotiated between the neoliberal economists and its more conservative allies, especially large landowners, who agreed on strengthening private property but not on taxing rights. The Water Code thus defined water rights as private property that was separate from land and could be freely traded, mortgaged or transferred; water rights were guaranteed by the state and could only be expropriated for compensation at market value, were governed by private law, and need not be used (it was argued that this would invade private liberty) (Bauer, 1997, 1998a, 1998b).

All flowing surface water and some groundwater resources were subject to water rights. The Water Code defined various types of rights, the most important of which are consumptive rights that allow for the use of water without replacing it (e.g. agriculture), and non-consumptive rights, which allow for use on the condition that water is returned to the source (e.g. hydroelectric power). Existing water rights were recognized and could be 'regularized'. Regularization required that water must have been used continuously for five years before the 1981 Water Code came into force and thereafter. New rights were allocated to land redistributed under agrarian reform and could be obtained by applying to the National Water Directorate (DGA), the government agency responsible for administering water rights. The DGA grants rights free of charge and, if water resources are available, cannot refuse to grant them. Once all rights are allocated, future transfers of water rights take place through the market. Local property registries record rights that are regularized, newly granted or formally traded. By the mid 1990s, however, many historic rights had still not been regularized, and many more had not been formally registered so that the registry was incomplete (Bauer, 1997).

Much Chilean writing on water markets contains positive claims and touts the Water Code in Chile as a symbol of neoliberal ideology (Bauer, 2004a). Such assessments of water markets contain little, if any, empirical evidence, yet produce sweeping claims that are framed as fact (Gazmuri, 1994; Donoso, 2003). One opines that 'perhaps the most important achievements of Chilean water policy are social benefits through redistribution of wealth and eradication of poverty' (Gazmuri, 1994, p76). Drawing on such material, the World Bank has published a number of documents characterizing the implementation of Chilean water markets positively (Rosegrant and Binswanger, 1994; Rosegrant and Gazmuri, 1994; Hearne and Easter, 1995, 1997; Ríos and Quiroz, 1995; Thobani, 1995; Briscoe, 1996; Gazmuri and Rosegrant, 1996; Briscoe et al, 1997; Simpson and Ringskog, 1998). Much of this material blurs the distinction between how water markets should work in theory and how they have actually worked in practice. Many documents neither provide nor cite field evidence, and are replete with inaccuracies and misleading statements about the functioning and outcomes of water markets (e.g. Rosegrant and Binswanger, 1994; Rosegrant and Gazmuri, 1994; Thobani, 1995; Gazmuri and Rosegrant, 1996).

Non-empirical studies also dismiss the existence of negative social impacts. For example one states that: 'After the turmoil caused by expropriation of land and water in the 1960s, the establishment of tradable water rights and redistribution of a large portion of these water rights to former landless labourers was seen as *a substantial improvement in equity* [author's emphasis]' (Rosegrant and Gazmuri, 1994, p19). Another says: 'Even though some specific equity problems might be involved with the initial implementation of a private water right market, it seems to be a *non-issue* in the case of Chile given the traditional operation of a water market among farmers [author's emphasis]' (Ríos and Quiroz, 1995, p27). This literature's claims of success were first challenged by the United Nations Economic Commission for Latin America and the Caribbean (ECLAC) in Santiago (ECLAC, 1995). ECLAC produced detailed conceptual critiques of the water markets system and called for investigation of alleged problems with the system (e.g. speculation, hoarding, market power, externalities and market inactivity), as well as non-economic concerns such as social equity (e.g. Dourojeanni, 1999; Dourojeanni and Jouravlev, 1999).

The first in-depth field research on Chilean water markets was conducted by Carl Bauer over two years (Bauer, 1997, 1998a, 1998b). This work examined the practical functioning of the water law in the agricultural and hydroelectric sectors and identified a series of problems, limitations and contradictions that fundamentally challenged many of the assertions made by World Bank-sponsored work. In particular, Bauer observed that the water market was largely inactive, even when users economized water, due to a number of barriers related to geography, infrastructure, the legal and administrative systems, cultural attitudes and pricing (Bauer, 1997). Subsequent field studies also confirmed that water markets were largely inactive (Hearne and Easter, 1995, 1997). However, this 'problem' was then reframed as the 'optimal' original allocation of water rights that resulted in few sales (Ríos and Quiroz, 1995). One exception is the Limarí Basin in central-northern Chile, which has an active market – but largely in water *volumes* rather than water *rights* (the 'spot market') – due to the presence of large irrigation reservoirs that enable users to calculate water availability in advance. The little work on social equity done in this basin suggests that water trading largely occurs from small to large farmers, especially in times of hardship (Hearne and Easter, 1995; Hadjigeorgalis, 2000; Romano and Leporati, 2002; León and Garay-Flühmann, 2005). For example, Romano and Leporati (2002, p56) found that 'the distribution of water resources is heavily unbalanced and it tended to worsen in the period of analysis [1981 to 1997]... Such a period is characterized by ... the accumulation of resources by the most powerful social-economic groups.'

An initial assessment of the impact of the Water Code on social equity among peasant farmers by Bauer (1998a) concluded that they were worse off than commercial farmers due to a lower proportion of legal rights, primitive or dilapidated infrastructure, limited influence in water users' organizations, a tendency to avoid both the legal system and government bureaucracy (because of cost, time and attitude towards them), and lack of socio-political power to undertake private bargaining. He also alleges that in some cases traditional rights were lost due to failure to register them before the water source became fully allocated. He concludes: 'These problems, of course, have deep historical roots and reflect peasants' overall poverty and social position more than recent changes in water law ... but the new Water Code seems to have made them worse off in several ways' (Bauer, 1997, p650).

In Chile, the DGA refutes serious social impacts, although the views of its staff are by no means consistent. According to the then director, Humberto Peña: 'There is nothing very serious because users have kept on using the water as they did before 1981 – nothing has changed' (Peña, pers comm, 4 August 2003). Peña proceeded to elaborate that there is no evidence of any concentration of rights among wealthier parties. He cited two examples of the more efficient and valuable reallocation of water through the market in practice: the spot market in the Limarí Valley, where in dry years peasant farmers sell their annual allocation to commercial farmers instead of producing crops themselves; and indigenous people, who sold their customary water rights to mining companies in northern Chile for high prices despite huge controversy surrounding these sales.

Two DGA colleagues, however, suggested that social impacts are more serious. In personal communications with this author, Pablo Jaeger highlighted the abandonment of traditional indigenous agriculture and migration to the mining town of Calama following water rights sales in northern Chile (Jaeger, pers comm, 6 March 2003). Jaime Muñoz explained his belief that only state subsidies can prevent monopolization through the transfer of water rights from poorer to richer water users (Muñoz, pers comm, 8 May 2003). These different arguments are understandable when considered in relation to the DGA's dilemma between publicly supporting the Water Code, while simultaneously trying to pass legislative modifications on the grounds that the market is not functioning correctly (e.g. Peña, 1996). Peña (1996) thus concludes that: 'The problem is not the Water Code itself, but the social context within which it is implemented.' Other authors disagree, however, on the basis that markets are *social relations* rather than *neutral and technical mechanisms*. Galaz (2004) used anecdotal evidence from Chile to show how socio-economic power presents significant barriers to both the market and the legal system for peasant farmers and indigenous groups in particular. He writes that 'the second [deficiency] is the failure fully to acknowledge the way in which natural resource markets are embedded in asymmetries of money and social power among users' (Galaz, 2004, p415). This stance concurs with Bauer's overall conclusion of water markets, that 'market mechanisms can do some things well and others poorly – that markets are not automatic or self-regulating but depend on their social, institutional and geographic contexts' (Bauer, 1998a, p119).

The view that the Water Code is not directly responsible for social impacts may underestimate their severity in terms of the lack of access to legal water rights or to justice in order to defend water rights violations. This question was explored in earlier work on access to water rights for irrigation between peasant and commercial farmers in the La Ligua River Basin in semi-arid central-northern Chile (Budds, 2004). This study found that commercial farmers had the necessary awareness, financial resources, and access to the legal system to acquire new water rights, particularly groundwater rights; but peasant farmers were largely unable to access new rights precisely due to the lack of those advantages. Similar questions will now be explored in the context of the Atacama region of northern Chile.

Water, mining and indigenous groups in the Atacama

The Atacama Desert of northern Chile is extremely arid. It is characterized by a barren landscape, a hot and dry climate with extremely low levels of precipitation, and the presence of vegetation only along river channels or in small and sporadic marshy wetlands. The Atacama region is inhabited by several diverse indigenous groups collectively denominated as Atacameños under the 1993 Indigenous Law (Gundermann, 2000). The northern Atacama is also inhabited by the *Aymara* indigenous group; but this is a separate group from those denominated Atacameños.

Surface fresh water is scarce but present in small streams or large rivers flowing westward from the Andes. In Antofagasta state in the central Atacama, the River Loa forms the main river basin. Wetlands are fed by subsurface fresh water and springs. Saline lakes known as *salares* are also present, the largest of these being the Salar de Atacama. Finally, groundwater also exists in deep aquifers (100m or more). Due to the low rate of recharge, this is largely saline 'fossil' groundwater. Surface fresh water is in the highest demand for most economic uses. The scarcity of freshwater resources in the Atacama region has given rise to competing demands among different water users present in the region – primarily the mining industry and, to a lesser extent, the water supply and tourism industries, indigenous groups and endemic wetlands.

Antofagasta state is Chile's primary region for mineral extraction, primarily metals. It constitutes the core of Chile's copper mining industry, which provides the country's most important export. A large number of mines are present in the region, including some of the world's largest copper mines (e.g. Escondida and Chuquicamata). Most mines are foreign owned. Mining demands huge volumes of water for different stages of the mineral production process. For example, Chuquicamata requires 12 cubic metres of water per second. The mining industry has a strong preference for surface freshwater resources for two reasons. First, surface water is more easily captured while groundwater exploitation requires drilling deep wells and installing pumps that are expensive to install and operate. Second, groundwater is often too saline to be used directly and purification is also costly. The lack of fresh water in the region has induced some mines to reduce their demand by recycling water (e.g. Escondida estimates that its water is reused seven times). According to Jorge Zeballos of Minera Escondida, this mine has installed a desalination plant that produces purified seawater, which is then piped to the mining area (over 100km). The high cost of this option means that it is only feasible for the most productive mines (Zeballos, pers comm, December 2003).

A state-wide water utility, Empresa de Servicios Sanitarios de Antofagasta (ESSAN), supplies piped water to the larger urban centres of the region, notably Antofagasta and Calama, as well as to some industrial consumers (mostly mining related). ESSAN draws its water from surface sources. The Atacama region also has a growing tourist industry, which is increasing the demand for piped water to tourist towns, especially San Pedro de Atacama, and luxury hotels.

The Atacameños live in small rural communities and practise traditional Andean terraced agriculture for subsistence crops (e.g. maize, alfalfa, quinoa and tubers), petty commerce (e.g. vegetables for Antofagasta and Calama) and livestock grazing. They use surface freshwater streams and springs as sources of irrigation water. Due to water scarcity, they only irrigate once every two weeks, which is much less frequently than in less arid regions of Chile. Atacameños do not use deep groundwater resources.

Since the liberalization of the economy under the Pinochet regime, the growth of industry – especially mining – from the 1980s has brought significant changes that are reflected in the regional economy, landscape and social structure. Central to these changes has been control over water resources, which, in turn, has been shaped by the institutional framework introduced by the Water Code. The Water Code allowed mining companies to access water resources in two new ways. They could regularize existing water rights or request new water rights to 'available' water in the region. They could also purchase water rights from other users so long as those rights had been formally registered. Both of these options, and the control over large flows of water, while perfectly legal, quickly became contentious. Soon after the introduction of the Water Code, mining companies (and other industrial users) were swift to claim as much water as possible for current and future ventures, which appeared to be 'available' even if other users were using it and had rights that were eligible for regularization (Bauer, 1998a).

Once surface water had been fully allocated, purchases from other users, including members of indigenous groups, increased. Buyers tended to be intermediaries rather than the companies themselves and they sought to resell the water rights at a profit. Indigenous people, especially in the town of Quillagua, were coerced into selling water rights at well below the market price.

Irrespective of the monetary value of the rights transferred, the wider debate revolves around economic efficiency and informed consent. Peña used the case of Quillagua to illustrate the *effective* functioning of the water market: to transfer water from low-value uses (subsistence agriculture or petty market gardening) to high-value uses (copper production). Peña also sought to legitimize this type of transaction by asserting that those indigenous peoples who sold water rights did so with full knowledge of their actions and made a rational decision based on the respective benefits of either keeping their water to practise traditional agriculture or selling their water and using the (significant) revenue to invest in a better life. However, the National Corporation of Indigenous Development (CONADI) alleges that some indigenous people sold their water by signing papers that they could not read and accepting money without knowing the market value. Others, it claims, did not fully understand the implications of the 'sale' because Atacameño people view water resources as a holistic and inseparable part of nature, and the concept of 'selling' their water is unfamiliar. The fact that a number of Atacameño people continued to use the water they had sold, apparently not understanding that having sold the rights meant that they could no longer use it, may support this claim. In contrast, the fact that the director of CONADI (herself an Atacameña) felt obliged to visit potential indigenous sellers to try to persuade them not to part with their water rights indicates that the large sums of money to be gained from selling water rights is often too hard to resist by people who are used to a largely subsistence lifestyle.

Following the publication of the Indigenous Law in 1993, and in response to the problems confronted by indigenous people, CONADI embarked upon a programme to regularize indigenous water rights in the same manner as for peasant farmers throughout Chile (CONADI, 1999; Cuadra, 2000b). In most cases, the regularizations proceeded smoothly; however, some cases inevitably meant reclaiming existing rights that had been registered by other users, and so some regularizations were challenged.

An emblematic example is the case of Toconce village in the upper Loa Basin. During the 1950s, the government granted the water supply company, now ESSAN, the right to extract 470 litres per second from the Río Toconce, which had been the source of irrigation water for the village. Of a flow of approximately 490 litres per second, the company left the village with only about 20 litres per second, which was insufficient to maintain agricultural production, and the terraces dried up. As a result, many villagers migrated to Calama to work in the mines. The lack of water and, thus, agriculture also meant that government agricultural support programmes were not extended to Toconce. CONADI supported Toconce to reclaim (regularize) 100 litres of water per second from ESSAN – not the original flow, but just the amount needed for the remaining villagers to restart agriculture. Following a field inspection, the judge awarded the community an even lower flow, a decision that was appealed not only by CONADI, which demanded 100 litres per second, but also by ESSAN, which argued that it owned the water rights. With a market value of approximately US$30,000 per litre per second for industrial use, ESSAN had a strong commercial interest in appealing. A Supreme Court decision later awarded 100 litres per second to Toconce (Cuadra, 2000a).

Importantly, the Indigenous Law allowed regularized rights to be registered as communal rights rather than individual rights. For cases in which members of indigenous groups had legally sold their water rights, CONADI implemented a programme to repurchase these. In accordance with the Water Code, CONADI was obliged to pay the market price for these rights, irrespective of the original sale price. The repurchased rights were registered as communal rights and could not be sold for 25 years and thereafter only if the entire community agreed. At least one major foreign mining company, Minera Escondida, now refuses to consider purchasing indigenous rights largely to avoid controversy and negative publicity in Chile and internationally.

It is difficult to value indigenous water rights because water supports uses that are not susceptible to a simple economic cost-benefit analysis. For instance, water feeds native wetlands that have environmental value as endemic natural habitats and are also socially important to Atacameños, who use them as pasture for their livestock. Tension has also arisen between pressure to preserve these wetlands and demands for the subsurface fresh water that sustains them for other economic uses. Therefore, the Chilean government has been reluctant to classify these areas as 'wetlands' under the 1971 Ramsar Convention, which would necessitate protection measures of the wetlands and their water sources. Indeed, allegations have already been made that many wetland areas in the region have either receded or dried up due to the exploitation of subsurface water for mining.

Similar tensions have arisen over the demarcation of indigenous land and water resources under the Indigenous Law. One of the intentions of this law is to pass control over ancestral land and water resources to Atacameño communities. However, the demarcation process has continually suffered long delays, and the area to be allocated has successively been reduced because of the possibility that it may harbour important mineral and water resources that potentially could be used to foster regional economic development.

Furthermore, while the Indigenous Law clearly states that indigenous customary water rights must be regularized in order to be protected, it contains no specific mechanisms or procedures to accomplish this. The Water Code makes no explicit references to water resources pertaining to indigenous groups, although indigenous rights are eligible for regularization on the basis of historic use. However, a problem throughout Chile is that historic rights are often very difficult to prove.

The situation with groundwater is different. Because water is separate from land under the Water Code, landowners do not automatically own the rights to groundwater beneath their land. Furthermore, in order to obtain groundwater rights, the applicant must demonstrate that the flow of water being requested exists, and this requires installing a well. Given water scarcity in the region, the huge demand for water by mines, and the high monetary value of water, groundwater exploration was extensive in the region for a time and encroached on land inhabited by Atacameño communities and on land proposed for demarcation. The problem has abated because the mining companies lost interest in groundwater, realizing the difficulty of transportation and uncertainties about its quality. Groundwater extraction has caused much resentment among Atacameños. They perceive it to be part of holistic nature and consider taking it to be wrong even when a well is not actually on their land. Moreover, they have observed impacts on wetlands from groundwater extraction. Most importantly, they are powerless to claim groundwater rights themselves – to guarantee resource protection – because a claim requires costly well drilling.

Another problem is that the DGA only has limited resources and is unable to monitor hydrological data throughout the country, particularly in remote areas. In the Atacama region, the DGA only has one gauging station in 6000 square kilometres. Because it has limited powers it does not, in any event, supervise levels of extraction. Instead, companies are expected to undertake their own monitoring and submit the data to the DGA. This gives them a greater influence and control than would be ideal.

Atacameños also struggle for control of water to which they have rights. Even though indigenous water rights are being regularized and applications of others to extract water from land set aside for demarcation are being rejected, the Water Code has nevertheless permitted non-indigenous actors to control the majority of the region's water resources. This has caused immense resentment among Atacameño communities, especially as these groups are receiving few benefits from the regional economic development generated by the mining industry, while they suffer the brunt of the impacts.

Some of the foreign mining companies are keen to reposition themselves as socially responsible and environmentally sustainable – in particular, Minera

Escondida (owned by Australian mining company BHP Billiton). Escondida has established the Fundación Minera Escondida in San Pedro de Atacama to promote local development around themes such as youth, health, education and handicrafts, but not water. Compared with the profits generated by the mine and the value of the freshwater rights obtained by Escondida in the region, the level of investment in these programmes, however, is minimal. Yet, it compromises the possibility of indigenous communities opposing the mine or its activities should they wish. The regional development benefits arising from mining remain contentious, especially in Chile where revenue to the government is low and environmental impacts are often significant. Many San Pedro de Atacama indigenous water users are of the opinion that, rather than having the foundation assist them, they would prefer to have ownership of the water rights on the land that they consider to be theirs.

Conclusions

Recent studies on the social outcomes of the implementation of Chile's Water Code increasingly suggest that the supposed social and developmental benefits of neoliberal mechanisms have not materialized in practice. Indeed, emerging empirical research, including that presented here, has identified negative social outcomes for lower-income water users. The Atacama region is particularly emblematic for these issues due to the presence of a powerful commercial sector, relatively weak social actors and the presence of fragile natural areas. Here, the introduction of the Water Code established conditions that enabled commercial interests to take control over the majority of water resources in the region, while the defence of traditional rights and water sources sustaining wetlands has been precarious. Indeed, during the long-running debate over the modification of the Water Code (1992 to 2005), the mining sector was among those that opposed any significant changes to the law. However, it is not just the market mechanisms embodied within the Water Code that have produced these changes, but rather the wider range of neoliberal measures introduced, including the regulation of water rights by private law and the reduction of DGA regulation. Together with new modes for allocating water in the Atacama, these measures have reshaped the economy, the landscape and the social structure of the region. As mineral extraction has increased, some Atacameño communities have suffered reductions in the availability of surface water and have been forced to migrate to urban areas; in addition, some native wetlands may have dried up.

The relationship between private property rights and security is a double-edged sword. On the one hand, private rights can be understood as offering security in terms of permanent duration, legal defensibility and a potential for using them as collateral. On the other hand, their tradable nature also brings them into the realm of the market where they can be simply and legitimately acquired by other economically more powerful users. Although the possibility introduced under the Indigenous Law of registering communal indigenous rights does increase the security of these rights, it also exposes them to unfamiliar free-market forces.

Supporters of the Water Code emphasize its efficiency benefits assessed through economic valuations of production. However, an economic cost-benefit analysis cannot incorporate either any cultural costs arising from water rights sales if they result in loss of traditional agriculture, lifestyles and social structure, or environmental (and cultural) losses through wetland degradation. These arguments highlight economic tensions between indigenous communities using high-value water to irrigate low-value subsistence crops and demanding that groundwater resources are not used. This reflects a fundamentally neoliberal utilitarian argument that rights should reflect economically efficient use. Indeed, conflicts over water in the Atacama revolve around the difficulties encountered by non-commercial groups to justify allocating water to other uses and to control water resources for such ends.

The social equity dimensions of the Water Code are often expressed as 'the law applies to all Chileans alike'. While this may be true in theory, it is not the case in practice. The statement is based on a conceptualization of legal mechanisms as technical and *neutral*, whereas, in practice, they depend on different types of social power, capital and relations, particularly access to information, knowledge, money and the legal system. Therefore, the social outcomes of the Water Code should not be understood as the effects of a policy *per se*, but as the result of a wider set of social relations (the law, the market and socio-economic status) that favour stronger social actors and disadvantage weaker ones (see also Chapter 2 in this volume).

The principal difficulty with making any changes to the Water Code is the fact that water rights enjoy status as private property, which is protected under the 1980 Constitution. This essentially means that existing water rights cannot be modified in any way without compensation at market value. It has only been in extreme cases – such as CONADI's programme to repurchase indigenous rights – that the latter mechanism has been employed. Nevertheless, some of the negative effects of the Water Code, such as the potential for hoarding water rights for speculative purposes (i.e. not using the water) and environmental and social concerns (e.g. stricter procedures for assessing groundwater availability), led to a debate in Chile over the need to reduce some of the law's most extreme free-market features. Following 12 years of debate and various draft bills, some modifications were finally passed in 2005. The principal changes include the introduction of fees for non-use (after five years for most rights), the need to justify the use and quantity of new rights, greater DGA powers for water resource assessments, and simpler mechanisms to regularize small wells. Given that the majority of surface water and many groundwater rights have already been allocated, especially in arid regions, these measures will have minimal impacts. The simpler regularization of small-scale wells principally applies to non-indigenous peasant farmers. The reforms are therefore unlikely to bring significant benefits to Atacameños.

References

Alegría, M. (2004) 'History of the water law in Chile and a theoretical and empiric [sic] analysis of the current water rights market', Paper presented to the American Water Resources Association International Specialty Conference, Dundee, 29 August–2 September 2004

Angell, A. and B. Pollack (1990) 'The Chilean elections of 1989 and the politics of the transition to democracy', *Bulletin of Latin American Research*, vol 9, no 1, pp1–23

Assies, W. (2003) 'David versus Goliath in Cochabamba: Water rights, neoliberalism, and the revival of social protest in Bolivia', *Latin American Perspectives*, vol 30, no 3, pp14–36

Bauer, C. (1997) 'Bringing water markets down to earth: The political economy of water rights in Chile, 1976–95', *World Development*, vol 25, no 5, pp639–656

Bauer, C. (1998a) *Against the Current? Privatization, Water Markets and the State in Chile,* Kluwer, Boston, MA

Bauer, C. (1998b) 'Slippery property rights: Multiple water uses and the neoliberal model in Chile 1981–1995', *Natural Resources Journal*, vol 38, no 1, pp109–155

Bauer, C. (2004a) *Siren Song: Chilean Water Law as a Model for International Reform*, Resources for the Future, Washington, DC

Bauer, C. (2004b) 'Results of Chilean water markets: Empirical research since 1990', *Water Resources Research*, vol 40, no 9, doi:10.1029/2003WR002838

Briscoe, J. (1996) *Water Resources Management in Chile: Lessons from a World Bank Study Tour*, World Bank, Washington, DC

Briscoe, J., P. Anguita and H. Peña (1997) *Managing Water as an Economic Resource: Reflection on the Chilean Experience*, World Bank, Washington, DC

Budds, J. (2004) 'Power, nature and neoliberalism: The political ecology of water in Chile', *Singapore Journal of Tropical Geography*, vol 25, no 3, pp322–342

Budds, J. and G. McGranahan (2003) 'Do the debates on water privatization miss the point? Experiences from Africa, Asia and Latin America', *Environment and Urbanization*, vol 15, no 2, pp87–113

CONADI (1999) *Proyecto regularización de derechos de agua Provincia El Loa – II Región: Informe Final, Tomo I*, CONADI, Calama

Cuadra, M. (2000a) *Caso regularización derechos de agua comunidad atacameña de Toconce*, Purifica Consultores, Calama

Cuadra, M. (2000b) 'Teoría y práctica de los derechos ancestrales de agua de las comunidades atacameñas, *Estudios Atacameños*, vol 19, pp93–112

Donoso, G. (2003) *Water Markets: Case Study of Chile's 1981 Código de Aguas*, Catholic University of Chile, Santiago, Chile

Dourojeanni, A. (1999) *Debate sobre el Código de Aguas de Chile*, ECLAC, Santiago, Chile

Dourojeanni, A. and A. Jouravlev (1999) *El Código de Aguas de Chile: Entre la ideología y la realidad, ECLAC*, Santiago, Chile

ECLAC (1995) *Mercados de derechos de aguas: Entorno legal, ECLAC*, Santiago, Chile

Galaz, V. (2004) 'Stealing from the poor? Game theory and the politics of water markets in Chile', *Environmental Politics*, vol 13, no 2, pp414–437

Gazmuri, R. (1994) 'Chile's market-oriented water policy: Institutional aspects and achievements' in G. Le Moigne, K. W. Easter, W. J. Ochs and S. Giltner (eds) *Water Policy and Water Markets*, World Bank, Washington, DC, pp65–78

Gazmuri, R. and M. Rosegrant (1996) 'Chilean water policy: The role of water rights, institutions and markets', *Water Resources Development*, vol 12, no 1, pp33–48

Grant, R. (2001) *A Future for Tradable Water Rights*, 2nd edition, The Free Market Foundation, Cape Town and Sandton, South Africa

Gundermann, H. (2000) 'Las organizaciones étnicas y el discurso de la identidad en el norte de Chile, 1980–2000', *Estudios Atacameños*, vol 19, pp75–91

Hadjigeorgalis, E. (2000) *Private Water Markets in Agriculture and the Effects of Risk, Uncertainty and Institutional Constraints*, PhD thesis, University of California-Davis, Davis, CA

Hearne, R. and K. Easter (1995) *Water Allocation and Water Markets: An Analysis of Gains-from-Trade in Chile*, World Bank, Washington, DC

Hearne, R. and K. Easter (1997) 'The economic and financial gains from water markets in Chile', *Agricultural Economics*, vol 15, no 3, pp187–199

León, A. and R. Garay-Flühmann (2005) 'Public policy as a means to increase human security in agriculture in a drought-prone area of northern Chile', Paper presented to the Human Security and Climate Change International Workshop, 22–23 June, Asker, Norway

Muchnik, E., M. Luraschi and F. Maldini (1997) *Comercialización de los Derechos de Aguas en Chile*, ECLAC, Santiago, Chile

Peña, H. (1996) 'Water markets in Chile: What they are, how they have worked, and what needs to be done to strengthen them?', Paper presented to the Fourth Annual World Bank Conference on Environmentally Sustainable Development, 25–27 September, Washington, DC

Ríos, M. and J. Quiroz (1995) *The Market for Water Rights in Chile: Major Issues*, World Bank, Washington, DC

Romano, D. and M. Leporati (2002) 'The distributive impact of the water market in Chile: A case study in Limarí Province, 1981–1997', *Quarterly Journal of International Agriculture*, vol 41, no 1–2, pp41–58

Rosegrant, M. and H. Binswanger (1994) 'Markets in tradable water rights: Potential for efficiency gains in developing-country water-resource allocation', *World Development*, vol 22, no 11, pp1613–1625

Rosegrant, M. and R. Gazmuri (1994) *Reforming Water Allocation Policy through Markets in Tradable Water Rights: Lessons from Chile, Mexico and California*, International Food Policy Research Institute, Washington, DC

Simpson, L. and K. Ringskog (1998) *Water Markets in the Americas*, World Bank, Washington, DC

Thobani, M. (1995) *Tradable Property Rights to Water: How to Improve Water Use and Resolve Water Conflicts*, World Bank, Washington, DC

Tietenberg, T. (2000) *Environmental and Natural Resource Economics*, 5th edition, HarperCollins, New York, NY

Trawick, P. (2003) 'Against the privatization of water: An indigenous model for improving existing laws and successfully governing the commons', *World Development*, vol 31, no 6, pp977–996

World Bank (1992) *Water Resources Management*, World Bank Policy Paper, World Bank, Washington, DC

11

Indian Water Rights in Conflict with State Water Rights: The Case of the Pyramid Lake Paiute Tribe in Nevada, US

Charles F. Wilkinson

Introduction: Indian reserved water rights

The story of the Pyramid Lake Paiute Tribe in Nevada, US, illustrates the difficulty of enforcing legal rights to water by indigenous peoples who compete with non-Indian water users and must use the legal system of the dominant society. The struggle of these people has resulted in success, but only after decades of legal and political manoeuvring and nearly a century of being deprived of water. It shows that being armed with legal rights is not enough. A sustained battle with the aid of lawyers and experts has been necessary to give meaning to the tribe's water rights. In the time that it has taken to vindicate the rights of the tribe, the once-thriving fishery on which the tribe depended for sustenance and culture has been driven to the brink of extinction for lack of water.

In the US, Indian tribes have the advantage of water rights that, according to a 19th-century Supreme Court decision, are superior to other water users on a waterway. The doctrine of 'reserved water rights', also known as the historic Winters Doctrine stemming from the decision in *Winters versus United States*, has come to be the legal precedent and source of legal rights to water claimed by tribes in the US.

In 1908, the court announced the Reserved Rights Doctrine in a conflict over use of the limited waters of the Milk River in Montana. The Fort Belknap

Reservation was a vestige of a huge area in the north-central area of the US near Canada that has been occupied by several tribes. After the tribes gave up all their other lands through treaties, they were confined to the small reservation. The surrounding area was being occupied by non-Indian settlers induced to come to the rugged area by government promises of land that could support farming. The formerly Indian land was parcelled out to the settlers and the government sponsored a project to divert and deliver water from the Milk River to the settlers. The tribe's reservation was set up with the ostensible purpose of turning the Indians into farmers and getting them to put aside their old way of life, which depended on hunting over an expansive area. But, of course, farming by the Indians in the dry area, with a short growing season, demanded water. Faced with the possibility that either the purpose of the Indian reservation or the purpose of the programme to bring settlers to the area so that they could take up homesteads would be defeated, the Supreme Court ruled for the Indians. It said:

> *We realize that there is a conflict of implications, but that which makes for the retention of the waters is of greater force than that which makes for their cession. The Indians had command of the lands and the waters – command of all their beneficial use, whether kept for hunting, 'and grazing roving herds of stock', or turned to agriculture and the arts of civilization. Did they give up all this? Did they reduce the area of their occupation and give up the waters which made it valuable or adequate?... By a rule of interpretation of agreements and treaties with the Indians, ambiguities occurring will be resolved from the standpoint of the Indians. And the rule should certainly be applied to determine between two inferences, one of which would support the purpose of the agreement and the other impair or defeat it.* (*Winters versus United States*, 1908)

For tribes throughout the US, this decision announcing that Indian reservations have reserved rights to water sufficient to fulfil the purposes for creating those reservations has ostensibly secured their ability to live in perpetuity in their homelands. But the theoretical right to water has not always been realized in practice. Often, non-Indian settlers proceed to develop water unchallenged by the neighbouring tribes who lack the financial and technical ability to develop the water themselves. The story of the Pyramid Lake Tribe is about tribal attempts to protect water for fishing, which was historically the source of their livelihood and culture and ostensibly the purpose of the reservation. Whether a reservation is established for agriculture, fishing or other purposes, tribes can find themselves struggling with non-Indian neighbours for a limited supply of water, as they did in the Winters case. In that historic case and at Pyramid Lake the tribes in question had to take their claims to court. And in both cases the government financed the non-Indians' development of water that caused the shortage of water for the Indians. As did the Indians of the Fort Belknap Reservation, the Pyramid Lake Paiute Tribe relied on the government to protect their water rights because the government is supposed to be in a fiduciary relationship with tribes with respect to their lands. The Pyramid Lake Tribe did not fare as well. Years after the government failed adequately to protect the tribe's rights, the tribe went to court to force action. By the time the US Supreme Court got the case, too many non-Indians had relied for too long on the water destined for the tribe's lake and the court denied relief.

A fishery in 'exclusive possession' of the Indians

Pyramid Lake is about 65km north-east of Reno, Nevada. People expect wild rivers in central Idaho, but not broad, deep natural lakes in the sagebrush and juniper high desert country of Nevada. And so, whether it is your first visit to Pyramid Lake or your tenth or hundredth, you catch your breath as you come over a ridge and gaze out over this aquamarine sheet that stretches more than 50km off in the distance. Certainly that was the reaction of John Charles Frémont, whose expedition reached the lake in January 1844, making it the first known visit by white people: 'It broke upon our eyes like the ocean... The waves were curling in the breeze and their dark-green colour showed it to be a body of deep water... It was set like a gem in the mountains, which, from our position, seemed to enclose it almost entirely.' Frémont also gave the lake the name that it bears today: 'we encamped on the shore, opposite a very remarkable rock in the lake, which had attracted our attention for many miles. It rose, according to our estimate, 600 feet above the water; and, from the point we viewed it, presented a pretty exact outline of the great pyramid of Cheops' (cited in Jackson and Spence, 1970, pp604–607).

Frémont was a newcomer to Pyramid Lake. Paiute people have lived there since at least 2000 years before Christ was born. Their name for themselves is *cui-ui-dakado*, meaning 'cui-ui eaters'. The cui-ui fish, which lives nowhere in the world except Pyramid Lake, is ugly to the uninitiated – a plump brownish-grey sucker; but the juicy flesh of these easily caught, good-sized fish (adults range from 2 to 6 pounds) has always been a staple in the diet of the Paiute people. Their other key food source was the Lahontan cutthroat trout, a strain that is native to Pyramid Lake. These fish, when smoked and dried, once formed the basis of a bustling commercial trade for the tribe throughout the Great Basin and over to the Pacific Coast. The large salmon-like Lahontans almost certainly approached 60 pounds in size; we know that they exceeded 40 pounds, because a tribal member named John Skimmerhorn caught a 41-pounder in 1925 (Knack and Stewart, 1984).

Other than a few small springs and ephemeral streams, the Truckee River is the only source of water for Pyramid Lake. The Truckee rises high in the Sierra Nevada as the outflow of Pyramid's sister lake, Tahoe, which is comparable in area and some 730m higher in elevation. The Truckee River winds down the steep eastern flank of the Sierra through what is now downtown Reno to its

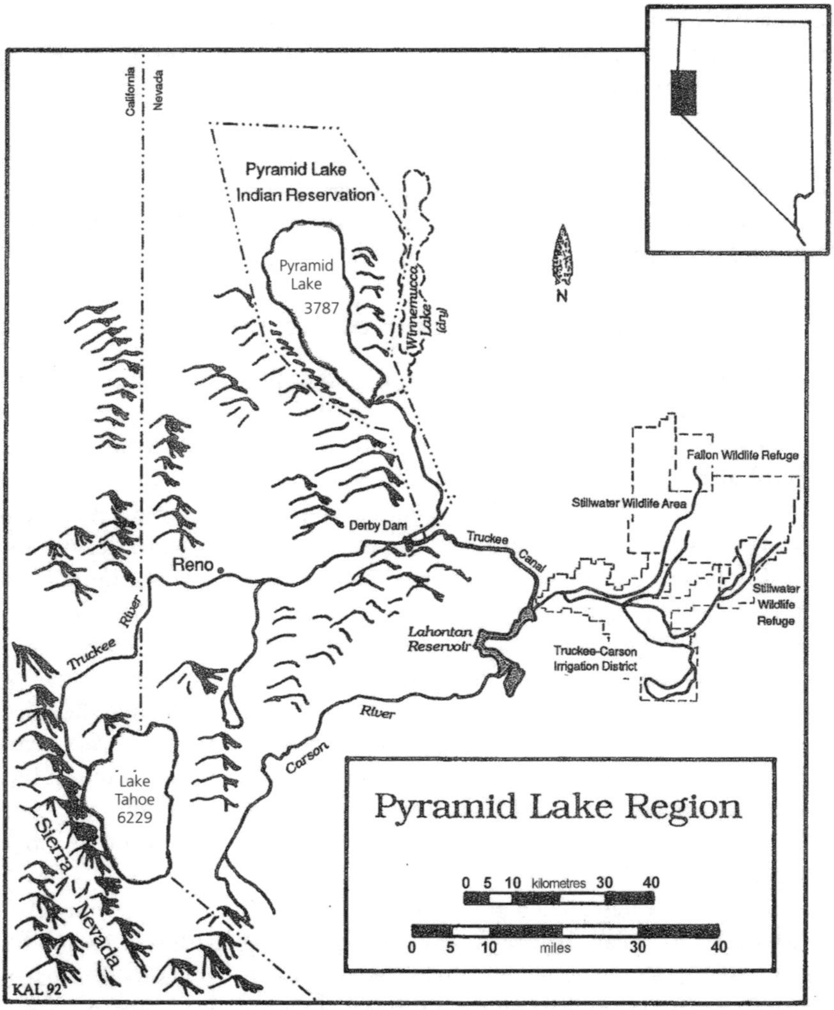

Source: Wilkinson, 1992, cartography by Karen Lewtosky

Figure 11.1 *Pyramid Lake, Nevada*

terminus at the southern end of Pyramid Lake. The Paiutes placed their villages at the mouth of the Truckee because both the Lahontans and the cui-ui were migrating fish, living most of their lives in Pyramid Lake but moving up through the Truckee River system to spawn. The mouth of the Truckee served to funnel the fish to the Paiutes, with their nets, traps and spears. Frémont's journal was expansive about this bounty:

> *An Indian brought in a large fish to trade, which we had the inexpressible satisfaction to find was a salmon trout... Their flavour was excellent – superior, in fact, to that of any fish I have ever known. They were of extraordinary size – about as large as*

> *the Columbia River salmon – generally from two to four feet in length... They doubtless formed the subsistence of these people, who hold the fishery in exclusive possession... These Indians were very fat and appeared to live an easy and happy life.* (cited in Jackson and Spence, 1970, p609)

The discovery of gold in California four years after Frémont's visit led to a spillover of population to Nevada, the opening of the fabulous Comstock Lode in 1859 and early statehood for Nevada in 1864. Conflicts over land arose between Indians and the new settlers of western Nevada. To alleviate tensions, in 1859 the US Department of the Interior set aside the territory of the Pyramid Lake Band of Paiutes from settlement; this administrative action was confirmed by President Grant's executive order in 1874, which reserved 1922 square kilometres for the tribe. In recognition of the central place of Pyramid Lake in the existence of these Paiutes, the reservation is dominated by Pyramid Lake itself, with a strip of Indian land surrounding the lake and a 30km arm of land reaching up the lower end of the Truckee River (see Figure 11.1). The late 19th century found the Paiutes' traditional way of life substantially unchanged, their millennia-old dependence on the lake's fishery still adequate to meet their needs. Indeed, the invention of an effective canning process allowed the Paiutes to expand their commercial use of the salmon trout: by the 1870s, shipments of Lahontan cutthroat trout to restaurants and grocery stores across the country amounted to 25 to 50 tonnes each year.

Water for non-Indian neighbours

Meanwhile, other Nevadans were undergoing frustrations of a kind found across the American West. Most sectors of the West, even the high desert land of Nevada, can support agriculture, but water must be brought in: beyond the 100th meridian (the north–south 'dry line' running through the middle of the Dakotas and then through Nebraska, Kansas, Oklahoma and Texas), annual rainfall by itself is insufficient to grow crops. In a few valleys, irrigation came fairly easily – water could be diverted from streams and transported through hand-cut ditches to nearby fields. More typically, however, potentially fertile farmlands were located on bench lands high above deep-canyoned rivers or in remote areas far from any watercourse. Furthermore, rainfall and high river flows come at the wrong time of year for successful irrigation of crops. The big runoff from the mountains arrives in April, May or June, at the front end of the irrigation season, leaving low flows for the key agricultural months of July, August and September. Individual settlers were unable to raise private capital to build the dams needed for storing the spring floods for the dry summer months and to construct canals for transporting the water to faraway fields. So, too, did the small western states lack the wherewithal to fund water development.

Finally, westerners succeeded in persuading the federal government to support irrigation on arid lands. The Reclamation Act of 1902, one of the two or three decisive laws in the history of the American West, was the chosen means to provide funding for construction of grand-scale water projects. The 1902 act was sponsored by Congressman Francis Newlands of Nevada, who, according to congressional custom, then as now, was entitled to have the first project in his home district. The Truckee-Carson Irrigation District, also called the Newlands Project, was begun promptly in 1903. It was designed to encompass some 142,000ha of land near Fallon, within the drainage area of the Carson River. Like all other reclamation projects, Newlands would be heavily subsidized by the US government: the non-Indian irrigators would receive free land and would pay only about 8 per cent of the more than US$10 million (in turn-of-the-century dollars) necessary to build facilities for storing and delivering water.

The Newlands Project was such an ambitious undertaking that water from the Carson River would be insufficient. To make up the deficit, Derby Dam was built on the Truckee River, 55km upstream of Pyramid Lake. The squat, unimposing structure diverted 50 per cent of the Truckee's flow out of the river channel. The water was transported south-east, via the Truckee Canal, to the Carson watershed and the farmers in the Newlands Project.

Pyramid Lake fishery imperilled

The Paiutes at Pyramid Lake have a name for Derby Dam. They call it 'the Killer'. Before Derby Dam, Pyramid Lake was the second largest natural lake in the west, after the Great Salt Lake. The diversions reduced the lake's surface areas by 25 per cent, from 572 square kilometres to 432 square kilometres. The lake level dropped 21.5m. Vast amounts of former lake bed were left exposed, especially at the southern end, where the Truckee River feeds in. During most years, fish could not enter the Truckee River to spawn because its mouth had been transformed into a delta of mud flats and scattered trickles of shallow, warm water. Those fish that could work their way up into the Truckee River were confronted by Derby Dam, which had no operable fish ladders. The big Lahontans slammed up against the concrete wall again and again, but to no avail. Unable to procreate, the native Pyramid Lake strain of Lahontan cutthroat trout died out, succeeded today by smaller, hatchery-reared fish with a similar, but not identical, genetic makeup that have been introduced into the river by the government.

The cui-ui is a hardier, more adaptable species and has done better. Some cui-ui have spawned in the few gravel stretches in the Truckee River below Derby Dam. In addition, it is a long-lived fish, having a life span of up to 40 years. Thus, cui-ui can survive if they can gain access to the spawning beds in the occasional years of high flows, when the fish can navigate the delta at the mouth of the Truckee River. In 1985, for example, 95 per cent of the spawning cui-ui population was composed of fish born in 1969, a year of exceptionally high runoff, and the high water of 1986 permitted the species to revive itself by natural regeneration once again. The cui-ui is also able to survive in degraded water. This is an important asset in today's Pyramid Lake, which is more heavily saline than in years past because the reduced flow of water does not sufficiently dilute the minerals that have built up over millions of years in this lake with no outlet (State of California, 1991; US Department of the Interior, 1991).

The Newlands Project has wrought many other changes. Anaho Island within the lake was declared a national wildlife refuge in 1913; if the lake level continues to drop, a land bridge soon will connect the island to the shore, allowing predators such as coyotes to reach eggs and fledglings on one of the nation's few nesting colonies for the great white pelican. In its natural state, the Truckee River sometimes overflowed into a level valley floor to the east, creating wetlands called Lake Winnemucca. This was prime habitat for migrating birds on the Great Western Flyway. Lake Winnemucca was declared a national wildlife refuge, but the designation was lifted after it became clear that there would be no more overflows from Pyramid Lake and that Winnemucca would dry up.

Ironically, the Newlands Project has itself served as a kind of substitute for Lake Winnemucca in the Carson watershed. Originally, the Carson River spread out on the floor of the Great Basin, creating Carson Sink and associated wetlands. The flow of the Carson River is now used for irrigation and the original wetland system has been drastically reduced. But unlike Lake Winnemucca, the area has not gone completely dry. Irrigation at Newlands has been notoriously wasteful – water has seeped through outmoded earthen transportation canals, flowed off improperly levelled fields and been applied excessively by flooding of fields rather than by use of more efficient sprinkler or drip systems. Today, Truckee River water once bound for Pyramid and Winnemucca lakes is mixed with Carson River water, and much of it runs off the Newlands Project as wastewater into the Stillwater National Wildlife Refuge.

But Stillwater, seemingly a place of refuge for birds and other wildlife, has taken a grotesque turn. The irrigation tail water from the Newlands Project carries highly toxic materials leached out of the soils and Stillwater has seen a build-up of many pollutants, including selenium, mercury and arsenic. The situation has worsened since federal officials forced farmers in the Newlands Project to conserve water. These conservation measures were taken so that diversions from the Truckee River could be reduced to provide more water and better habitat for the cui-ui and Lahontan cutthroat trout; yet water conservation at Newlands, made in the name of wildlife protection in the Truckee, has helped to create a death trap for wildlife in Stillwater. With less irrigation outflow from Newlands, the toxins have become more concentrated. Blue herons have dropped dead in flight and white pelicans have developed deformities. In 1987, Tom Harris of the Sacramento *Fresno Bee* newspaper visited Stillwater and gave this description:

> *A yard-wide band of death rings the massive, shallow and shrinking lake they call the Carson Sink, overwhelming evidence that*

> *the ecological system here is in complete collapse. Dead fish by the uncountable millions are washing up along the gooey shoreline, bobbing across the surface or decaying on the bottom, where bloating gases soon will pop great fetid masses more of them to the surface. Duck carcasses dot the shore ... herons, egrets, grebes, geese, cormorants – all are represented among the carcasses. It is a wretching, reeking sight.* (Harris, 1987)

The causes for this breakdown are complex and are not fully understood; but the main contributing factor seems plainly to be the chain reaction of events resulting from water development in the Truckee and Carson watersheds: a case of dramatic and large-scale water, land, economic and social engineering. All the while, as hardworking non-Indian farmers in another basin irrigate with Truckee River water, the Pyramid Lake Paiutes face an unemployment rate exceeding 40 per cent and are left with an uncertain supply of water, either for the fish and the lake that sustained them for 4000 years or for the irrigation that opened the West for non-Indian settlers.

The legal struggle

It is clear that the movement of water away from the Truckee River and Pyramid Lake was undertaken by the federal government for the benefit of irrigators. Federal money was used to build facilities to divert and deliver water, all to the detriment of the tribe. When it occurred, the US went to court and obtained water rights for the water taken away. Not only did the US represent itself as the sponsor of the Newlands Reclamation Project, but it also represented the tribe. It did not try to claim any water rights for the tribe to maintain the fishery. That was in 1913. The tribe finally got access to lawyers during the 1970s and began litigating to try to get water rights to enough water to save the fishery. After difficult litigation, one court said that because the US is a trustee for Indian tribes and their property, and because the Supreme Court of the US long ago had ruled that Indian tribes have 'reserved rights' to sufficient water to fulfil the purposes of their reservations, the US could not simply continue operating the Newlands project without regard to the tribe's rights (*Pyramid Lake Paiute Tribe versus Morton*, 1974).

Then the US – pressed by the tribe and moved by this 1974 court decision – attempted to claim rights for the tribe's fishery. But the Supreme Court said it was too late. In *Nevada versus United States* the court ruled that the opportunity of the tribe to claim a reserved water right, even if valid, was barred by the doctrine of *res judicata* because the US failed to raise the tribal right in litigation begun in 1913. The Supreme Court declined to correct the underlying inequity caused by the government's early disregard for the water rights of the same tribe by refusing to allow the tribe itself to reopen the adjudication. It cited the value of not disturbing non-Indian expectations implicit in notions of finality and *res judicata* in water rights decisions. So the very significant water rights ostensibly available to tribes under the reserved water rights doctrine were beyond the tribe's reach. The US, after failing to claim the tribe's rights and years later being prodded to do so by a court decision, was told by the nation's highest court that it had waited too long.

In 1990, the US Congress at last enacted promising legislation based on a settlement agreed to by almost all of the many parties. The Fallon Paiute Shoshone Indian Tribes Water Rights Settlement Act is designed to protect Pyramid Lake and its fish and to bring some justice to the Paiutes by improving the flow regime in the Truckee River and establishing recovery programmes for the Lahontan cutthroat trout and the cui-ui. The dedication of more water to maintaining the Pyramid Lake fishery has forced the Truckee–Carson Irrigation District to be more efficient. Not only is less water available to the district, but return flows to the wildlife refuge have been decreased. Tribal, federal and state wildlife biologists are hard at work to make the recovery programmes on the Truckee succeed, and the same is true at Stillwater National Wildlife Refuge with biologists from the US Fish and Wildlife Service. In the Newlands Project, conservation measures continue to be implemented so that less water will be drawn out of the Truckee; at the same time, The Nature Conservancy has headed up a water purchase plan so that farmers, if willing, can sell water rights and allow fresh Carson River water to flow into the refuge and prevent further toxic horrors.

Yet there is still a very long way to go. The Settlement Act for the Truckee and Carson watersheds is contingent upon further negotiations and actions involving the US, Nevada and California, the city of Reno, the Sierra Pacific Power Company and the tribe. Furthermore, it is still unclear whether the conservation and water purchase programmes will be fully implemented or whether the threatened Lahontan cutthroat trout and endangered cui-ui – even given a first-rate recovery effort – will have the staying power to eke out an existence from their depleted, battered habitat. The same is true with the birds and fish on the contaminant-laden flats of the Stillwater National Wildlife Refuge. Because the cultural survival of the Pyramid Lake Paiutes depends on the restoration and long-term survival of this fragile ecosystem, much is at stake.

Conclusions

The case of Pyramid Lake illustrates the culture clash between traditional water uses by indigenous peoples and the competing uses that are embraced by the dominant society. The US Congress enthusiastically created a programme to construct irrigation projects to bring irrigation water to desert lands. Federal lawyers were single-minded in their efforts to establish water rights for the Newlands Irrigation Project coming from the Truckee River. They did not pause to consider the devastating effect that this would have on the fishing culture of the Pyramid Lake Paiute Tribe, whose only source of water was the Truckee. When the tribe, long without lawyers or resources to conduct their own legal fight, finally got to court with the support of the government, their claim was found to be too disruptive to be asserted at such a late date, long after non-Indians had established their rights. On the one hand, the legal system had announced in the Winters case a formidable water right for tribes. On the other, the Pyramid Lake Paiute Tribe's historical lack of access to the system became the basis for denying them the right itself.

References

Fallon Paiute Shoshone Indian Tribes Water Rights Settlement Act of 1990 (1990) 104 Statute 3289

Harris, T (1987) 'Scientists try to find out what is wiping out life at Carson Sink', *Fresno Bee*, 15 February, pAI

Jackson, D. and M. Spence (eds) (1970) *The Expeditions of John Charles Frémont: Travels from 1838 to 1844*, vol I, University of Illinois Press, Urbana, IL, pp604–609

Knack, M. and O. Stewart (1984) *As Long as the River Shall Run: An Ethnohistory of Pyramid Lake Indian Reservation*, University of California Press, Berkeley, CA

Nevada versus United States (1983) 463 U.S. 110

Pyramid Lake Paiute Tribe of Indians versus Morton (1974) 354 F.Supp. 252 (D.D.C.1972)

Reclamation Act of 1902 (1902) 43 U.S.C. § 371 et seq

State of California, Department of Water Resources (1991) *Truckee River Atlas*, Department of Water Resources, Sacramento, CA

Truckee–Carson–Pyramid Lake Water Rights Settlement Act of 1990 (1990) Title II of Pub. L. No 101–618, 104 Statute 3294

US Department of the Interior, Fish and Wildlife Service, Cui-ui Recovery Team (1991) *Technical Agency Draft Cui-ui Recovery Plan*, 2nd revision, US Fish and Wildlife Service, Reno, NV

Wilkinson, C. (1992) *Crossing the Next Meridien*, Island Press, Covelo, CA

Winters versus United States (1908) 207 U.S. 564

12

Acequias of the South-Western US in Tension with State Water Laws

Gregory A. Hicks

Introduction

This chapter introduces the history of the problematic relationship between state-centred governance of water in Colorado and New Mexico, US, and the historic Hispano irrigation institutions that have operated there since before the area became part of the US. This history has much to say about the complex relationship between local cultures of water and the imperatives of state-centred agendas for the use and allocation of water. It is a history that speaks to the challenges of achieving pluralism in water institutions, particularly to the challenge of carrying forward a collective community-based understanding of water when the prevailing legal and administrative order prioritizes water's commodity value. This narrative is presented in the hopes that it may be useful to the current struggles for collective water control in the Andean region, offering additional comparative evidence that water policymakers ought not to be indifferent to the fate of well-adapted water institutions grounded in communities of use.

Foundations of the South-west: *Acequias*

The Hispano irrigation communities of present-day Colorado and New Mexico were originally established as frontier settlements under Spanish and Mexican law between the 17th and 19th centuries, and based on the *acequia* (Hutchins, 1928; Meyer, 1984; Rivera, 1998). The term refers to the ditches that distribute water as well as to the local institutions that govern that distribution. Relying on irrigation technologies adapted to their locales and on an evolving practical knowledge of the movement of water and the properties of soils, communities of smallholders built and maintained effective irrigation systems in the river valleys of an arid landscape (Carlson, 1967; Clark, 1987; Nostrand, 1987; Baxter, 1997; Peña, 1999). The pattern of land apportionment was designed to ensure that each landholding would be served by a communal irrigation system, with farms laid out as narrow strips running perpendicular to the principal streams and to the ditch network (Carlson, 1967, pp111, 115–119). Thus, neighbours lived in close proximity to each other and shared the *acequia*.

The structure of the built landscape reflected and reinforced the dependence of neighbours on each other and on their irrigation commons (Van Ness, 1987; Peña, 1999, pp107–132). Their mutual commitments were expressed not only in the construction of the *acequia* and in the ongoing expectation of shared labour in maintaining the ditch network. Water was allocated on the basis of need and local equity perceptions, and not simply on the basis of temporal priority of use, further expressing the community's bonds of mutuality (Meyer, 1984; Baxter, 1997; Ebright, 2001; Hicks and Peña, 2003). Another defining expression of equality of circumstances, shared fate and mutuality of commitment among *acequia parciantes* was the rule giving all members equal votes in the election of ditch officers and in decision-making, irrespective of the size of their landholdings (Hicks and Peña, 2003, p393).

Practices of communal governance also lay at the heart of the *acequia* system, stressing the participation of all user families in maintaining the network of ditches and emphasizing the sharing of water in times of scarcity. Essential to the fixed settlement of the arid regions of Hispano North America, in the US *acequias* have survived as vital institutions only in localities in New Mexico and Colorado, where they anchor rural communities and provide continuing and extraordinary arrays of environmental services in their watersheds (Peña, 1999; Hicks and Peña, 2003).

Western US expansion and Mexican water institutions

Following cession through treaty at the end of the US–Mexico War of 1846, the area now comprising Colorado and New Mexico became part of the US. The change of dominion initiated a hectic phase of settlement and commercial development throughout the American West, shaped in large part by the availability or scarcity of water (Pisani, 1992). The development of new water institutions was a central concern of this period of build-up and settlement. The new American legal regime included the doctrine of appropriative water rights, which allocates water priority in conditions of scarcity to the first-in-time beneficial users (Pisani, 1992). By the early 1870s, Colorado had moved decisively towards a uniform law of water, adopting the rule of prior appropriation and abandoning an earlier toleration of other rules of allocation, including those of the *acequias* (Hicks and Peña, 2003, pp418–425). New Mexico, too, moved towards the system of prior appropriation and the marginalization of *acequia* institutions, although the process of change there was more fitful (Clark, 1987). In the end, however, the imperative for uniform principles of water allocation, based on priority in time and a concept of beneficial use, prevailed there as well.

Each state's water law, committed to the logic of prior appropriation and to centralized administrative authority over water, generated ownership rules and governance principles at odds with *acequia* practices. Chief among these were the rules that priority in time decides who shall have water in periods of scarcity, that decision-making by groups of water users shall be based on how much or how little water rights they own rather than on equality of voting, and the doctrine that ownership of a water right might carry with it the right to transfer that right to another place or use. This set of legal principles undermined *acequia* norms of equitable sharing and the understanding of water as an in-place resource for the community. Even though individual *parciantes* might well continue to feel bound to communal norms and to an understanding of water as a feature of a physical and social landscape of work and production, the law's characterization of water as individually owned, severed from social context and moveable, created pressures on longstanding social commitments and made those commitments vulnerable (Hicks and Peña, 2003, pp399–404).

The differing trajectories of the two states in their handling of *acequia* rights have had lasting consequences for the modern-day status of *acequia* systems and for the relationship between those systems and contemporary state water administration. New Mexico from the beginning chose to recognize *acequias* and aspects of Mexican water administration, while Colorado, after a brief early period in which variant water institutions were accommodated, did not. That difference may be attributable to the relative visibility and sense of importance of traditional Hispano water institutions in the development of water resources in the two states. *Acequias* were far more integrated within the overall water culture of New Mexico (Clark, 1987; Pisani, 1992; Rivera, 1998), but present in Colorado only in the Hispano regions of the southern part of the state (Hicks and Peña, 2003).

In New Mexico, *acequias* are known as community distribution systems and as governance structures of a particular sort, though the scope of their possible authority is periodically contested (Clark, 1987; Hall, 2000; Hicks and Peña, 2003). In Colorado, they have, in general, been viewed, when seen at all, as rather quaint and historical, highly local and peripheral to the real business of water law (Hicks and Peña, 2003, pp462–466). Those differing responses have been significant in defining the place that *acequias* have in the imaginations of water policy-makers. They create distinct challenges for achieving an institutional pluralism in each state. In New Mexico, *acequias* are recognized as legal entities and as political subdivisions of the state; but they face continuing resistance when they attempt to give substance to *acequia* values in the regulation of water rights (New Mexico Court of Appeals, 2008). In Colorado, formal legal status was never obtained, and the prevailing structure of water administration has found it easy to marginalize *acequia* institutions as relics of an earlier era.

In the first decades following New Mexico's absorption into the US, territorial water law there demonstrated a striking receptiveness to Mexican institutions (Clark, 1987; Baxter, 1997). Territorial water law seems to have functioned as a structure for enabling and legitimating local law and custom. The result was a form of pluralism of administration calculated to support practices of allocation and administration familiar and legitimate in their locales (Clark, 1987; Hall, 2000). There were great local variations in the administration of water rights from *acequia* to *acequia* (Clark, 1987), and the decisions of the courts suggest both indefiniteness in the content of fundamental law and a tolerance of local variation in the statement of water rights. In this, there may be evidence of a characteristic New Mexico exceptionalism, demonstrated through deference to local institutions. That fluidity was also characteristic of frontier legal regimes under Mexican and Spanish dominion in New Mexico (Clark, 1987, pp30, 41–42; Ebright, 2001).

However, with the coming of the system of prior appropriation to New Mexico, the nature of the legal accommodation of *acequia* institutions shifted dramatically. Territorial and state law changed from being a statutory and judicial expression of Mexican and Spanish law and custom and began instead to reflect the emerging American law of prior appropriation (Clark, 1987; Baxter, 1997). During this period, laws relating to the special circumstances of *acequias* continued to be adopted, but in a way that nested *acequias*, as local water institutions, in the more general framework of governance of water by a centralized administrative structure: the state engineer's office (Clark, 1987; Hicks and Peña, 2003). In addition, the scope of local authority was subordinated to the prior appropriation system's commitment to individual appropriation for beneficial use, and to the understanding that the state alone could confirm water rights (Baxter, 1997, pp83–93). For example, the questions whether *acequias* retained any authority to create or define water rights or to reserve water within their basins for future use was answered firmly in the negative (Hall, 2000). The role of *acequias* was thus redefined as administrative and managerial, and in support of rights viewed by the law as held by individual rights holders, not by the *acequias* as collectives.

An important factor leading to the recasting of *acequia* rights and to the limitation of *acequia* authority in New Mexico was a self-described progressive reform agenda characteristic of natural resource management in the US at the end of the 19th century and the beginning of the 20th century. That agenda focused on conservation of natural resources and the efficient management of watersheds. The regulation of water under centralized and technocratic management systems and the elimination of competing systems were central imperatives of the reform agendas. It was a time of national irrigation congresses, of model statutes, and of enlisting the talents of engineers, agronomists and hydrologists, all in aid of rationalizing the management of a scarce resource (Worster, 1985; Clark, 1987; Pisani, 1992; see also Chapter 4 in this book). Those technocratic imperatives, driven in part by confidence that the methods and knowledge of those who deployed them were correct, were complemented by a dismissive attitude towards the methods and technologies of older systems, including *acequias* (Newell, 1890, pp193–199). The marginalizing of local water norms and practices in the American West was a by-product of public policies that insisted, first, on an administration of water responsive to uniform rules and central authority, and, second, on a particular vision of water rights: one that facilitated the movement of water from wherever it might be to wherever it might be used. Thus an earlier period of community-based irrigation projects gave way in the latter part of the 19th century to larger systems and more controlling structures of state governance (Pisani, 1992).

In consequence, public policy imposed on traditional community water practices a regime with two aspects: one, administrative and legal, and the other, functional. There was to be a single source of water rights, a single set of criteria for the creation and maintaining of rights, and, in addition, a substantive definition of the content of those rights that allowed water to be severed from its immediate surroundings and moved to support beneficial uses elsewhere. Competing structures that either claimed command of the waters or the capacity to define rights, or that proposed different understandings of how water could be used, simply were not to be tolerated. The dismantling of the authority of New Mexico's community ditches was a direct reflection of these policies (Hall, 2000). There was, unhappily, a racialist cast to some end of the 19th-century commentary on Hispano water institutions, emphasizing their wastefulness, their apparently casual arrangements, and the crudeness of their diversion structures (Pisani, 1992, p43; Hicks and Peña, 2003, p405). Oddly, such commentary at times expressed a grudging admiration for the aptness of *acequias* to their locales and for the farmers who used them.

In Colorado, the reformist attack on local *acequia* systems was direct and harmful, particularly in the San Luis Valley. The most senior water rights in the state, those of the *acequia* ditches centred on the town of San Luis, were cut in half, and the water formerly distributed by the *acequias* was made available to a major entrepreneurial concern seeking to direct Northern European settlers to the San Luis Valley (Hicks and Peña, 2003; Hicks, 2005). A leading historian of water institutions of the American West has pointed out the significant variation in the willingness of state water administrators to attack or tolerate older ditch systems, depending on the depth of commitment of policy-makers to pre-existing allocations (Pisani, 1992). In the case of the San Luis *acequias*, the easy tolerance that state water administrators had often demonstrated for older ditch systems was nowhere in evidence, although the efficiency of the *acequias* was well within a range the law had protected in other disputes (Hicks and Peña, 2003).

Loss of watershed autonomy; loss of *acequia* landscape

The loss of water rights has not been the only factor undermining the effectiveness of *acequia* institutions in Colorado's San Luis Valley. The opportunities for a more vigorous local agricultural economy have also been constrained by shortages of labour, a rigorous high-altitude climate, and by the loss of arable land to environmental damage (Peña, 1999). With the coming of the railroads, the forest and steppe uplands constituting the watershed for the San Luis *acequias* were developed by their owners as sources of forage and timber for regional and national cattle and lumber markets. The loss of this upland commons to private enclosure deprived *acequia* farmers of access to grazing and timber resources; but the more enduring injury was loss of control over the watershed on which their irrigation systems and their farms depended (Hicks and Peña, 2003). A period of agricultural decline followed episodes of significant soil erosion caused by stock grazing, commercial logging and mineral prospecting by other owners beginning in the late 1870s and growing in intensity into the 20th century (Lantis, 1950; Hicks and Peña, 2003). The loss of bottomland farms to erosion and flooding was caused precisely by overcutting of the timberlands, speeding up snowmelt, and by overgrazing of the upland steppes, leaving the soil open to erosion. The surface geology of the area, characterized by relatively light and gravelly soils atop a shallow impermeable layer of clays, is typical of many of the *acequia* lands of Colorado and New Mexico (Lantis, 1950; Carlson, 1967; Hicks and Peña, 2003). In intact condition, that soil structure can produce highly effective sub-irrigation of croplands (Hicks and Peña, 2003). However, the soil is also vulnerable, being susceptible to erosion when disturbed. On a quotidian level, this vulnerability shows itself each year when the *acequia* farmers have to backhoe the gravel and sand out of their ditches, and shovel it out of their head gates.

With the opportunities for a more vigorous local agricultural economy limited by factors of environment and population, the work of maintaining high levels of commitment to local water institutions has been challenging in San Luis (Hicks and Peña, 2003). One cannot know for certain whether retention of the uplands by the Hispano farmers would have produced different patterns of use on those lands, or if the *acequia* farmers, too, would have overgrazed the pasturage and overcut the timber, producing the destruction of the farmlands on which their subsistence so directly depended. However, there is a strong ideology favouring collective action, mutual aid and the cherishing of the *acequias*, and those commitments might have produced approaches to managing the uplands that would have protected both the *acequias* and the watershed.

A commitment to *acequia* norms as a structure of cultural identity has come down through time. People prize neighbourliness and the annual round of *acequia* tasks, including meetings and ditch cleanings. These mutual commitments are strengthened by the lack of material resources in the communities. People have less equipment and money. They must help each other, and they do. But the scarcity of working farmers, relatively low crop values and uncertain supplies of water discourage larger investments for any but holders of the most senior water rights. There are some countervailing developments, chiefly a growing market for organic specialty crops – landrace varieties of corn and beans and orchard fruits (Hicks and Peña, 2003); but the constraints of climate and of labour availability continue to limit more substantial commitments to building up local agricultural capacity. Sincerely held norms of solidarity, collective action and commitment to place and territory fail to produce lasting participation in building up the effectiveness of *acequias* as institutions or as water distribution systems.

To offer this critique is not to devalue the strength of commitment to *acequias* in Colorado, only to make clear that before *acequia* institutions can flourish, they must be seen by their users as capable of supporting substantial economic opportunities for *parciantes* and for their communities. State actors and *acequia* organizations have begun projects to strengthen the functioning of *acequias*, and on their own terms; but the success of those initiatives has been limited by the difficulty of persuading *acequia* farmers to do their parts. Dwellers along the *acequias* value their ditches as cultural expressions and as water distribution systems; but constraints on agricultural economic opportunity have diminished investment by irrigators in their own systems. Most *acequia parciantes* have had to develop a way of remaining on their land as a tie to cultural and familial values and as a form of social insurance, without making substantial new investments of labour or money.

This convergence of limited economic prospects from farming and strong cultural commitment to a place and a way of life has produced deep paradoxes. People are acutely aware of the potential value of their land to new buyers, especially those seeking recreational property in the West. One of the most pressing threats to the *acequia* lands of the San Luis Valley is that they may be absorbed and their culture diminished, not only by modes of agricultural production and water use that are foreign to them, but through colonization by newcomers attracted to the physical and social landscape (the arrangements of space, the scale of holdings, the vegetative patterns, and even an aesthetic of agro-pastoral production) that the *acequia* system has created and conserved (Hicks and Peña, 2003). Rumours circulate of the sales prices that land in the area is commanding, and landowners are increasingly aware that their *acequia* farms are very valuable economic assets. Yet, fascination with the possibility of an irresistible offer coexists with anger about the loss of the uniqueness of the tie between people and place.

Restoration of the *acequia* commons: Landowner commitment and state action

The marginalization and the supplanting of *acequia* rules of water allocation would matter less if there were no continuing commitment to *acequia* principles or if the *acequia* systems were not highly functional as a sustainable mode of irrigation and a potent source of social and natural resource capital fostering mutual commitments to communities. Traditions of communally maintained ditches and the norm that water ought to be a shared resource shapes landowners' conduct in using their land, as well as their understandings of place and community.

Boelens and de Vos (2006) note that local property rights systems are often characterized by high levels of engagement by the people who rely on those systems. Where those systems are expressive of culture and shared experience, they are reported to produce in their users a sense of autonomy and of self-authoring, which in its turn supports the possibility of creativity and self-generated action in response to changing economic and environmental circumstances affecting resource use (Endter-Wada et al, 1998; Folke et al, 2005). Those features are all characteristic of *acequia* communities (Peña, 2003a, 2003b).

Such a culture of natural resource use can project itself forward, supporting continuing adaptations over time, grounded in known institutions and continually able to embody commitments that people have made to each other and to their home places in constructing natural resource landscapes and in building social capital. In the case of the Colorado and New Mexico *acequia* communities, there has been a remarkable tenacity of commitment by water users to the values and practices of the older Hispano water regime. Yet, those commitments have been made precarious, dependent on the willingness of people to hold themselves to conduct and commitments that the law no longer requires and that produce modest economic returns.

Part of the challenge for state actors in providing the institutional validation that *acequias* need to flourish is simply accepting the premise that variant approaches to water management, and even variant understandings of the worth of water, should thrive. That willingness may come from a commitment to the autonomy of the communities who wish to manage their water in a particular way; but more typically it will be grounded in the conclusion that some combination of social welfare and environmental functioning is to be achieved though locally appropriate alternatives, or perhaps that there is no harm in allowing the variation. In any case, formal recognition of the lawfulness of local governance institutions and rules – and in the case of water governance, of principles of allocation, rules of ownership, methods of diversion and distribution, and even understandings of efficiency – is essential for *acequias* to become more secure.

New Mexico law has gone furthest in that regard. It recognizes *acequias* as a form of community water distribution organization, has a highly developed body of statutory law governing *acequias*, and now in section 73-2-21(E) of its laws even grants *acequias* the power to restrict water transfers in order to protect the *acequia* against private choice (New Mexico Statutes Annotated, 2003). But its courts continue to struggle with the tension between private and *acequia* interests in water (New Mexico Court of Appeals, 2008). This difficulty reflects essential conceptual differences in thinking about water: whether it is to be viewed as a constitutionally defined and protected individual property right, transferable and alienable, or whether it can be viewed as an in-place natural resource in service to a community of users and available to a particular user on terms consistent with the best interests of the *acequia*. So, while New Mexico *acequias* have regained limited authority to regulate water transfers, they have learned to exercise that authority with delicacy and meticulous adherence to legal process. They exist in a water culture long since reshaped by the imperatives of the prior appropriation system, where water is generally thought to be owned by its user.

Colorado is now beginning to acknowledge and respond to the tensions between *acequia* practices and the generality of the state's water administration. A bill introduced in the 2009 session of the state legislature was signed into law in April of that year, allowing *acequias* to organize themselves as associations of water users enjoying a right of first refusal with respect to any proposed transfers of water rights outside of the *acequia*; to allocate water among *parciantes* on the basis of rules other than temporal priority; and to adopt a one-user, one-vote electoral structure to formalize the historical practice of voting equality among all *acequia* irrigator families, irrespective of size of landholding (Vigil, 2009). Similar to New Mexico's statute, the Colorado legislation would require that water users voluntarily bind themselves to *acequia* rules, a necessary concession to the constitutional protection of individual water rights under existing state law.

These movements in the law are rather modest in themselves; but they represent significant developments after the long hegemony of both the law of prior appropriation and the practice of centralized administration of uniform water rules. The revival under New Mexico and Colorado law of the possibility of local option, based on a community's commitment to its water practices, represents a recasting of a once powerful idea important in the formation of state-sponsored irrigation organizations in the American West. During the building of government-funded irrigation systems, a key strategy was educating farmers not only in technologies of water use and in particular understandings of efficiency, but also in the new content of their political and social relations to each other and to their water (Worster, 1985, pp186–187). Those new irrigation systems, crossing property lines and depending on the recognition of a new form of social commons, required users to understand water management as a foundation for community. It was understood that an irrigation system ought to be an expression of the community that depended on it, or at least that it was necessary that the new community of farmers be naturalized to the irrigation system. The New Mexico and Colorado laws are giving a new face to this model of intimate relation between irrigators and their water resource in the context of the ancient *acequia* system. Rather than recruiting users to a freshly minted ideology of community, they would instead use existing social capital based on valued relations to neighbours and to landscape in order to create a structure of water administration responsive to local circumstances. Perhaps the greatest challenge for converting these potentialities into significant improvements in water management and landscape functioning within *acequia* communities is the continuing legacy of the loss of commons and water rights, and the contraction of economic opportunities for *acequia* farmers. Those realities weigh heavily on *acequias* as working landscapes and as communities of water use. The effect of reformed laws will depend on their interaction with continually evolving social practices and economic circumstances within these communities. Those interactions are a necessary topic of continuing study and are being investigated by the recently established Justicia Hídrica alliance launched in Cusco, Peru, for the study and promotion of fairness in resource access and allocation (Hicks and Peña, 2010).

Conclusions

There is little risk that if traditional irrigation communities were allowed to manage water with more autonomy, they would not be responsive to management practices that are adapted to their localities and needs. A traditional water culture may well be deep seated, but its very success in winning allegiance by its practically minded users over time means that it will also have evolved continuously against a background of social, economic and environmental change. A traditionally grounded system may, in fact, be a highly adaptive and flexible system, suitable for dynamic rural environments where changes in patterns of land occupancy, labour methods, community life and market opportunities put constant pressure on the soundness of established practices (Peña, 2003a, 2003b; Boelens and de Vos, 2006). One form of that adaptability may, indeed, be a culture of successful change in the face of past challenges, a form of social capital that can give users of the system a sense of confidence and orientation as they face new challenges.

The opposite of this empowering of local capacity has too often been the sad reality. This is evinced by the difficult relationship between western water law and policy and the rules and rights practised by *acequias* communities. *Acequia* natural resource landscapes have been under constant pressure as a result of incentives for commercial development of land and water resources. The favouring of irrigation technologies and modes of agricultural production requiring investments beyond the economic reach of *acequia* communities, and the adoption of policies easing the movement of water towards higher investments, have not only reduced the availability of water to *acequias*, but also supported an easy dismissiveness of their value as water institutions.

There is great risk in undermining social and natural resource capital created by long and successful engagement between people and place. Existing relations between people and landscapes are typically constructed at great cost and often embody highly adaptive and effective technologies. The history of the engagement between the *acequia* communities of New Mexico and Colorado and the dominant system of water law and administration has largely been a history of the certainties of the dominant water regime, including the certainty that *acequias* were not much needed as part of the future of water management. There is now a stirring of more pluralist impulses. They should be encouraged because of what they offer for the health of watersheds and the flourishing of water communities.

References

Baxter, J. (1997) *Dividing New Mexico's Waters 1700–1912,* University of New Mexico, Albuquerque, NM

Boelens, R. and H. de Vos (2006) 'Water law and indigenous rights in the Andes', *Cultural Survival Quarterly*, vol 29, no 4, pp18–21

Carlson, A. (1967) 'Rural settlement patterns in the San Luis Valley: A comparative study', *The Colorado Magazine*, vol 44, no 2, pp111–128

Clark, I. (1987) *Water in New Mexico: A History of its Management and Use,* University of New Mexico, Albuquerque, NM

Ebright, M. (2001) 'Sharing the shortages: Water litigation and regulation in Hispanic New Mexico, 1600–1850', *New Mexico Historical Review*, vol 76, no 1, pp3–45

Endter-Wada, J., D. Blahna, R. Krannich and M. Brunson (1998) 'A framework for understanding social science contributions to ecosystem management', *Ecological Applications*, vol 8, no 3, pp891–904

Folke, C., T. Hahn, P. Olsson and J. Norberg (2005) 'Adaptive governance of social-ecological systems', *Annual Review of Environment and Resources*, vol 30, pp441–473

Hall, G. (2000) 'Tularosa and the dismantling of New Mexico community ditches', *New Mexico Historical Review*, vol 75, no 1, pp77–106

Hicks, G. (2005) 'Memory and pluralism on a property law frontier: The contested landscape of New Mexico's Costilla Valley', *American Journal of Legal History*, vol 47, no 4, pp383–411

Hicks, G. and D. Peña (2003) 'Community *acequias* in Colorado's Rio Culebra watershed: A customary commons in the domain of prior appropriation', *Colorado Law Review*, vol 74, no 2, pp387–486

Hicks, G. and Peña, D. (2010) 'Conflicting norms in the struggle for communitary governance in the *acequias* of the Upper Rio Grande watershed', paper presented at the First seminar of the programme Justicia Hídrica: Acumulación de Agua, Conflictos, y Acción de la Sociedad Civil (coordinated by Wageningen University, The Netherlands), 22–28 November 2009, Cusco, Peru

Hutchins, W. (1928) 'The community *acequia*: Its origins and development', *Southwestern Historical Quarterly*, vol 31, no 2, pp261–284

Lantis, D. (1950) *The San Luis Valley, Colorado: Sequent Rural Occupancy in an Intermontane Basin*, PhD thesis, Ohio State University, Ohio

Meyer, M. (1984) *Water in the Hispanic Southwest: A Social and Legal History 1550–1850,* University of Arizona, Tucson, AZ

Newell, F. (1890) *Report on Agriculture by Irrigation in the Western Part of the United States at the Eleventh Census*, United States Government, Washington, DC

New Mexico Court of Appeals (2008) *Peña Blanca Partnership v. San Jose de Hernandez Community Ditch*, Opinion no 2009-NMCA-016, Docket no 28,005, Court of Appeals of the State of New Mexico, Santa Fe, NM

New Mexico Statutes Annotated 1978 (2003) Section 73-2-21(E), The State of New Mexico, Santa Fe, NM

Nostrand, R. (1987) 'The century of Hispano expansion', *New Mexico Historical Review,* vol 62, no 4, pp361–386

Peña, D. (1999) 'Cultural landscapes and biodiversity: The ethnoecology of an Upper Rio Grande watershed commons', in V. Nazarea (ed) *Ethnoecology: Situated Knowledge/Located Lives,* University of Arizona, Tucson, AZ

Peña, D. (2003a) 'The watershed commonwealth of the Upper Rio Grande', in J. Boyce and B. Shelley (eds) *Natural Assets: Democratizing Environmental Ownership*, Island Press, Washington, DC

Peña, D. (2003b) 'Identity, place, and communities of resistance', in J. Agyeman, R. Bullard and B. Evans (eds) *Just Sustainabilities: Development in an Unequal World*, MIT Press, Cambridge, MA

Pisani, D. (1992) *To Reclaim a Divided West: Water, Law, and Public Policy 1848–1902,* University of New Mexico, Albuquerque, NM

Rivera, J. (1998) *Acequia Culture: Water, Land and Community in the Southwest,* University of New Mexico, Albuquerque, NM

Van Ness, J. (1987) 'Hispanic land grants, ecology and subsistence in the uplands of northern New Mexico and southern Colorado', in C. Briggs and J. Van Ness (eds) *Land, Water, and Culture: New Perspectives on Hispanic Land Grants*, University of New Mexico, Albuquerque, NM

Vigil, E. (2009) *Reengrossed House Bill 09-1233: A Bill for an Act Concerning the Recognition of Acequias, and, in Connection Therewith, Authorizing Acequia Water Districts*, Introduced to the first regular session of the sixty-seventh General Assembly, General Assembly, State of Colorado, Denver, CO

Worster, D. (1985) *Rivers of Empire: Water, Aridity and the Growth of the American West*, Pantheon, New York, NY

13

Community-Controlled Codification of Local Resource Tenure: An Effective Tool for Defending Local Rights?

André J. Hoekema

Introduction

In many parts of the world, indigenous peoples have achieved legal recognition of community rights to exploit local natural resources such as land, water and forests under their own rules and practices. This legal doctrine of recognizing rights that were never specifically given up by indigenous peoples when they ceded their lands to colonizing nations in the 18th and 19th centuries began in North America. Now, commonly subject to diverse restrictions, it is being followed in Latin America, the Philippines and Australia. It may hold promise as a means of preserving local community control of water in the Andes.

A Canadian case is illustrative of the struggle to preserve traditional rights through contemporary legal procedures. In its 1999 Marshall decision, the Supreme Court of Canada confirmed that Mr Marshall, a descendant of the indigenous Mi'kmaq people in Nova Scotia, had the right to fish in waters that once formed the traditional territory of the Mi'kmaqs, but now are supposed to be under the full regulatory power of federal and provincial authorities. A right of access to resources such as fish, wildlife and other traditionally gathered things was enshrined in 1760 and 1761 by a treaty between the British Crown and various indigenous peoples of that area.[1] The Marshall decision reaffirmed this treaty as valid law and dismissed a criminal conviction of Mr Marshall, finding that a government licensing scheme was in conflict with the treaty right. The recognition of the Mi'kmaq's treaty fishing right had a price, however. The right to fish is hedged in with important restrictions. The court qualified the right by allowing the government to enforce regulations that it proves to be necessary to conserve and protect the resource and limits the overall catch by construing the Mi'kmaqs' right of trading in fish for 'necessaries' in the treaty to mean only enough fish to provide 'a moderate livelihood'.

The qualifications on rights imposed by the court require the indigenous leadership to justify to state officials how their communities regulate the resource. This demands an extended and soul-searching local dialogue aimed at retrieving and, if need be, reinventing what the tribe considers to be 'our fundamental laws' and 'the way we are'. Eventually these fundamental laws will be put into writing: a form of bottom-up indigenous codification. Local tenure, which usually is a form of 'communal resource tenure', can be restored in a codified form after years of neglect or official suppression. Communal resource tenure, 'community based' as it is sometimes aptly called (Lynch and Talbott, 1995), refers to the rules or practices of access, use, control, conservation, transfer and 'sale', etc. of specific resources such as land, grazing areas, forests, wells, fishing grounds and waters that are developed, sustained or enforced by non-state (and sub-state) community organizations and derive their effectiveness and legitimacy from that community. Communal tenure regimes can control irrigation systems or water use directly from rivers, ponds or wells. This kind of codified local law tends to be a hybrid form that is the product of inter-legality, rather than being a restoration of certain historically static 'customary law' norms that fulfil a romantic notion of making community life 'more harmonious' or 'better adapted to the circumstances'. This chapter explores the tendency towards codification and discusses some consequences for local people in the context of water management.

The following section shows how negotiations concerning territorial rights provoke the writing down of indigenous tenure rules, and also discusses so-called co-management schemes. These schemes represent a minor devolution of power over their territory to indigenous people. The third section elaborates upon and analyses this trend by placing the reorganization and codification of local tenure institutions in the broader context of ethnic reconstruction using two case studies, the Navajo (Dine') Nation in the US and Quechua-speaking Indian communities in the Andes of Bolivia. The fourth section discusses the strong tendency in current neoliberal land reform programmes to look for a 'third way' rather than either imposing a 'one-size-fits-all' system of Western-style private, individual ownership, or simply embracing uncritically communal tenure systems. The third way borrows the strong elements of both systems, bringing together elements of 'modern' and 'customary' systems of land tenure that are then codified. Sometimes this emulates the old colonial ways of transforming local tenure into a Western dominant form by 'incorporation'. Or, by 'recognition', it can provide indigenous leaders the opportunity to maintain control over the development of their own local law.[2] These processes of incorporation and recognition are dealt with in the fifth section. The final section draws some conclusions.

Territorial rights negotiations and codification of local tenure

A fairly typical case of land negotiations and writing down of local law is reported by Peter Grant (1988), who acted as the lawyer for the Gitksan-Wet'suwet'en Tribe of British Columbia in Canada. During the 1980s, this community was engaged in a protracted struggle for recognition of the right to traditional fishing and harvesting in its territory. In the process of wrenching from the state government a 'comprehensive claims agreement' – a means of restoring local self-rule – the community sought to restate and codify its own fishery and harvesting laws in order to regulate access by people to parts of the territory, and to regulate the management of stock, water and so on. The community even codified practices such as the rules for adopting members of one house or clan into another. This required lengthy consultation among the clans and houses.

This process is not strictly codification; but, as Grant (1988, p262) observes: 'The best way of setting out the laws would be through a description of examples of their application in different circumstances, a method similar to the development of the common law system.' By contrast, the typical European-continental style of codification involves breaking down common-sense principles, rules of thumb, exemplary decisions, wisdoms, etc. into underlying analytical elements, reconnecting these in more abstract norms, reformulating other norms as the more concrete manifestations of the abstract ones, and thereby reconstructing law as a more or less logically coherent whole. Here, we use the term 'codification' to mean writing down important normative principles or rules of a community or society as a binding or at least orienting guide for authorities dealing with social conflicts. Codification, therefore, creates a written normative guide for authorities. Such codification is usually done on the basis of a broad process of consultation with the respective community members.

Local law is often codified ostensibly 'for the advancement of Indian self-government' (Grant, 1988, p266). This is a typical motive in the indigenous land rights negotiations that have abounded in recent decades in Canada. Seeking to succeed in land claims, a First Nation may feel pressure to present itself as a 'civilized' and 'modern' government and, hence, to show that it has rules, courts and constables. There is a second, even stronger, practical reason in land rights cases to promote codification of local law. In the Canadian cases, granting specific land rights implies some degree of self-rule regarding use of the land, access and some management decisions. Canadian court decisions recognize that indigenous rights to land imply a right to self-government (*Campbell versus A.G.B.C., 2000 BCSC 1123*). Moreover, broader arrangements for self-government can arise from land rights cases. In the case of the Nisga'a in British Columbia (Canada), Nisga'a law is emerging as the dominant law in this territory and Nisga'a authorities are able to administer this law through their own courts. Even if only grudgingly accepted by the dominant society, such an arrangement goes farther than a scheme of co-management in which, for instance, local ways of regulating access to and use of a resource merely play a role in the overall policies regarding the management of this resource. Although the Nisga'a approach leads to fuller recognition of indigenous identity and law, it requires the development of jurisdictional competence and creating new institutions.

In the old days, codification was often forced on indigenous peoples from the top down; but in the case of the Gitksan, the initiative came from the leadership itself. They determined how to consult with elders and the population at large, and it appears that they controlled the terms used to write down their fundamental laws. One problem that the leaders encountered was that the laws they sought to retrieve were highly dynamic and hybrid, not expressed in their 'pure, traditional form'. If local law is usually the outcome of a dynamic process of interaction, including elements of rule imposition from other societies, the codification task cannot rely only on restating the old traditional law. To succeed in practice, codifying local laws governing use and management of resources would be a patchwork of traditional and alien elements. Next, sorting out what must be kept as 'fundamental' and what modern practices should be included and excluded is further complicated by the tendency of people 'producing local law' to choose from diverse legal and normative repertoires whatever suits their interests.

Another hurdle for some populations is establishing that a traditional community with identifiable norms still exists. It may be that such a community has fallen apart or has become mixed with newcomers and 'modern elements' to such a degree that a reference to 'fundamental laws' is inappropriate. In this situation, it could serve to divide people in the present community to introduce an essentialized history and strained notions of identity. This problem arises in several Latin American contexts. For example, as Guevara-Gil explains in Chapter 9 of this book, the irrigation system of the Achamayo River in the Mantaro Valley in the Peruvian Andes is run by people who no longer self-identify as a single community distinct from the surrounding world. Nevertheless, some participants in the system seek to apply traditional norms, invoking the centuries-long efforts of their ancestors to construct and maintain the structures.

Although the practice is becoming widespread, scholars rarely deal with the difficulties of codifying indigenous law. One difficulty is that politicians and bureaucrats can easily manipulate the process by coining its complexities in straightforward, simplistic concepts. Land rights negotiations, for example, may not offer an appropriate environment to initiate broad, time-consuming internal reflections about 'our fundamental laws' and how to mix the Western with the non-Western. General wisdoms and vague references are difficult for state representatives to understand and they may be labelled as 'not law'. State bureaucracies may resist accepting 'unknown' and distinct patterns of handling affairs, particularly when they implicate sensitive areas such as the use of natural resources. So, in the negotiating process, representatives of the state may insist on clear-cut rules and refuse to respect indigenous rules that are not codified or at least written. Another issue is the role of non-governmental organizations (NGOs) in promoting projects in Latin American, Asian and African countries. They have financial and networking power and may introduce concepts and norms from international law or from NGO philosophies that have been called 'project law' and may displace local law (von Benda-Beckmann, 2001, p38). Protracted negotiations and sensitivity to cultural differences, therefore, may be needed to make a complete inventory of a group's way of life and the pertinent laws and rules affecting it.

Co-management regimes

The process of retrieval and codification of local law is also used for the expanding practice of co-management. Under co-management schemes, the official authority that controls water or another natural resource shares its powers with local groups or allows official decisions to be influenced by local interests. The fierce battle for indigenous territorial rights often ends in a compromise between local peoples and national authorities, establishing a joint committee or similar body that has some advisory and some decision-making power over the way in which national authorities regulate and manage wildlife, fisheries, water and forests, and even such issues as housing, medical care and education. Except in a few Canadian cases, broad powers of indigenous governance over the territories (self-rule) are usually not provided for and these committees do not replace official governmental authority.

An illuminating example of the gap between being a so-called partner in a co-management regime and having collective territorial rights with recognized governing authority is the case of the Sámi in Norway (see, for example, Svensson, 2003, 2006). The Sámi struggle for social and cultural survival started as a land rights case claiming that the Sámi were a component of Norway and had territorial and self-governing rights. The proposed regime for resolving the claim was, however, a weak variety of co-management. Svensson (2006) says that the Norwegian Finnmark Act that passed the Norwegian Legislature in 2005 was presumed to recognize Sámi land rights in this northernmost part of Norway, but now appears to be no more than a co-management arrangement treating all groups and persons living in the area as having equal interests in resources. 'The Sámi are offered no distinctive rights, but are placed on equal footing with other regional inhabitants' (Svensson, 2006, p71). A board was proposed to regulate and administer resources in the north, with membership from various segments of society, including the Sámi members, but not as a majority. The law reaffirms specific Sámi rights to reindeer pasture and related rights that were already recognized by the Norwegian Supreme Court. It recognizes a Sámi court that can apply Sámi customs, but as a system that complements the national court system – local custom and law are not stated to be on par with national law: 'there are no Sámi aboriginal rights to be codified. In a similar way the customary law is completely invisible in the law' (Svensson, 2006, p72).

A co-management regime, however, may offer to local peoples important group rights to manage water collectively, to hunt or to graze animals on claimed territory. An example of such a broader regime is the James Bay and Northern Quebec Agreement (JBNQA), struck between the government of Québec and of Canada and the Cree and other indigenous peoples. The 1975 agreement resulted from a long legal fight against a huge hydroelectric project on what the Cree saw as their territory. They won a bold temporary legal prohibition of the works by a Canadian judge. The Cree leadership then entered negotiations in which they accepted the abolition of all their claims to ownership or governmental authority in the expansive territory in exchange for firmly grounded and legally entrenched rights that allowed them to go on living as they were accustomed. They also received a large cash settlement. The Cree obtained the right to hunt and fish in the whole territory – for some species an exclusive right and for other species to be shared with non-native hunters. Canadian officials are not allowed to interfere with these uses except in emergencies involving urgent conservation goals. Local law, in the form of the Cree communal tenure institutions, was recognized to some extent, not repressed, ignored or declared backward and barbaric. If state regulations restrict, for example, harvesting, the Cree will have hunting priority as long as feasible. Moreover, any state regulation has to be presented to the Hunting–Fishing–Trapping Coordinating Committee, a joint committee of which half of the members are Cree and the other half national and regional officials. The minister cannot ignore the committee's advice (Scott, 2005).

The concept of 'co-management' has become attractive as a means of achieving the sustainable management of resources as the effectiveness of top-down public regulation is drawn into question. 'These indigenous institutions are now viewed by development theorists and practitioners as having qualities that make them valuable resources for achieving development goals' (Watson, 2003, p288). This comports with current tendencies to integrate user groups or 'stakeholders', not specifically indigenous groups, within resource management. Empowerment of local stakeholders, participation and co-management are discussed in policy documents and publications on development theory. In Western societies the same phenomenon is commonly labelled as 'horizontal', 'negotiated' or 'interactive' governance (Tache and Erwin, 2003, pp5–6). Often these projects only lead to some informal and temporary arrangements between an NGO and a local people on how to develop jointly some development programmes and organize its implementation. Sometimes, ambitions go further: towards a more durable and formally guaranteed position of the local tenure regimes. In the course of such more durable co-management designs, one may expect a strong pressure on local peoples to participate in a process of reinvention, or at least reformulation of their local tenure system. Possibly this leads to attempts at codification, too.

Although the greater attention being paid to local law provides opportunities for indigenous groups to express their identity and to pursue self-determination, in most cases, such as land claims, recognition of local law may effectively be a substitute for fuller rights to territorial control. The state and NGOs may embrace local law neither for its own sake nor for the sake of promoting indigenous self-rule, but rather for the sake of effective public resource management.

Ethnic reconstruction and inter-legality

Local water rights are often part of a larger body of institutions through which local communities regulate and control aspects of their world. Indigenous communities generally want to control the use and management of land and other natural resources. However, they are often confronted with state laws and policies that negate the existence and legitimacy of local institutions. These state policies aid powerful economic actors – as well as poor landless farmers – in penetrating local communities. These forces can subject them to the process of assimilation.

On the other hand, many local peoples, particularly many indigenous peoples, have strongly resisted assimilation. This resistance is not manifested by a retreat and retrenchment into isolation or an illusory form of purity, but by actively developing hybrid forms of institutions in which elements of the dominant law and policies are reconstructed within what is considered 'our traditional way of life'. This is a process of ethnic reconstruction, a conscious effort by indigenous peoples to articulate what they see as their fundamental values and institutions, thereby avoiding oppression and marginalization, while asserting that 'we are all here to stay'.[3] Responding to the demands of the dominant culture by reorganizing part of the social structure is also called 'ethnic reorganization' (Nagel and Snipp, 1993, p204). In this process new hybrid laws serve as weapons to ward off assimilative pressures. It is, as stated in the British Columbia case, a weapon 'to advance Indian self-government'. Ethnic reconstruction has enabled many territorial rights negotiations and co-management regimes, and it helps to explain why the bottom-up codification of local tenure does not necessarily produce the death of local living law or threaten the survival of distinct communities.[4] Two examples illustrate what codification of local law can mean for indigenous peoples.

The fundamental principles of the Dine'

The Navajo of the American South-West call themselves the Dine' (Zion, 2002). Like many other US tribes, they have long possessed a formally recognized system of justice with a hierarchy of courts, including a supreme court to hear appeals. The Navajo Nation has its own law enforcement agencies, with a local police force, and enacts its own laws within certain limits. The limiting rules are part of federal law that defines the jurisdiction retained by the Indian nations within US territory. These rules have been enacted in federal statutes and developed in court decisions, forming a patchwork of restrictions on tribal law, especially in terms of competence over non-indigenous persons and over Indian non-tribal members. In some cases, federal or tribal laws apply exclusively and in others they may apply concurrently with state laws or with one another (Krakoff, 2004). These rules affect the reach and effectiveness of any scheme of granting or recognizing rights to land, water and territory, as well as schemes reassigning jurisdiction to local courts and authorities.

The Navajo Nation developed its own institutions and substantive and procedural laws. In conducting court sessions, the tribe more or less copies the demeanour of 'white justice'. In the realm of substantive norms, Navajo legal principles mingle with copies of US federal or state law. The courts regularly apply Navajo common law and other unwritten principles. This mixing of the Western and the indigenous norms can be called 'inter-legality', which is generally a synthesis of two different legal orders or a hybrid legal order. Of course, indigenous laws and culture are not static; but the mixing process is now being initiated and more consciously and broadly assumed by the Dine'.

Some Navajo think the practice of hybridization gives away too much of the Navajo culture. Recently, a group of leaders, medicine men and other concerned people took the initiative to stop this process of creeping assimilation into Anglo culture (Zion, 2002). Beginning in about 1981, elders and their friends attempted to revitalize Navajo values and to sort out which elements of the US system could or should be retained and which elements rejected. These advocates claim that fundamental principles distinguish the Dine' from other peoples, particularly those of the dominant society. Many traditional values can be applied in Navajo peace-making courts that are often used as alternative forums from the formal courts in order to decide disputes among individuals. Eventually, after lengthy grassroots consultations, discussions and reflections, a document was produced containing a statement of principles of Dine' traditional law – *Dine'Bi Beenahaz'aanii* – which is the Navajo word for law, referring to the basics (Yazzie and Zion, 2003). The Dine' declaration is a document of wisdoms, proverbs, poetic reminders of fundamental duties, and other fairly broad principles, not rules. It resembles the biblical Ten Commandments more than Western codes. For instance, Article F, para 4 states: 'It is the right and freedom of the people that our children are provided with education to absorb wisdom, self-knowledge and knowledge to empower them to make a living and participate in the growth of the Navajo Nation', and Article G, para 5 states: 'It is the duty and responsibility of the Dine' to protect and preserve the beauty of the natural world for future generations.'

These fundamental principles are intended to transform Navajo politics, law-making and the administration of civil and criminal justice, as well as 'constitutional law' matters pertaining to relations between the tribal government and tribal members. Whether this process of mobilization (ethnic reconstruction) is effective against creeping assimilation will depend on the Navajo judicial personnel and on the degree to which the people are modernists or traditionalists. Some say 'the white way is the right way' and cherish material welfare coming from exploitation of resources. Others fight for a revival of centuries old traditions. Finally, successful resistance depends on the extent to which US economic interests will tolerate indigenous self-rule, particularly over natural resources, given the power of the US Congress to alter tribal governmental powers.

The task of integrating Anglo cultural elements within the Navajo legal system seems daunting. The fundamental laws of the Dine' are written to express traditional and longstanding values. Mixing Western and Dine' law and principles occurs in the daily practices of Navajo administrators, mediators and judges. Sceptics may doubt whether inter-legality is possible at all in such cases where the legal orders involved are so far apart.

Inter-legality describes both a process and an outcome. It is a process of adopting elements of a dominant legal order, national and international, into the practices of a local legal order. The outcome of such a process is a hybrid new legal order.

Inter-legality can also occur as the adoption and adaptation of local elements into national and international law. The Canadian scholar Craig Proulx (2006) writes about such reverse inter-legality, asking whether dominant legal actors can learn from and take over concepts and values from practices of indigenous communitarian justice; while Weatherford (1988, pp133–150) highlights the influence of the Iroquois League in the set-up of US federalism and even the League of Nations. This way of thinking may seem a bit optimistic, but it avoids the common notion that indigenous peoples are only the victims of absolute and overwhelming power.

Self-rule or autonomy

Indigenous leaders know that the road to cultural survival is to gain a lasting and respected position within a multinational society and state. During the 1950s, the Navajo leadership reorganized the tribal courts after the model of the Anglo system 'in order to preserve control over the system' (Krakoff, 2004, p1129). They wanted to produce respectability in the eyes of powerful interests eager to deny Navajos the capacity to rule themselves or to conduct independent court sessions. This required compromises with modernity rather than insisting on formalizing traditional norms. Such concessions may seem more reasonable when one realizes that Western society has penetrated most indigenous societies and that members of local communities enjoy some of the benefits of modern economies.

Human rights principles, generally defined by Western institutions, may also be extended over indigenous governments, especially when they apply their laws to others. These principles are sometimes accused of ethnocentrically undermining indigenous societies. The impact has been limited. The Colombian Constitutional Court only applies very basic human rights when reviewing indigenous court decisions. An Ecuadorian draft law calls for an 'intercultural interpretation' of human rights when local indigenous decisions are found to have violated human rights. In the US, the 1968 Indian Civil Rights Act is written to respect tribal sovereignty, although it places certain limits on tribal courts. In Canada, the Canadian Charter of Rights and Freedoms is less respectful of First Nations' self-rule (Morse, 2004).

I believe that the application of human rights norms may be appropriate in some cases. Many indigenous women would like to 'step out of the normative structures of their local society' and liberate themselves of male-dominated marriage arrangements, or get a fair share of their father's estate. Someone accused of being a witch and risking a death penalty would accept the 'encroachment' on their societies by the so-called Western-only human rights. For example, in 1999, the highest authorities of a self-governing Indian nation in Colombia (Jambaló) overruled a village decision to punish someone only very leniently who had murdered a supposed witch. 'It is unacceptable that a community condones homicide on the single argument that the victim was a witch, since this violates the right to life and human rights' (Assies, 2003, p183). Because the mixing of cultural and of legal institutions has been under way for a long time, a distinction between outside and inside is misleading. Indigenous leadership is aware of pressures within their own society to incorporate and appropriate many 'foreign' elements. This requires inter-legality and mixing the old and the new.

Internal versus external codification

The process of Navajo codification shows features that could be employed in similar projects around the world, including effective ethnic reconstruction and the stress on general principles as opposed to hard and detailed rules. In addition, the Navajo codification is an example of internal codification in contrast to post-colonial 'external codification' of customary law. The reconstruction of local law was a bottom-up process initiated and controlled by Navajo leadership, including respected citizens and elders, and not imposed by the dominant society.

During the 1950s and 1960s, top-down external codification in other places responded to state interests and the interests of the colonial or post-colonial legal elite. Moreover, evolutionist theories or sentiments dominated the scene. Modernization was expected to cause a gradual withering away of customary law. Eugen Cotran was an enlightened jurist who advised the Kenyan government on one of the restatement of African law projects of the 1960s (Allott, 1995). He did not share the modernization zeal of others who believed in a rapid assimilation process of customary law into national (i.e. British) law and stressed that customary law would continue to play an important role in the lives of large portions of the population (Cotran, 1966). He argued that local institutions must not be forced into Western legal concepts, citing customary marriage with gifts of chattel as an example. During the 1960s, British-educated judges considered this a category of a contract and deemed repugnant 'the purchase of a wife'. Cotran scorned the socio-cultural ignorance of such judges. 'This is going a bit too far', he wrote in typically English understatement.

Cotran held the firm opinion that unification and integration of folk law will be brought about by codification, and eventually 'one can also foresee a gradual integration not only as between the different customary laws, but also as between customary law and statutory law ... by bringing those different legal systems under one regulatory enactment' (Cotran, 1966, p91). He believed, consistent with the premises of the law and development movement of the time, that codification (external or internal) fulfilled the potential of legal enactments to make and change society and that the integration of the local into the national law will serve 'to meet modern conditions' in the 'not too distant future' (Cotran, 1966, p92).

Later, legal anthropologists eschewed even talking about codification. But as the Navajo show, almost 40 years after Cotran, codification is accepted in practice and is happening on the initiative of local leaders. It is motivated not to encapsulate local law in the dominant Western legal order, but to free it from that dominance and to gain space for an autonomous position and for locally controlled development.

Inter-legality and autonomy in the Bolivian Andes[5]

The Bolivian scholar and current Minister of Water Resources René Orellana studied two Indian Bolivian highland territories, each made up of various villages, together comprising about 12,000 people. He participated in many conflict-solving sessions involving cattle theft, aggression, land boundaries, succession, killings and marriage. In addition, these sessions resolved civil matters such as debts and contractual problems, and matters of governance such as contempt of leaders and disobedience by the members. The character of the proceedings resembled the elements of restorative and conciliatory justice well described in Laura Nader's studies of justice and control in a Zapotec mountain village. There, disputes are settled by uncovering the underlying causes and by taking into account an array of social relations besides those between the parties. Restoration of good relations is valued more than finding of 'the truth' (Nader, 1990). Similarly, the Bolivian Andean villages practise law and order in this spirit of reaching consensus.

In one of the areas studied, the territory of Raqaypampa, written rules are rare and proceedings are long and full of rhetoric. The public participates, and sometimes the leading official, the *secretario de justicias*, calls for the public to vent opinions and suggestions. Meetings go on indefinitely until reconciliation is reached. However, decisions or, rather, the commitments that parties make (either two parties in civil, land tenure or marriage matters, or the accused in criminal cases) are written down and documented. Sometimes heavy fines are stipulated in case the parties do not live up to these commitments. Orellana Halkyer (2004) observed that authorities presiding over conflicts entered into extensive, if not always consistent, mixing of legal elements in their decisions.

In one case the official bringing the charges urged the court to follow the 'correct procedure' (i.e. to hear witnesses), which is a practice adopted from state law. He also suggested that the accused had to be condemned because his behaviour infringed a recently enacted rule, booked officially and sent to all villages (*el acta*), although other lower leaders opposed its validity. The reviewing official chastised the village authority who had failed to deal with the case in a satisfactory way. The way to dispense justice, he said, is to look into state law and into local law and then come up with some form of combination. He said: 'If you don't pay attention to the national laws and completely go for our own ways, you are going to fall' (Orellana Halkyer, 2004, p150).

Other researchers have observed that in conflict-solving at the village level, indigenous authorities ornate their speech with Western-style legal jargon and cling to new procedural rules. In the Andes, components of the oral, relational and participatory style are mixed with elements of the written, case-oriented, authoritative approach. Those who are concerned with indigenous survival often deliberately embrace this kind of inter-legality as a strategic move. But often there is more to it than strategy. Some indigenous leaders might sincerely desire to develop systems of law and justice that borrow from Western ones, while at the same time preserving core features of their own. For example, they may want to incorporate concepts of individual human rights.

Is incorporation a sound strategy or does it merely help local officials who want to consolidate their power? On the one hand, clinging stubbornly to old forms is likely to annoy the political elite. Indeed, those who are sceptical about the viability of traditional norms and systems point to corporal punishments and the sometimes harsh treatment of witchcraft and other 'barbaric' features to discredit any scheme of recognition of indigenous justice. One perspective is that First Nations have to show themselves as unique, distinct and in possession of 'interesting' features and knowledge to be taken seriously, but not so traditional as to require 'development' (Tsing, 1999, cited in Wiber, 2002).

Specifying jurisdiction for dispute resolution is another area in which Andean communities have tried to accommodate but limit inter-legality. In terms of state law, many of the practices of Bolivian Indian communities are formally illegal. To prevent creeping assimilation, the leadership tries to define the 'normal' way of conduct and to induce the members, the *comuneros*, to pass first through the village institutions and then go to the central indigenous conflict-solving bodies, and only after having received permission, to go to town to submit the conflict to state tribunals. They rarely give permission. The courts or councils actively try to attract disputes 'from which they expect to gain politically' (von Benda-Beckmann, 1984, p37). Whether or not this method of reconstruction of Raqaypampa society will be successful remains to be seen; but there are indications of a considerable success. This is not the case in another territory that Orellana Halkyer (2004) studied (Rinconada). There the local leaders acted passively and shunned taking responsibility, let alone pursuing ethnic reconstruction. They did not intervene to prevent *comuneros* from turning to the provincial capital, leaving conflicts largely to official state authorities.

Assuming that reconstructions of traditional approaches will eventually be codified as they have been in the Navajo Nation, will this transform the oral, reconciliatory and participatory style of doing justice into a more formal and bureaucratic style? Santos (2002, p435) described this as the biblical style giving way to a more Homeric approach. Is this process of making customary law causing an abyss between what goes on in 'court' and 'real life' (or 'living law')?

There are ongoing tensions in communities who have embraced codification. The Navajo Nation grapples with how to retain conciliatory elements of their system such as the institution of peace-making. In Canada, indigenous communities seek to retain the sentencing circles and the concept of, and institutional process for, healing. In the Andes of Bolivia, communities actually pursue changes that, in fact, may lead to a more differentiated and bureaucratized society. Despite old fears of external codification in colonial and early post-colonial times, the tendency towards internally motivated codification today need not lead to recolonization. Instead, it could be a conscious step towards recognition of indigenous communities as distinct, yet equal, partners in the national society if indigenous peoples themselves control the transformation.

Water rights and inter-legality

As water in many places becomes scarcer every year, state governments attempt to control it more fully. Water departments and other official bodies may dispute ownership of, for example, an irrigation system against traditional owners in a local community. In this struggle for control, local communities often engage in processes of inter-legality by using the language and legal tools of the dominant society to defend themselves. Pradhan and Pradhan (1996) analyse such struggles in an area located in the mid-western plains of Nepal. Similar struggles are encountered by Guevara-Gil (see Chapter 9 in this volume) in the Achamayo irrigation system in the Mantaro Valley, Peru. The two cases show a similar pattern. Traditional users from the villages consider themselves to be the exclusive rights holders to a system built and maintained by their ancestors and now kept in good order by their own effort. Yet, the state now claims legal ownership of all water resources. An administration imposes a levy for water use to be paid by users. Moreover, the state asserts authority, for example, by giving permission to outsiders to use the water, thereby reducing the flow of water available to the original users. After various skirmishes with officials, these users have organized themselves and resisted state interference. They have borrowed norms and forms of the state apparatus.

Both in Nepal and Peru the villagers mimicked a state-sanctioned organizational form but changed it to serve their own purposes (for an analysis of such 'mimicry strategies' in water control, see Boelens, 2008). In Nepal, no state representatives were nominated to the board of the association and no official registration with the Irrigation Department was pursued. In Peru, local users took one of the new organizational forms allowed by the state and organized as a committee of irrigators (*comité de regantes*). The community used this entity to develop and enforce their own system management. Moreover, at times they borrowed the state's way of measuring water input for each user by volume instead of duration of water flow into a field. They did so strategically to counter the state-permitted taking of much water by a competing fish farm. In both cases, elements of state administrative regulation are absorbed into local practices by conscious bottom-up manoeuvring.

Such processes of inter-legality may lead to assimilation into the dominant culture. But in these examples, at least in Nepal, the community went through a process of self-organization and mustered its strength to counteract the state claims. It remains to be seen whether or not, in the Peruvian case, the inter-legality will strengthen ethnic identity. In a community so near busy centres of commerce and transport and already so diverse in its population, perhaps the degree of 'indigenousness' and ethnic cohesion will be more difficult to maintain.

Ethnic reorganization through inter-legality is shown in another Peruvian area also studied by Guevara-Gil (2007). People living at the shores of Lago Titicaca use water resources that they see as their own. They borrow state legal forms such as 'nature reserve', 'indigenous people' and 'community', and use them to ward off state claims so they can continue to use the water according to their own practices.

In Nepal, the community continued their water tenure arrangement as before. In Peru the struggle goes on. In these processes two different worlds clash. Communities foster notions of long-time local autonomy, take pride in an ancestral arrangement of community-based water tenure and nurture a vision of water resources as 'natural' and not fit for privatization and market transactions. On the other hand, the state promotes neoliberal development notions based on the supremacy of markets, private transactions and individualized property title to land and water. There may, however, be a 'third way' to resolve this tension.

The third way in matters of land tenure reform

Until recently, in Kenya and elsewhere, post-colonial land reform typically ignored local communal tenure institutions and formally enacted Western-style private ownership and administrative regimes (Shipton, 1988). Now, the thesis that only uniform Western-style property rights pave the way to economic development has been largely abandoned. Present land reform projects reveal a bewildering variety of ways to take local tenure law into account. Communal land tenure is now 'recognized' in many places in Africa. The reappraisal of local tenure institutions and apparent recognition of a kind of legal pluralism, however, reflect regulatory goals of neoliberal state policies rather than being the result of appreciating the diversity of cultures.

State authorities and experts and the World Bank are now looking for ways 'to bridge the gap between customary and formal rules' (the subtitle of a brochure by Dubois, 1997), or how 'to integrate statutory and customary tenure' (Mwebaza, 1999). Exemplified by Uganda, Mozambique, Tanzania, Niger and Namibia, there is 'a new wave of land tenure reforms ... [in which] many recent laws protect customary land rights and provide or allow for their registration' (Cotula et al, 2004, p5). This, in turn, often provokes the codification of local rules or practices. In Latin American countries such as Bolivia, Colombia, Ecuador, Mexico, Panama and others, land and water tenure laws, as well as forest management schemes, have integrated some local tenure institutions. This new political and legal interest in local law avoids absorption of local law into the national legal order but does not necessarily lead to uncritical acceptance of local tenure either. Instead, it is a mixing process – the quest for a third way. 'Rather than suppressing legal pluralism by absorbing one system into another, the aim is to retain the most dynamic aspects of each' (Lavigne Delville, 2000, p116). Formal hybridization or mixing of elements is utilized in land reform by 'taking into account' local law and strictly testing it against elements of human rights, gender issues, 'barbaric features' and so on. To evaluate the extent to which indigenous peoples can capture the codification process for their own benefit and promote ethnic reconstruction, one must distinguish between incorporation and recognition.

Incorporation and recognition of local resource tenure: Blessing or curse for indigenous communities?

Incorporation is the most common way of dealing with local law. It refers to the legal adoption of a set of local institutions, such as communal resource tenure, as part of national law. Local laws are integrated within national law in land reform projects, co-management of natural resources and related policies. Often, new legal categories are invented to accommodate unknown and distinct local institutions. For instance, in Uganda's Land Act of 1998, *common land associations* were created to accommodate pastoralists' communal tenure of water sources and grazing. Another example is the new legal concept of 'Finnmark property', introduced in the draft bill in Norway as an attempt, albeit inadequate, to incorporate traditions of the large Sámi majority for holding and administering land in Norway's northernmost province.

Although occasionally incorporation policies are called recognition of local customs, this term is somewhat misleading as neither the dominant motive nor the institutional set-up of these policies fully recognizes the distinctiveness of local communities or indigenous peoples, or a legal right to be different and conduct their own affairs. Nor is this 'recognition' about the right of indigenous peoples to hold or use traditional land and resources. Incorporation usually relates to a single activity such as herding and grazing; it is therefore a 'one-law' incorporation. It is not like recognition of a complete local legal order, which would be an important feature of a grant of self-rule. Moreover, the dominant state policy is to create conditions that are believed to allow for more successful management of resources or agrarian development. In a sense, incorporating local law is doing legal business as usual. In Western-style legal orders, it is normal to give some weight to customs of a trade, to internal rules made by associations and the like as a way of increasing the local acceptance and overall effectiveness of the law.

Full recognition would entail the adoption of a more or less complete legal order of a distinct community as an equally valuable and legally valid part of the national law. In practice, recognition of local law can fulfil the idea that society is a cooperative effort between various distinct peoples or communities. This notion runs against the grain of liberal political thinking in that it implies acceptance of First Nations as a distinct but essential part of the society as a whole and values the preservation of distinct identities within a larger political whole, effectively instituting a form of socio-cultural (often ethnic) pluralism. It tries to strike a balance between self-rule by some specific communities or groups within a larger society and shared rules regarding matters of common concern. Such recognition, however, does not convey state-like sovereignty to an indigenous society. Indeed, it is usually the national government who authorizes a distinct community to rule itself and to live under its own laws, while the state dictates the conditions. The domain and scope of autonomy in law and governance is always restricted, and sometimes heavily so, by conflict rules mostly made unilaterally. This may result in confining indigenous communities to a specific way of life. Sometimes, however, such 'federal' regimes are the product of more equitable negotiations between partners, leading to an agreement entrenched in constitutional or other formally binding law.

In practice, recognition proposals can end up furthering policies of incorporation, hedged in with restrictions motivated by assimilative tendencies so that local authorities lack the opportunity to take control over developing their own law and a bottom-up inter-legality. The earlier mentioned Marshall case in Canada restricted the scope of self-rule in matters of fishing by allowing traditional practices and uses only, banning commercial and new enterprises and precluding the development of new tenure institutions (Hoekema, 2003; Wiber, 2003).

Recognition of local legal orders offers a better chance for bottom-up ethnic reconstruction, although achieving such recognition requires considerable struggle, and other struggles continue after the formal granting of self-rule. Recognition, thus, is a rare feat. For a while it seemed as if the drive for multinational federalism was gaining ground, particularly in international law. Today, it seems to have fallen into disfavour and evokes references to 'ethnic cleansing', 'ethnic strife' and disintegration of society.

Formal recognition is accepted in Colombia, Panama, Greenland, Nicaragua and, to some extent, Canada and the US. But stumbling blocks on the road to a real partnership between the distinct communities and the dominant government are many. In order to analyse recognition proposals, it is important to consider the extent to which the arrangement defines jurisdiction of the local self-ruling entity over activities, people and territory as exclusive – and, if not, the circumstances in which state laws have validity in the indigenous territory. In cases of conflicts, what court or other body will have jurisdiction and to what extent do local representatives participate in these courts? Often there are provisions that might limit local autonomy by specifying that local laws and decisions should not violate 'the constitution and the national laws'.

Under recognition regimes, more than under incorporation, the application of local law can promote inter-legality. There are examples, such as the Colombian experiment, where broad powers of self-rule are attributed to indigenous nations in matters of legislation and administration of justice. Indian self-government in the US is another example of such a regime. The Colombian Constitutional Court has held that the local system of administration of justice is obligatory and members cannot opt out. Using a narrow human rights test, the court did away with the usual requirement that local decisions and local law must not contravene many individual human rights: 'In a nation where cultural diversity is recognized, no world view can prevail over the other, let alone attempt to dominate' (Colombia Constitutional Court, Verdict T 496, 26 September 1996). The Colombian court's approach exemplifies a rare acceptance of the diversity of constituent nations and their right to be different (Assies, 2003; Hoekema, 2003). However, the court maintains a core of essential human rights which have to be respected by local authorities: the right to life, the right not to be enslaved, the right not to be tortured and the right not to be judged in ways that are unforeseeable for the accused. Compare this short list with the extensive list of human rights that are often imposed on First Nations, even in cases where a special human rights regime for such nations is instituted, as in the US (Morse, 2004).

Unlike the Colombian court's approach, in other cases granting recognition to indigenous peoples 'repugnancy clauses' are typically introduced, curtailing the blending of new and old practices well beyond the limitation on invading certain human rights. Restricting land and water rights to 'traditional use' is one example of the stringent restrictions.

A more important barrier to full recognition resides in financial and economic policies and circumstances that often thwart or cripple systems of self-rule. Van de Sandt (2003) sketched some of these circumstances applying to Colombian self-governing nations: the *resguardos*. Financial assistance to such communities from the national government may be conditioned on rules preferring local governments that mimic national systems of governance and law by adopting elected representative bodies, Western-style police forces and the like (van de Sandt, 2003). Attempts to hold self-governing authorities (judges included) accountable by withholding aid could coerce them to adopt Western practices.

Recognition encourages codification because if courts encounter local law, such as when testing local decisions against human rights, they need to examine local standards. In order to know what due process entails, for instance, the court will seek a written record so that it can learn about the community, what the laws say and how they are applied. The fact that such territories and their laws are officially part of the constitutional and legal structure of the national society implies an extensive traffic of written documents, statements, justifications of proper policies and formalization, and codification of the local law.

Conclusions

Modernization policies today, even of a neoliberal type, open opportunities for local (particularly indigenous) distinct communities to develop alternatives to assimilation into mainstream cultures and economies. Local resource tenure regimes, including rights to water, are being accepted as part of national policies of land and water reform, development projects and environmental policies. There are also instances of relatively successful indigenous land rights claims leading to forms of official legal pluralism: local tenure institutions become part of the law of the land, or at least indirectly get a more respected place in managing resources. These developments provoke a reconstruction, sometimes reinvention, of 'our fundamental laws' and eventually lead to a process of bottom-up codification of local law. The regimes of local law are going through intense processes of mixing proper ('old') elements with elements coming from national and other dominant orders, producing interlegality. Codification is not necessarily like the old colonial and post-colonial way of forcing local communities to assimilate into the dominant society. Anthropologists should revise their usual abhorrence of codification of indigenous law. Whether or not indigenous communities can resist the negative effects of codification depends on the presence and strength of local mobilization efforts promoting 'ethnic reconstruction'. Indeed, codification of local land and water tenure regimes can be used as a weapon in the struggle for self-rule. Perhaps under those conditions indigenous peoples may see their strong desires for controlling their own destiny fulfilled.

Notes

1 Originally, the treaty protected indigenous peoples' access to their territorial resources, but was also used by colonizers to exact title to lands occupied by the Mi'kmaqs that the newcomers desired for settlement.

2 'Local law' (von Benda-Beckmann et al, 1996, p89) refers to the typical melange of 'traditional' and 'modern' ways of organizing and ordering social life, which is characteristic of the laws of many distinct communities and nations. It critiques the concept of customary law as a distinct, homogeneous and more or less static ensemble of rules and practices. It is 'the locally dominant mixture of interpretations and transformations of the surrounding universe of plural legal repertoires'.

3 The phrase 'we are all here to stay' stems from Chief Justice Lamer in the Canadian Delgamuukw decision, 11 December 1997.

4 Several authors have analysed customary law as codified in Africa to be a colonial invention (Snyder, 1981; Woodman, 1988; Chanock, 1992).

5 Adapted from Orellana Halkyer (2004) and Hoekema (2003).

References

Allott, A. N. (1995) 'International development in customary law: The Restatement of African Law Project and thereafter', in T. W. Bennett and M. Rünger (eds) *The Ascertainment of Customary Law and the Methodological Aspects of Research into Customary Law*, Proceedings of a workshop on Law Reform and Development Commission, Windhoek, February–March 1995, pp31–35

Assies, W. (2003) 'Indian justice in the Andes: Re-rooting or re-routing?', in T. Salman and A. Zoomers (eds) *Imaging the Andes: Shifting Margins of a Marginal World*, Aksant Academic Publishers, Amsterdam, The Netherlands, pp167–186

Boelens, R. (2008) *The Rules of the Game and the Game of the Rules: Normalization and Resistance in Andean Water Control*, Wageningen University, Wageningen, The Netherlands

Chanock, M. (1992) 'The law market: The legal encounter in British East and Central Africa', in W. J. Mommsen and J. A. de Moor (eds) *European Expansion and Law*, Berg, Oxford and New York, p279

Cotran, E. (1966) 'The future and place of customary law in Africa', in *East African Law Today*, British Institute of International and Comparative Law, Commonwealth Law Series no 5, Stevens and Sons Ltd, London, pp72–92

Cotula L., C. Toulmin and C. Hesse (2004), *Land Tenure and Administration in Africa: Lessons of Experience and Emerging Issues,* International Institute for Environment and Development (IIED), London

Dubois, O. (1997) *Rights and Wrongs of Rights to Land and Forest Resources in Sub-Saharan Africa: Bridging the Gap between Customary and Formal Rules*, Forest Participation Studies no 10, International Institute for Environment and Development (IIED), London

Grant, P. R. (1988) 'Recognition of traditional laws in state courts and the formulation of state legislation', in B. W. Morse and G. R. Woodman (eds) *Indigenous Law and the State*, Dordrecht/Providence, pp259–268

Guevara-Gil, A. (2007) 'Espejismos desarrollistas y autonomía comunal: El impacto de los proyectos de desarrollo en el lago Titicaca (1930–2006)', *Global Jurist*, vol 8, no 3, www.bepress.com/gj/vol8/iss3/art5, accessed 2 April 2009

Hoekema, A. J. (2003) 'A new beginning of law among indigenous peoples: Some observations from anthropology of law', in F. J. M. Feldbrugge (ed) *The Law's Beginnings*, Brill Academic Publishers, Leiden, pp181–220

Hoekema, A. J. (ed) (2006) *Multicultural Interlegality: Special Issue of Journal of Legal Pluralism and Unofficial Law*, no 51, Lit Verlag, Berlin

Krakoff, S. (2004) 'A narrative of sovereignty: Illuminating the paradox of the domestic dependent nation', *Oregon Law Review*, vol 83, no 4, pp1109–1202

Lavigne Delville, P. H. (2000) 'Harmonising formal law and customary land rights in French-speaking West Africa', in C. Toulmin and J. F. Quan (eds) *Evolving Land Rights, Policy and Tenure in Africa*, DFIOD/IEED/NRI, London, pp97–121

Lynch, O. W. and K. Talbott (1995) *Balancing Acts: Community-Based Forest Management and National Law in Asia and the Pacific*, World Resources Institute, Washington, DC

Morse, B. W. (2004) 'The scope of human rights law concerning indigenous and minority peoples', Readings for a seminar of the Commission on Folk Law and Legal Pluralism, August 2004, Fredericton, New Brunswick, Canada

Mwebaza, R. (1999) 'Integrating statutory and customary tenure systems in policy and legislation: The Uganda case', Paper presented at the DFID workshop on Land Rights and Sustainable Development in Sub-Saharan Africa, Sunningdale, UK

Nader, L. (1990) *Harmony Ideology: Justice and Control in a Zapotec Mountain Village*, Stanford University Press, Stanford, CA

Nagel, J. and C. M. Snipp (1993) 'Ethnic reorganization: American Indian social, economic, political and cultural strategies for survival', *Ethnic and Racial Studies*, vol 16, no 2, pp203–235

Orellana Halkyer, R. (2004) *Interlegalidad y campos jurídicos: Discurso y derecho en la configuración de órdenes semiautónomos en comunidades quechuas de Bolivia*, Private edition (salvador@supernet.com.bo), Cochabamba, Bolivia

Pradhan, R. and U. Pradhan (1996) 'Staking a claim: Law, politics and water rights in farmer managed irrigation systems in Nepal', in J. Spiertz and M. G. Wiber (eds) *The Role of Law in Natural Resource Management,* special issue of *Recht der Werkelijkheid,* VUGA, The Hague, pp61–76

Proulx, C. (2006) 'Blending justice: Interlegality and the incorporation of aboriginal justice into the formal Canadian justice system', in A. J. Hoekema (ed) *Multicultural Interlegality: Special Issue of Journal of Legal Pluralism and Unofficial Law*, no 51, Lit Verlag, Berlin, pp79–109

Santos, B. de Sousa (2002), *Towards a New Common Sense: Law, Globalization and Emancipation*, Butterworths, London and Edinburgh

Scott, C. H. (2005) 'Co-management and the politics of aboriginal consent to resource development: The agreement concerning a new relationship between the government of Québec and the Crees of Québec (2002)', in M. Murphy (ed) *Canada: The State of the Federation: Reconfiguring Aboriginal–State Relations*, McGill-Queen's University Press, Montreal and Kingston, Canada

Shipton, P. (1988) 'The Kenyan land tenure reform: Misunderstandings in the public creation of private property', in R. E. Downs and S. P. Reyna (eds) *Land and Society in Contemporary Africa*, University Press of New England, Hanover, NH, pp91–135

Snyder, F. G. (1981) 'Colonialism and legal form: The creation of 'customary law' in Senegal', *Journal of Legal Pluralism*, vol 19, pp49–90

Svensson, T. G. (2003) 'On customary law: Inquiry into an indigenous rights issue', *Acta Borealia*, vol 2, pp95–119

Svensson, T. G. (2006) 'Interlegality, a process towards strengthening indigenous peoples' autonomy: Case of the Sámi in Norway', in A. J. Hoekema (ed) *Multicultural Interlegality: Special Issue of Journal of Legal Pluralism and Unofficial Law*, no 51, Lit Verlag, Berlin, pp51–77

Tache, B. and B. Irwin (2003) *Traditional Institutions, Multiple Stakeholders and Modern Perspectives in Common Property: Accompanying Change within Barona Pastoral Systems*, International Institute for Environment and Development (IIED), Securing the Commons series no 4, London

Tsing, A. L. (1999) 'Becoming a tribal elder and other green development fantasies', in T. Murray Li (ed) *Transforming the Indonesian Uplands*, Harwood Academic Publications, Amsterdam, The Netherlands, pp159–202 van de Sandt, J. (2003) 'Communal resource tenure and the quest for autonomy: On state law and ethnic reorganization in two Colombian *resguardos*', *Journal of Legal Pluralism and Unofficial Law*, vol 48, pp125–162

von Benda-Beckmann, F., K. von Benda-Beckmann and H. L. J. Spiertz (1996) 'Water rights and policy', in J. Spiertz and M. G. Wiber (eds) *The Role of Law in Natural Resource Management,* VUGA, The Hague, pp77–99

von Benda-Beckmann, K. (1984) *The Broken Stairways to Consensus: Village Justice and State Courts in Minangkabau*, ICG Printing, Dordrecht, The Netherlands

von Benda-Beckmann, K. (2001) 'Transnational dimensions of legal pluralism', in W. Fikentscher (ed) *Begegnung und Konflikt – Eine kulturanthropologische Bestandsaufnahme*, Abhandlungen der Bayerischen Akademie der Wissenschaften, Philosophisch-Historische Klasse, Neue Folge Heft 120, C. H. Beck, München, pp33–48

Watson, E. E. (2003) 'Examining the potential of indigenous institutions for development: A perspective from Borana, Ethiopia', *Development and Change*, vol 34, no 2, pp287–309

Weatherford, J. (1988) *Indian Givers: How the Indians of the Americas Transformed the World*, Fawcett Columbine, New York, NY

Wiber, M. (2002) 'Immanent local law? Immanent culture? Review of Orin Starn, *Nightwatch, the Politics of Protest in the Andes* (Duke University Press, Durham and London, 1999)', *Newsletter of the Commission of Folk Law and Legal Pluralism*, vol xxxv, February, pp81–87

Wiber, M. (2003) 'Review of Elizabeth Povinelli, *The Cunning of Recognition, Indigenous Alterities and the Making of Australian Multiculturalism* (Duke University Press, Durham and London, 2002)', *Newsletter of the Commission on Folk Law and Legal Pluralism*, vol xxxvii, June, pp61–64

Woodman, G. R. (1988) 'How state courts create customary law in Ghana and Nigeria', in B. W. Morse and G. R. Woodman (eds) *Indigenous Law and the State*, Dordrecht/Providence, pp181–220

Yazzie, R. and J. W. Zion (2003) *A Practioner's Guide to the Fundamental Laws of the Dine'*, www.nativeweb.org, accessed 15 November 2003

Zion, J. W. (2002) 'Codifying traditional law', *Newsletter of the Commission of Folk Law and Legal Pluralism*, vol xxxv, February, pp75–80

Cases

Campbell versus A.G.B.C., 2000 BCSC 1123

R. versus Marshall S.C.R., 1999 [3] (17 September and 17 November)

Delgamuukw versus British Columbia (1997) 3 S.C.R. 1010, 11 December 1997, file no 23799

Part IV

Social Mobilization and Grassroots Strategies for Water Rights

14

Using International Law to Assert Indigenous Water Rights

David Getches

Introduction

The field of international law is expanding rapidly. Indigenous peoples – especially in Latin America – have begun to assert norms under this body of law to protect their lands and resources. Yet, it is difficult to find any express protection for water or even natural resources. Thus, indigenous peoples are looking to the established international law of human rights, cultural rights and emerging law regarding environmental protection. International agreements and norms gleaned from them and from the practices of individual nations offer possibilities for arguing that rights to land and natural resources, including water, must be protected as an element of physical and cultural survival, and the failure to do so is a violation of basic human rights.

This volume includes chapters studying several examples of indigenous peoples' struggles to gain access to water sources in the Andes and elsewhere in the Americas, and to apply their cultural norms to regulate the use of water. Indigenous peoples' ability to perform water-dependent vocations such as farming, to continue occupying their own territories and to perpetuate cultures and spiritual practices may depend on achieving success in limiting the encroachment of incompatible legal systems and practices embodied in the water law of host nations.

In the Andean countries and around the world indigenous peoples are mobilizing efforts to assert and protect their rights. As a result of these struggles, the domestic law of many nations purports to provide for protection of indigenous rights. Some of the chapters in this volume discuss such formal legal protections found in constitutions of Latin American countries; but none concludes that the experience of indigenous peoples under those laws has been entirely satisfactory or sufficient. Even in the United States, as introduced in Chapter 11 on the water rights struggle of the Pyramid Lake Tribe of Indians, the generous court-made doctrine of reserved water rights for Indian reservations has encountered problems in its enforcement. It is not surprising that indigenous peoples might, therefore, look to international law for a source of enforceable rights.

As this chapter concludes, indigenous peoples are only beginning to assert international norms to protect their natural resources rights and it is too early to conclude how useful they will be in the context of water. These international legal battles are part of a broader strategy of indigenous peoples. Other chapters in this book, such as Chapters 15 and 16, address the ways in which local, *campesino* and indigenous water-user groups have joined together in federations to defend their collective water interests, producing some promise of success but incomplete results. It remains to be seen whether formal 'legal mobilization' or such localized social mobilization will ultimately be more successful. Very likely it will be a combination of the two strategies.

International law has begun to embody respect for cultural integrity related to indigenous and customary practices, if not practices specifically related to water. This chapter identifies six types of rights that exemplify ways in which indigenous water rights claims could be framed and the various international law instruments and norms that could serve as the bases for claims of deprivation of indigenous water uses.[1]

International law

Extending rights to indigenous peoples

Although there is no body of international law that specifically protects indigenous peoples' water rights, potential indigenous claims of rights to water – and to land and natural resources – could fit within several categories of protection for human rights and for rights to property, environmental protection, subsistence, cultural preservation, racial discrimination and self-determination. The fact that water has not been specifically addressed in these accords is partly because international law only recently has begun to comprehend the unique nature of the natural resource claims of indigenous peoples.

A body of international human rights law as applied to the rights of indigenous peoples has emerged over the last 50 years (Anaya and Williams, 2001). In 1948, the Organization of American States General Assembly took the first step in accepting Article 39 of the Inter-American Charter of Social Guaranties. The charter required American states to take 'necessary measures' to protect indigenous peoples' lives and property, 'defending them from extermination, sheltering them from oppression and exploitation'. Since this regional recognition of indigenous rights in 1948, various multilateral and bilateral agreements have been adopted in an effort to protect indigenous peoples' rights.

As various chapters in this book have illustrated, one of the most important accords is International Labour Organization Convention No 169 on Indigenous and Tribal Peoples of 1989 (ILO No 169). The basic theme of ILO No 169 is embodied in the convention's preamble, which recognizes 'the aspirations of [indigenous] peoples to exercise control over their own institutions, ways of life and economic development and to ... develop their identities, languages and religions, within the framework of the states in which they live'. Anaya (1991, p7) explains the pluralistic intent behind this agreement: 'The convention includes extensive provisions advancing indigenous cultural integrity, land and resource rights and non-discrimination in social welfare spheres; in addition, it generally enjoins states to respect indigenous peoples' aspirations in all decisions affecting them.'

Several other international conventions primarily relating to human rights or, more recently, to environmental protection are potential sources of indigenous rights. International tribunals charged with carrying out international agreements and domestic courts of some countries have upheld the application of these international laws to protect various interests of indigenous peoples.

Besides formal binding agreements, a variety of norms that could apply to indigenous water rights have been accepted by the nations of the world. For instance, Agenda 21, adopted at the 1992 United Nations Conference on Environment and Development (UNCED) in Rio de Janeiro, provides a set of standards for countries in their use and conservation of natural resources. Nearly all the nations of the world have accepted these standards. One standard requires the full participation of the public, including water user groups and indigenous peoples and their communities.

Other potential sources of rights discussed in this chapter are merely draft documents that have not yet been presented to the United Nations or other international organizations for their consideration. Nevertheless, they remain helpful sources in attempting to find the norms that arguably should guide international conduct towards indigenous peoples.

Based largely on the content of international human rights conventions and customs apart from domestic laws, Cohan (2001/2002, p154) argues that 'the international community now regards indigenous peoples as having environmental rights that rise to the status of international norms' and that 'because indigenous peoples' way of life and very existence depends on their relationship with the land, their human rights are inextricable from environmental rights'.

International development agencies have also incorporated indigenous rights within their policies. The rules and policies of these agencies provide a further demonstration that such rights have increasingly been accepted as international norms.

Enforcement

Assuming that international agreements and norms include protections that could be invoked to support indigenous claims related to water, it nevertheless may be difficult to find a forum to hear such claims. Williams (1990, pp695–696) writes that 'the International Court of Justice and many other more effective and high-profile forums of international law are available only to states – a term which under present conceptions of international law does not include indigenous peoples'.

Some tribunals created to carry out agreements protecting human rights allow individuals and groups to bring claims before them. These include the United Nations Committee on Human Rights. Similar entities, such as the Inter-American Commission on Human Rights, exist to enforce and interpret regional agreements. The threshold issue is whether a claim is admissible according to the protections of the relevant instrument and the rules under which they operate. It is not always possible to have a claim heard by these bodies; but indigenous peoples have been successful in gaining access to pursue their natural resource claims in a number of recent cases (de Bolivar, 1998).

In July 2002, the Inter-American Commission took a substantial step towards providing for indigenous peoples' rights when it released findings that the US had violated international human rights by depriving the Western Shoshone Indians of ancestral lands claimed by the federal government where some tribal members continued to graze cattle (Sansani, 2004). Yet, the essentially advisory nature of such rulings was shown by the reaction of the US when the US Department of State rejected the commission's findings because it determined that the Indians' land claims, as asserted by certain tribal members, the Dann sisters, had been litigated to finality in US courts without success. Having rejected the commission's recommendations, the US government seized and confiscated livestock from the federal land, stating that the claim was 'fundamentally not a human rights claim, but an attempt by two individual Indians to reopen the question of collective Western Shoshone tribal property rights to land' (Getches et al, 2005, p294).

National courts and administrative tribunals may also apply international law. The utility of international agreements and norms for indigenous peoples can vary among countries based on the degree to which their courts accept international law as a source for their rules of decision. For instance, US courts rarely apply international law even when the US has subscribed to particular instruments (McFadden, 1995). The courts of other nations are more hospitable to claims brought under international law.

Free trade issues

Special treaties and tribunals to ensure free trade have emerged, creating a separate and currently influential body of international law. This raises a collateral concern about the enforceability of international law to protect indigenous rights to natural resources when to do so would inhibit international trade and investment. Some of the special trade courts interpret bilateral and multilateral agreements to elevate free trade above other values and interests. Thus, for example, the World Trade Organization (WTO) has been reluctant to allow signatories to the General Agreement on Tariffs and Trade

(GATT) to adopt measures that an individual country believes to be necessary to protect species covered by the Convention on International Trade in Endangered Species of Wild Fauna and Flora (CITES) if it would inhibit free trade (e.g. GATT Dispute Settlement Panel Report on US Restrictions on Imports of Tuna, 1991).

It is possible that the WTO and other entities, such as the tribunals enforcing the North American Free Trade Agreement (NAFTA), will subordinate the interests and rights of indigenous peoples to the promotion of free trade. The trade treaties have been said to reflect 'a disturbing lack of balance between the protection of private interests and the need to promote and protect the public welfare' (Runnalls and Fuller, 2001; Stiglitz, 2003).

Conflicts between indigenous rights and free trade could arise in a number of ways. For example, an international company could invest in the development and transfer of water that harms indigenous cultures and economies. If the national government attempted to restrict water development for the sake of affected peoples – particularly after the same government has granted a concession to the company allowing it to use the water, a trade court could see it as an expropriation in conflict with free trade principles protecting international investments. This arguably would entitle the company to compensation. It is possible that even if indigenous groups themselves persuaded a court to protect their water rights *vis-à-vis* such a company a trade court would perceive it as an expropriation at the hands of the court.

On the other hand, indigenous peoples and their allies could consider taking the initiative and using the free trade regime as a vehicle to enforce their rights. Some First Nations in Canada joined with environmental groups and with the US before the WTO to assert that Canada effectively subsidized exports of wood by failing to accommodate and compensate for the interests of indigenous peoples in the timber cut from their aboriginal lands. If Canada respected the economic and cultural rights of the indigenous peoples, the price of timber would be higher. Therefore, disregard of the tribes' land and resource rights resulted in an unfair price for timber, and the challengers argued that this would, in turn, enable importing countries to impose a tariff. In April 2004, the WTO decided the case without considering protections for indigenous peoples derived from other international agreements in calculating subsidies (WTO Panel Report, 2004). Nevertheless, an affirmative use of trade law to enforce indigenous resource rights may be an option for peoples affected by uncompensated resource exploitation in their territories (Barsh, 2002).

Potential indigenous water rights claims under international law

The many international agreements and potential sources of customary international law that could be sources of indigenous rights to water can be roughly divided into categories that correspond with the kinds of claims that indigenous peoples might assert when they are deprived of access to water.

Protection for lands and resources

Because indigenous populations, according to various international law definitions and regulations, are usually tied inextricably to their lands for sustenance, cultural identity and spirituality, their demands for the recognition of indigenous land rights are frequently closely linked to demands for human rights protections (Torres, 1991). Some international environmental agreements also mention indigenous rights to land and resources. Few mention water specifically; but this should not be a barrier given the close connection of land and water use.

The Inter-American Convention on Human Rights (IACHR) recognizes that indigenous peoples' rights to land and resources do not derive from formal state recognition, but from traditional use and occupation. This understanding is based in large part upon the realization that 'Certain indigenous peoples maintain special ties with their traditional lands and a close dependence upon the natural resources provided therein – respect for which is essential to their physical and cultural survival' (Wagner, 2001, p497).

Human rights agreements include strong protections for indigenous land rights. Article 13 of ILO No 169 provides that indigenous and tribal peoples are to enjoy full human rights that, in turn, include rights to use land. The land rights provisions of ILO No 169 are framed by Article 13, which refers to lands or territories that 'they occupy or otherwise use'. It goes on to say that 'The use of the term "lands" in Articles 15 and 16 shall include the concept of territories, which covers the total environment of the areas which the peoples concerned occupy or otherwise use'. Article 14 states that the possessory rights of indigenous peoples established by use and occupancy must be protected. The recognized rights include the right of indigenous peoples 'to participate in the use, management and conservation of these resources' (ILO No 169). Finally, the convention requires states to provide penalties for unauthorized intrusion upon or use of indigenous lands.

The convention only binds signatory states; but its impact may extend farther. A distinguished commentator argues that 'In addition to creating treaty obligations among ratifying states, Convention No 169 is properly viewed as reflecting a new and still developing body of customary international law' (Anaya, 1991, p8). According to this approach, whether or not a particular country has signed Convention No 169 is immaterial – it may be bound by customary international law's recognition of indigenous peoples' property rights in natural resources that they have traditionally used.

The Union of Huichol Indigenous Communities of Jalisco successfully used ILO No 169 to secure land title to land which had been illegally adjudicated during the 1960s by the Mexican government. Although the community brought the case under Articles 13 and 14, which recognize indigenous rights to land, including its use, occupancy and management, the ILO Committee of Experts did not ground its decision on these provisions. Instead, it found that Mexico had violated Article 2, which prevents governments from discriminating against indigenous peoples and requires them to 'promote the economic, social and cultural rights of those people'. In addition, the committee emphasized the importance of Article 4, which instructs governments to protect individuals and cultures of the people affected by a project.

Indigenous land and resource rights are also protected by various human rights instruments of the Organization of American States (OAS). The 1978 American Convention on Human Rights is binding upon the signatory states of the OAS. Although the convention does not specifically mention indigenous peoples, it includes 'general human rights provisions that protect traditional indigenous land and resource tenure', including 'provisions explicitly upholding the rights to property' (Anaya and Williams, 2001, p41).

The Awas Tingni Community successfully used the American Convention on Human Rights to defend their traditional lands in the Atlantic coast region of Nicaragua. Although Nicaraguan law generally recognized that indigenous peoples had certain rights in lands that they traditionally used and occupied, it did not recognize indigenous landownership and treated these untitled lands as state lands. The Nicaraguan government granted a foreign company rights to construct roads and to exploit timber on traditional Awas Tingni lands that threatened damage to the environment and to social and cultural integrity.

After failing to obtain protection for its rights in domestic courts, the Awas Tingni Community petitioned the Inter-American Commission on Human Rights, asserting rights to their communal lands and resources. Ultimately, the Inter-American Court considered the case and delivered its judgement in 2001. It held that the American Convention on Human Rights includes the right of indigenous peoples to hold their customary lands and resources under the protection of domestic governments. The court resolved that Nicaragua 'violated the right to property protected by Article 21 of the American Convention on Human Rights to the detriment of the members of the Mayagna (Sumo) Community of Awas Tingni in connection with Articles 1(1) and 2 of the convention'. It directed that Nicaragua 'must adopt the ... measures required ... for delimitation, demarcation and titling of the property of [the Awas Tingni and do so] ... in accordance with their customary law, values, customs and mores' (*Mayagna (Sumo) Awas Tingni Community versus Nicaragua*, 2001). This decision exemplifies the possibility of invoking the protections of individual rights found in international human rights instruments to sustain the collective land and resource rights of indigenous peoples when they are ignored by domestic law.

An earlier OAS instrument, the American Declaration on the Rights and Duties of Man, was accepted in 1948. The American Declaration is 'the principal instrument for determining the applicable substantive rights for those [OAS] countries [not signatories to the American Convention on Human Rights] in proceedings before the Inter-American [Court]' (Anaya and Williams, 2001, p41). Like the American Convention on Human Rights, the American Declaration does not specifically mention indigenous people, but rather provides general human rights provisions upholding the right to property and other basic human rights of individuals. The same principle adopted by the court in the Awas Tingni case arguably could be invoked to extend the individual rights contained in the declaration to secure collective rights based on indigenous customary use and occupancy.

The Proposed American Declaration on the Rights of Indigenous Peoples was approved by the Inter-American Commission on Human Rights at its 95th session on 26 February 1997. Article XVI provides that indigenous law should be made an integral part of a nation's legal system and Article XVII provides that indigenous practices are to be included in national organizational structures. Anaya and Williams (2001, p43) argue that:

> *Excluding indigenous property regimes from the property protected by the American Convention and American Declaration would perpetuate the long history of discrimination against indigenous peoples. Such discriminatory application of the right to property would be in tension with the principle of non-discrimination that is part of the Inter-American human rights system's foundation.*

Indigenous peoples will have difficulty protecting their property rights in lands and natural resources if the extent of those lands is uncertain. For this reason, the Inter-American Commission on Human Rights long ago cited the rights of indigenous peoples under the American Declaration to support its decision that Brazil should demarcate the lands of the Yanomami Indians (*Yanomami versus Brazil*, 1985). The commission found that the laws of Brazil recognized the rights of Indian communities to possess the lands they traditionally occupied and gave them the exclusive use of natural resources within those lands. When the government began a programme of building highways and allowing widespread mining, however, the occupation and development of lands in the area of the Yanomami caused profound social, economic, cultural and environmental harm to the people and land. The Yanomami had violent conflicts with miners and, in some cases, highway workers and miners drove the Indians out of their villages. Without demarcation of the Yanomami territory, their rights to protect lands and resources from illegal occupation or exploitation by others were mostly theoretical.

In 1988, Brazil revised its constitution and ordered the demarcation of all indigenous territories by 1993. However, in 1996 the Brazilian government issued a decree delaying the demarcation of new reserves and effectively impeded the indigenous rights that were guaranteed under the new constitution. The decree states that cities and non-Indians can challenge demarcation and suspend Indian property claims. Although the Yanomami have secured some victories, at present they have gained control of only one quarter of their original lands, lands the security of which remains threatened by mining interests, politicians and the Brazilian military.

Demarcation becomes important when others compete for rights to indigenous lands, waters or other resources. If a government has already granted rights to exploit indigenous lands to others, as in the Awas Tingni case, the lack of demarcation will make it more difficult for the indigenous peoples to challenge the government action.

Environmental protection

Three kinds of international agreements or norms potentially protect indigenous rights to environmental quality relating to use and control of water. First, the right to environmental protection is implicit in international instruments specifically acknowledging indigenous land rights, such as those discussed in the preceding section. Second, environmental rights can be derived from human rights guarantees because the way of life and the very existence of indigenous peoples often depend on their relationship with the land. Finally, the emergence of international law directly protecting the environment has special significance for indigenous peoples. Taken together, these norms provide support for the rights of indigenous peoples to protect, use and manage water resources.

One of the most important instruments protecting the human rights of indigenous peoples is ILO No 169. Article 15 and paragraph 4 of Article 7 require states to safeguard indigenous peoples' rights concerning the environment and natural resources of indigenous lands. Article 7 requires governments to 'take measures, in cooperation with the peoples concerned, to protect and preserve the environment of the territories they inhabit'. And Article 15 provides that 'The rights of the peoples concerned to the natural resources pertaining to their lands shall be specially safeguarded.'

The first major international agreement specifically addressing the environment was the Stockholm Declaration emanating from the 1972 United Nations Conference on the Human Environment. The declaration recognized in Principle 5 that 'The non-renewable resources of the earth must be employed in such a way as to guard against the danger of their future exhaustion and to ensure that benefits from such employment are shared by all mankind.' Although it did not deal specifically with the rights of indigenous peoples, the declaration can be read to imply that indigenous peoples have a right to an equitable share of a state's waters.

In 1992, the international environmental community adopted the Rio Declaration reaffirming and updating the principles set forth in the Stockholm Declaration; a new consensus among states was reached concerning international environmental policies to protect the world's ecosystems and biological diversity. Principle 22 of the declaration recognized that 'Indigenous people and their communities and other local communities, have a vital role in environmental management and development because of their knowledge and traditional practices. States should recognize and duly support their identity, culture and interests and enable their effective participation in the achievement of sustainable development.'

Agenda 21, also adopted at UNCED in Rio in 1992, was 'essentially a plan of action for carrying out the principles in the Rio Declaration'. Chapter 26 of section 3 of Agenda 21 deals with the role of indigenous peoples and their communities:

> *Indigenous people and their communities have a historical relationship with their lands and are generally descendants of the original inhabitants of such lands. In the context of this chapter the term 'lands' is understood to include the environment of the areas which the people concerned traditionally occupy... In view of the interrelationship between the natural environment and its sustainable development and the cultural, social, economic and physical well-being of indigenous people, national and international efforts to implement environmentally sound and sustainable development should recognize, accommodate, promote and strengthen the role of indigenous people and their communities.*

By recognizing that 'traditional and direct dependence on renewable resources and ecosystems, including sustainable harvesting, continues to be essential to the cultural, economic and physical well-being of indigenous people and their communities', Agenda 21 would support the rights of indigenous peoples to water needed to maintain indigenous culture (UNCED, 1992).

Chapter 26.4 of Agenda 21 recognizes the cultural importance of indigenous environmental controls:

> *Some indigenous people and their communities may require, in accordance with national legislation, greater control over their lands, self-management of their resources, participation in development decisions affecting them, including, where appropriate, participation in the establishment or management of protected areas.*

The Convention on Biological Diversity (CBD) is another instrument adopted by nearly all the nations of the world at the Rio conference in 1992. Its attention to indigenous peoples was largely directed at protecting their property interests in biological resources. One writer argues that the convention could assist indigenous peoples, at least indirectly, in resisting environmentally damaging dam construction in their territories. He cites an analogous 1995 ruling that plans to build a paved road in wild areas within Daisestuzan National Park would breach the Japanese government's obligation 'not to destroy biodiversity' under the CBD (Ichikawa, 2002).

Subsistence rights

Several international instruments on human and civil rights protect the rights of people to seek and acquire basic subsistence. Because access to water is vitally important to the quest for subsistence by most indigenous peoples, these instruments are another potential source of rights to use and control water resources.

The United Nations International Covenant on Civil and Political Rights (ICCPR) states in Article 1, section 2, that a people cannot be deprived of its own means of subsistence, and in Article 27 that 'ethnic minorities ... shall not be denied the right, in community with the other members of their group, to enjoy their own culture'. The Human Rights Committee was established to determine complaints made pursuant to the ICCPR. The committee's decisions recognize broad protections for traditional uses of natural resources because:

> *... indigenous peoples' subsistence and other traditional economic activities are an integral part of their culture and interference with those activities can be detrimental to their cultural integrity and survival. By necessity, the land, resource base and the environment thereof also require protection if subsistence activities are to be safeguarded ... [thus] necessary steps should be taken to restore and protect the titles and interests of indigenous persons in their native lands ... [and that] securing continuation and sustainability of traditional forms of economy of indigenous minorities (hunting, fishing and gathering) and protection of sites of religious or cultural significance for such minorities ... must be protected under Article 27.* (MacKay, 2002, pp596–597)

A case decided by the Human Rights Committee in 1996 illustrates the applicability, and the limits, of Article 27. In *Länsman versus Finland*, Sámi reindeer herders in northern Finland challenged plans of the national government to allow logging and road construction in a 300ha area used by the Muotkatunturi Herdsmen's Committee as winter pasture and spring calving grounds. The claim was rejected because the logging was not of a large enough scale to threaten the survival of traditional reindeer husbandry and therefore did not constitute a violation of Article 27. The decision stated that 'Measures that have a certain limited impact on the way of life and the livelihood of persons belonging to a minority will not necessarily amount to a denial of the rights under Article 27' (*Länsman versus Finland*, 1996).

Another vehicle for seeking access to water resources through international protections for subsistence uses is the United Nations International Covenant on Economic, Social and Cultural Rights (ICESCR). Although the ICESCR does not directly address indigenous peoples or land and natural resource rights, several of the articles can be applied to indigenous water rights. For example, Article 1 states that 'In no case may a people be deprived of its own means of subsistence.' Article 11 includes a right to an adequate standard of living and the right to share in efficient agrarian systems. Article 12 contains a right to a secure, healthy and ecologically sound environment, which arguably could include access to sufficient amounts of clean water.

Indigenous peoples could attempt to use these provisions to claim that they are entitled to water sufficient to irrigate crops upon which they rely for an adequate standard of living and to potable water and other environmental conditions related to their subsistence. The 'healthy environment' provision of the ICESCR is so sweeping and so general, however, that it is difficult to imagine its being enforced against a state in the absence of other, more specific, guarantees.

Cultural identity

Respect for indigenous culture is found in several international instruments. For instance, Principle 20 of the Vienna Declaration adopted by the 1993 United Nations World Conference on Human Rights 'recognizes the inherent dignity and the unique contribution of indigenous people ... and strongly reaffirms the commitment of the international community to their economic, social and cultural well-being'. Presumably, state actions that erode the cultural uniqueness of indigenous peoples would be contrary to the declaration.

For instance, ILO No 169 frames its requirements ensuring indigenous peoples' rights to ownership of traditionally used natural resources in terms of cultural integrity. Article 2 specifies that the mandated action should include measures aimed at 'Promoting the full realization of social, economic and cultural rights of these peoples with respect for their social and cultural identity, their customs and traditions and their institutions.'

Some international agreements speak primarily of indigenous peoples' rights to cultural integrity and prohibit government actions that would erode or destroy traditional culture. These prohibitions could be interpreted to extend to depletions or contamination of water sources that result in deprivation of traditional water uses such as agriculture or spiritual purposes, or that would render it impossible for indigenous peoples to continue community life in their historical territory. Scholars have observed that:

> *... under international law, the states' obligation to protect indigenous peoples' right to cultural integrity necessarily includes the obligation to protect traditional lands because of the inextricable link between land and culture in this context. Thus, rights to lands and resources are property rights that are prerequisites for the physical and cultural survival of indigenous communities.* (Anaya and Williams, 2001, p53)

In one decision, the Mexican National Human Rights Commission applied several articles of ILO No 169 and the Mexican Constitution to protect the cultural survival of the Cucupá people (Comisión Nacional de los Derechos Humanos, Recomendación, 008/2002). The Cucupá are an indigenous community living in the Colorado River Delta region whose population has declined from several thousand to fewer than 200 people. The government's management of the land, waters and fisheries of the delta over the last 50 years has led to the near-extinction of the Cucupá community and has damaged the delta's environmental health in an area with great biodiversity values that has been designated a biosphere reserve under the United Nations Man and the Biosphere Programme. This destructive management led Defenders of Wildlife to petition the commission on behalf of the Cucupá people and argue that the government owed a duty to ensure them a decent living through improved management.

The Mexican Commission cited several sections of ILO No 169 for the 'obligation of governments to recognize, protect and respect cultural values and practices of indigenous peoples, such as their environment, especially their spiritual and cultural relation with the lands' (Comisión Nacional de los Derechos Humanos, Recomendación, 008/2002). It then ordered Mexico's natural resources agency, Secretario de Medio Ambiente y Recursos Naturales (SEMARNAT), to update the biosphere reserve management plan to ensure that the cultural, ecological and economic needs of the Cucupá people are fulfilled. It also ordered SEMARNAT to develop a social development programme for the Cucupá and required the agricultural agency, Secretario de Agricultura, Ganadería, Desarrollo Rural, Pesca y Alimentación (SAGARPA), to grant additional fishing permits to the Cucupá. This decision is an example of an indigenous people using international human rights law in a domestic tribunal to force government institutions to promote the economic and cultural survival of indigenous communities through improved management of, and access to, natural resources. Mexican government agencies, however, failed to respond fully to the commission's recommendations.

Article 27 of the ICCPR guarantees indigenous peoples the right to enjoy their cultures. The United Nations Human Rights Committee has interpreted Article 27 of the ICCPR to include the protection of cultural integrity. In the Lubicon Lake Band case the provincial government of Alberta, Canada, granted leases for mineral exploration and timber harvesting within the aboriginal territory of the Lubicon Lake Band (McGoldrick, 1991). The Human Rights Committee found that this violated the cultural integrity guarantees of Article 27 of the ICCPR. Extensive extractive resource development by outsiders, combined with the government's historical failure to assure the band a land base, had threatened the band's way of life and culture (Kingsbury, 1998).

Another case involved the indigenous Ainu of Japan. When the Japanese government constructed a dam in an area where Ainu ceremonial places had been, two Ainu sued. The Sapporo District Court held that the construction impaired the Ainu culture and that the government had violated the Ainu's right to enjoy their indigenous culture under the ICCPR (*Kayano versus Hokkaido Expropriation Comm*, 1997). Ichikawa (2002) argues that it is possible to construe Article 27 of ICCPR as securing Ainu fishing rights and water rights as a part of their culture.

Indigenous peoples can also benefit indirectly from national cultural protection laws that do not refer to indigenous culture. For instance, Australia enacted a statute prohibiting dam construction planned by the State of Tasmania in an area listed as a World Cultural and National Heritage site under the World Heritage Convention. The dam's proponents challenged this national law; but the Australian High Court upheld the legislation, saying that the convention imposed an obligation on Australia to take appropriate measures for the preservation of the site (*Commonwealth versus Tasmania*, 1983).

The United Nations Declaration on the Rights of Indigenous Peoples was adopted by the General Assembly in September 2007. Article 20 says that 'Indigenous peoples have the right to maintain and develop their political, economic and social systems or institutions, to be secure in the enjoyment of their own means of subsistence and development, and to engage freely in all their traditional and other economic activities.' The declaration was negotiated and debated for 22 years and, while declarations are technically not legally binding, scholars believe that its provisions demanding respect for traditional practices, cultural integrity and economic uses of lands fairly reflect accepted international norms that should be useful in asserting indigenous rights to natural resources (see Williams, 1990, pp689–690).

Several declarations arising out of international conferences are also useful in determining the attitudes and norms of the international community concerning indigenous cultural rights. While they are not 'hard law' in the sense that they are likely to be applied by international or domestic tribunals to require or prevent particular actions of a country, they can assist in interpreting and applying principles found in formally adopted instruments and can provide evidence of the acceptance of norms which are themselves 'soft law' or customary law, which is an accepted source of international law.

The Declaration of Principles of Indigenous Rights adopted by the Fourth General Assembly of the World Council on Indigenous Peoples called for legal recognition of indigenous cultural rights and related them to water by declaring in Principles 10 and 11 that 'Indigenous people have inalienable rights over their traditional lands and resources', including rights to 'coastal economic zones'. And Principle 13 states that 'No action or process shall be implemented which directly and/or indirectly would result in the destruction of land, air, water, glaciers, animal life, environment or natural resources, without the free and well informed consent of the affected indigenous people.'

In March 2003, the Third World Water Forum was held in Kyoto, Japan, bringing participants from 182 countries. The United Nations Educational, Scientific and Cultural Organization (UNESCO) and the Water Law and

Indigenous Rights (WALIR) programme assembled indigenous leaders from all continents to present their cases, claims and visions, and to formulate jointly the Indigenous Peoples Water Declaration through which indigenous participants also committed to forming a network to encourage local communities in protecting their water rights. Principle 11 states that 'Self-determination includes the practice of our cultural and spiritual relationships with water and the exercise of authority to govern, use, manage, regulate, recover, conserve, enhance and renew our water sources, without interference' (Indigenous Peoples Water Declaration, cited in Boelens et al, 2006).

Racial discrimination

Destroying or inhibiting access to indigenous natural resources, including water, can be a form of racial discrimination. The United Nations International Convention on the Elimination of All Forms of Racial Discrimination of 1966 (CERD) has been widely accepted by the nations of the world. The convention recognizes the right of communal property ownership for indigenous peoples, and so the failure of a government to respect communal ownership can therefore violate its provisions. MacKay (2002, p594) explains that:

> *... under Article 5 of the CERD, for instance, states-parties are obligated to recognize, respect and guarantee the right 'to own property alone as well as in association with others' and the right to inherit property, without discrimination. Failure to recognize and protect indigenous property ownership and inheritance systems and rights is discriminatory and denies equal protection of the law.*

In Australia, indigenous peoples' rights received little legal recognition until the last decade. The 1992 Mabo decision of the Australian High Court recognized the existence of native title to land after finding that the doctrine used to deny such title in the past was racially discriminatory. In 1975 the Australian Parliament passed the Racial Discrimination Act in order to implement CERD, and after that date state extinguishment of native title was subject to legal challenge as racial discrimination.

Rights under native title include certain traditional uses such as hunting, fishing and gathering. Although rights can be lost when indigenous peoples sever their connection with the land or by extinguishment when the government grants the land to others, it seems clear that Aborigines have interests in the lands and waters still held by the Crown. The Native Title Act of 1993 confirmed that native title could be claimed in waters as well as lands. This opens the possibility that Aborigines could claim rights to use the waters of Crown lands for everything from irrigation to fishing to cultural and spiritual uses.

A recent case supports the conclusion that Aboriginal communities can assert rights to traditional uses of water. In the Spinifex case the federal court confirmed an agreement between the parties that gave native title holders a right to take water for the purposes of satisfying their personal, domestic, social, cultural, religious, spiritual or non-commercial communal needs, including the observance of traditional laws and customs (*Spinifex People versus Western Australia*, 2000). However, a claim to water for irrigation might not fare as well. There is a common assumption that historical practices of Aboriginal communities did not include irrigation, and it is possible that the state will deny water rights for non-traditional uses. In contexts other than water, however, the courts have said that they will not limit the rights under native title strictly to the customary uses that prevailed in ancient times, but will allow, at least, 'the maintenance of the ways of the past in changed circumstances' (*Members of the Yorta Yorta Aboriginal Community versus State of Victoria*, 2001).

A serious problem with native rights to water in Australia, even for demonstrably traditional uses, is that they are subordinate to all of the rights that have been granted to others by the state under water legislation. The courts have not determined whether this is racially discriminatory. Furthermore, the very enactment of legislation allocating water rights to others may pose problems for indigenous peoples seeking to assert rights to water.

Right of self-determination

Ultimately, the most important right for indigenous peoples may be to govern the allocation, use and protection of natural resources. State systems of water law often allow for uses that are damaging to indigenous cultures and that compete with indigenous values, ranging from economic to spiritual. By controlling the waters in their territories, indigenous peoples can affect decisions that ensure that their own needs are met. Moreover, protecting the dignity of indigenous institutions strengthens indigenous cultures. Maintaining indigenous community control of water is often difficult, however, because states may consider it a threat to their sovereignty, as well as to their ability to promote national economic development.

Some international law sources appear to support the self-determination of indigenous peoples, including their right to control water and other natural resources and the obligation of states to respect indigenous laws and customs concerning water use. In addition, multiple provisions of international law insist that states consult with affected indigenous communities when they make decisions. Nevertheless, provisions relating to self-determination are fewer and less specific than the international law provisions protecting the other potential sources of rights related to water.

An important example of an instrument validating indigenous self-determination is, again, ILO No 169. The preamble frames the agreement in terms of indigenous peoples' right 'to exercise control over their own institutions, ways of life and economic development'. The agreement goes on to provide that 'Governments shall have the responsibility for developing, with the participation of the peoples concerned, coordinated and systematic action to protect the rights of these peoples and to guarantee respect for their integrity.'

Another instrument, the International Covenant on Economic, Social and Cultural Rights (ICESCR), states in Article 1 that 'All peoples have the right of self-determination. By virtue of that right they freely determine their political status and freely pursue their economic, social and cultural development.'

The Proposed American Declaration on the Rights of Indigenous Peoples (approved by the Inter-American Commission on Human Rights), like ILO No 169, contains more elaborate and specific protections for indigenous land rights than are found in human rights accords. Although it remains a proposed agreement, the American Declaration includes the fullest elaboration of the important elements of indigenous control of resources. Several articles support the rights of indigenous peoples to self-government and to have their laws recognized within the state legal system. Article XVI says that:

1 *Indigenous law shall be recognized as a part of the states' legal system and of the framework in which the social and economic development of the states takes place.*
2 *Indigenous peoples have the right to maintain and reinforce their indigenous legal systems and also to apply them to matters within their communities...*
3 *In the jurisdiction of any state, procedures concerning indigenous peoples or their interests shall ... include observance of indigenous law and custom and, where necessary, use of their language.*

Among the provisions related to natural resources, Article XVIII states that:

1 *Indigenous peoples have the right to the legal recognition of their varied and specific forms and modalities of their control, ownership, use and enjoyment of territories and property.*
2 *Indigenous peoples have the right to the recognition of their property and ownership rights with respect to lands, territories and resources they have historically occupied, as well as to the use of those to which they have historically had access for their traditional activities and livelihood...*
4 *Indigenous peoples have the right to an effective legal framework for the protection of their rights with respect to the natural resources on their lands.*

As mentioned above, the United Nations General Assembly adopted a Declaration on the Rights of Indigenous Peoples in 2007. Article 26 includes language especially protective of indigenous peoples' right to apply customary law in managing their lands and waters:

1. *Indigenous peoples have the right to the lands, territories and resources which they have traditionally owned, occupied or otherwise used or acquired.*
2. *Indigenous peoples have the right to own, use, develop and control the lands, territories and resources that they possess by reason of traditional ownership or other traditional occupation or use, as well as those which they have otherwise acquired.*
3. *States shall give legal recognition and protection to these lands, territories and resources. Such recognition shall be conducted with due respect to the customs, traditions and land tenure systems of the indigenous peoples concerned.*

Finally, Principle 5 of the Declaration of Principles of Indigenous Rights of the World Council on Indigenous Peoples states that 'The customs and usages of the indigenous peoples must be respected by the nation-states and recognized as a legitimate source of rights.'

Conclusions

International law provides several grounds for asserting indigenous rights to water. Although there is a dearth of precedent for using international law to pursue indigenous water rights claims, there are a small but growing number of cases in which indigenous peoples have used international law to protect their rights to other natural resources such as timber or minerals.

The six types of rights identified in this chapter illustrate various ways in which such claims could be framed and some of the international law instruments and norms that could be invoked by indigenous peoples when a government takes action that deprives them of quantities of water or pollutes water within their traditional territories. Claims could also arise from denial of access to water for traditional uses or commercial uses by tribes and indigenous communities.

Indigenous groups asserting international law claims need not confine them to rights under a single instrument. It is likely that they will need to allege several grounds under various instruments. For instance, the Carrier Sekani Tribal Council of Canada submitted a claim to the Inter-American Human Rights Commission arising out of the provincial government's decision to allow companies to cut the timber resources on their traditional lands, claiming that this violated its rights to property, cultural integrity, self-determination and consultation, equality under the law and to have their rights secured by the state (MacKay, 2001).

This chapter has concentrated on international law remedies for violation of indigenous water rights, but pursuit of remedies under domestic laws should not be ignored if individual countries have statutes or constitutional provisions that speak to indigenous rights to land and water. Moreover, domestic laws in

Latin American countries may include rights of customary use of resources pertaining to all *campesinos* and not limited to indigenous groups. When they seek the aid of the law in asserting water rights in domestic courts and administrative tribunals, indigenous peoples and their lawyers should consider citing both domestic and international law sources.

The unexplored possibilities for using international law as well as domestic law to establish and protect indigenous water rights are many. The most promising legal approaches and the most appropriate legal forums for particular indigenous groups to protect their water rights will vary with the nature of each claim. It would be wise for indigenous groups from several countries to coordinate regional or international efforts to find the best cases to advance the development of international law as a tool for securing indigenous water rights.

Note

1 This chapter is an abridged and revised version of a 2005 article published by the author in the *Colorado Journal of International Environmental Law and Policy*, vol 16, pp259–294 that was written in connection with the Water Law and Indigenous Rights (WALIR) project coordinated by Wageningen University and the United Nations Economic Commission for Latin America and the Caribbean (ECLAC).

References

Anaya, S. J. (1991) 'Indigenous rights norms in contemporary international law', *Arizona Journal of International and Comparative Law*, vol 8, pp1–39

Anaya, S. J. and R. Williams (2001) 'The protection of indigenous peoples' rights over lands and natural resources under the Inter-American Human Rights System', *Harvard Human Rights Journal*, vol 14, pp33–86

Barsh, R. (2002) 'A social theory of fair trade, with special reference to indigenous peoples', *American Society of International Law Proceedings*, vol 96, pp279–290

Boelens, R., M. Chiba and D. Nakashima (2006) *Water and Indigenous Peoples (including Indigenous Interventions at the Third World Water Forum, Kyoto, Japan, 2003, and Indigenous Peoples Kyoto Water Declaration)*, UNESCO and WALIR, Paris, www.indigenouswater.org

Cohan, J. (2001/2002) 'Environmental rights of indigenous peoples under the Alien Tort Claims Act, the public trust doctrine and corporate ethics and environmental dispute resolution', *UCLA Journal of Environmental Law and Policy*, vol 20, pp133–185

Comisión Nacional de los Derechos Humanos, Recomendación (008/2002) 'Decision Letter', www.cndh.org.mx/recomen/2002/008.htm

de Bolivar, M. (1998) 'A comparison of protecting the environmental interests of Latin-American indigenous communities from transnational corporations under international human rights and environmental law', *Journal of Transnational Law and Policy*, vol 8, pp105–149

Getches, D., C. Wilkinson and R. Williams (2005) *Cases and Materials on Federal Indian Law*, 5th edition, Westgroup, St Paul, MN

Ichikawa, M. (2002) 'Protection of ecological and cultural values of watersheds under the Convention on Biological Diversity and the International Covenant on Civil and Political Rights', Paper presented to the Conference on Allocating and Managing Water for a Sustainable Future: Lessons from Around the World, Natural Resources Law Center, University of Colorado School of Law, Boulder, CO

Kingsbury, B. (1998) 'Indigenous peoples in international law: A constructivist approach to the Asian controversy', *American Journal of International Law*, vol 92, pp414–457

MacKay, F. (2001) *Forest Peoples Programme: A Guide to Indigenous Peoples' Rights in the Inter-American Human Rights System*, www.forestpeoples.org/documents/law_hr/iachr_briefing_oct_01_eng.pdf

MacKay, F. (2002) 'Universal rights or a universe unto itself? Indigenous peoples' human rights and the World Bank's Draft Operational Policy 4.10 on Indigenous Peoples', *American University International Law Review*, vol 17, pp 527–624

McFadden, P. (1995) 'Provincialism in United States courts', *Cornell Law Review*, vol 81, pp4–65

McGoldrick, D. (1991) 'Canadian Indians, cultural rights and the Human Rights Committee', *International and Comparative Law Quarterly*, vol 40, pp658–669

Runnalls, D. and K. Fuller (2001) 'Preface to Howard Mann, International Institute of Sustainable Development and the World Wildlife Fund', in *Private Rights, Public Problems: A Guide to NAFTA's Controversial Chapter on Investor Rights*, www.iisd.org/pdf/trade_citizensguide.pdf

Sansani, I. (2004) 'American Indian land rights in the Inter-American system: *Dann v. United States*', *Human Rights Brief*, vol 10, no 2, pp2–5

Stiglitz, J. (2003) *Globalization and its Discontents*, W. W. Norton, New York, NY

Torres, R. (1991) 'The rights of indigenous populations: The emerging international norm', *Yale Journal of International Law*, vol 16, pp127, 137–140

Wagner, M. (2001) 'The international legal rights of indigenous peoples affected by natural resource exploitation: A brief case study', *Hastings International and Comparative Law Review*, vol 24, pp491–502

Williams, R. (1990) 'Encounters on the frontiers of international human rights law: Redefining the terms of indigenous peoples' survival in the world', *Duke Law Journal*, no 4, pp660–704

World Council of Indigenous Peoples (1984) *Declaration of Principles*, www.cwis.org/fwdp/Resolutions/WCIP/wcip.txt

WTO Panel Report (2004) *US–Final Dumping Determination on Softwood Lumber from Canada*, WT/DS264

Treaties and agreements

American Convention on Human Rights (1978) *Pact of San José, Costa Rica*, OAS Treaty Series No 36, 1144 U.N.T.S. 123, www.cidh.org/Basicos/English/Basic3.American%20Convention.htm

Convention on Biological Diversity (1992) 1760 U.N.T.S. 79

Inter-American Charter of Social Guaranties (1948) Reprinted in E. J. Osmanczyk (ed) *Encyclopedia of the United Nations and International Relations* (2003) pp763, 1012

International Covenant on Civil and Political Rights (1966) G.A. Res. 2200(XXI), Article 1(2), 999 U.N.T.S. 171, www2.ohchr.org/english/law/ccpr.htm

International Covenant on Economic, Social and Cultural Rights (1966) G.A. Res. 2200(XXI), 993 U.N.T.S. 3, www2.ohchr.org/english/law/cescr.htm

International Labour Organization Convention No 169 (1989) *Convention Concerning Indigenous and Tribal Peoples in Independent Countries*, 28 I.L.M. 1382

Proposed American Declaration on the Rights of Indigenous Peoples (1997) Inter-American Commission on Human Rights, 95th Session, OAS Doc OEA/Ser/L/V/. II.95 Doc 6, www.cidh.oas.org/Indigenous.htm

Rio Declaration on Environment and Development (1992) U.N. GAOR, 46th Session, UN Doc A/CONF.151/5/Rev.1, reprinted in 31 I.L.M. 874

Statute of the International Court of Justice (1945) 59 Statute 1055, 1060, T.S. No 993, http://avalon.law.yale.edu/20th_century/decad026.asp

Stockholm Declaration of the United Nations Conference on the Human Environment (1972) G.A. Res. 2997, UN GAOR, 27th Session, UN Doc A/ Conf.48/14/Rev/1 (1972), reprinted in 11 I.L.M. 1416

United Nations (1966) *International Convention on the Elimination of All Forms of Racial Discrimination*, 660 U.N.T.S. 195

UNCED (United Nations Conference on Environment and Development) (1992) *Agenda 21*, UN Doc A/CONF.151/26 (vol II), www.un.org/esa/sustdev/documents/agenda21

United Nations Declaration on the Rights of Indigenous Peoples adopted by General Assembly Resolution 61/295, 17 September 2007

Vienna Declaration and Programme of Action (1993) UN GAOR, 48th Session

Other

Amended Petition and Response to the Inter-American Human Rights Commission submitted by the Chiefs of the Member First Nations of the Carrier Sekani Tribal Council against Canada (2000) Case No 12.279

Commonwealth versus Tasmania (1983) 158 C.L.R. 1

GATT Dispute Settlement Panel Report on US Restrictions on Imports of Tuna (1991) 30 I.L.M. 1594, 1623 (1991)

Kayano versus Hokkaido Expropriation Comm (1997) 1598 Hanrei Jiho 33 (Sapporo District Court)

Länsman versus Finland (1996) Communication No 671/1995, para 2.1, UN Doc CCPR/C58/D/671/1995, www.umn.edu/humanrts/undocs/html/ VWS67158.htm

Mabo versus Queensland (1992) 175 C.L.R. 1

Mayagna (Sumo) Awas Tingni Community versus Nicaragua (2001) Case 79, Inter-Am. C.H.R., ser. C, www.javier-leon-diaz.com/property/Mayagna%20case.pdf

Members of the Yorta Yorta Aboriginal Community versus State of Victoria (2001) 110 F.C.R. 244

Ominayak, Chief of the Lubicon Lake Band versus Canada (1990) Comm No 267/1984, Report of the Human Rights Committee, UN GOAR, 45th Session, Supp No 40, UN Doc A/45/40

Racial Discrimination Act of Australia (1975)

Spinifex People versus Western Australia (2000) F.C.A. 1717

World Heritage Properties Conservation Act (1983) Australia

Yanomami versus Brazil (1985) Case 7615, Inter-Am. C.H.R. 24, O.E.A./ser. L./ V./II.66, doc 10 rev 1

15

Networking Strategies and Struggles for Water Control: From Water Wars to Mobilizations for Day-to-Day Water Rights Defence

Rutgerd Boelens, Rocío Bustamante and Tom Perreault

Introduction

This chapter examines struggles for water and water rights in the Andes where water plays a key role in the livelihoods of peasant and indigenous communities. These local user collectives face growing threats from powerful extra-local interest groups who claim or undermine their water rights for their own economic, political or military objectives (e.g. Gelles, 2000; de Vos et al, 2006; Boelens, 2009). Given that these intervening actors, such as state agents, agro-commercial enterprises, mining companies, hydropower stations and other dominant players, use strategic tools and governance techniques that sprout from national frameworks and global networks, 'local' user groups also look for responses that extend beyond their home domains. Thus, threats to their territories and water resources require user collectives to organize not just within their local, mostly common property institutions, but increasingly to pursue their objectives, organizations and defence mechanisms at a variety of scales.

This chapter concentrates, in particular, on the multi-scalar networks within which local user communities and organizations are embedded and through which they mobilize to advance and defend their water interests. It looks at their 'scalar capabilities' and alliance strategies since these are critical to their capacity to defend and promote local water access and control rights (Hendriks, 2002; Perreault, 2005; Assies and Gundermann, 2007). As Swyngedouw (2004, pp26–27) rightly argues: 'the success or effectiveness of social and political strategies for empowerment is related to the ways in which geographical scale is actively considered and mobilized in struggles for social, political, or economic resistance or change'. The politics of dominant players (trying to align user communities to their frames, rules and regulations or to take over their water access rights), as well as the resistance strategies of local user groups (aiming for the localization of water access and decision-making power), are fundamentally related to their power to compose or manipulate patterns of multiple scales (Perreault, 2005; Boelens, 2008a).

Water rights struggles on every scale involve a fundamental conflict between control-externalizing governance networks and control-localizing networks that foster local dominion over resources, rule-making and authority (Boelens, 2008a). These struggles also take place within and among networks whose actors are not just striving to get access to water. They also battle over the material control of water use systems and over the right to define culturally, organize politically, and shape these systems and their governance frameworks (Boelens and Gelles, 2005; Zwarteveen et al, 2005; Perreault, 2008).

This chapter aims to deepen insights into water rights conflicts and multi-scale network strategies through analytical elaboration and case illustration. The section that follows elaborates on struggles over livelihoods in the Andes, and the ways in which transformational movements strategize spatial scales and networks to secure access to resources. The third section examines the role of scale and network in what has become known as the Bolivian Water Wars. It focuses on the ways in which claims and strategies were broadened beyond just local community and Cochabamba city scales, and new national policies and legislation have emerged through an irrigators' movement. The fourth section analyses the scalar strategies of peasant and indigenous irrigation and drinking water organizations in Chimborazo Province, Ecuador. It analyses why and how local users have federated at supra-local scales to be able to struggle for their water interests *vis-à-vis* state agencies and powerful private holdings. Elaborating on the foregoing concepts and case illustrations, section five examines the dialectical relationship between rural people's social movements and translocal networks, and the scalar strategies to influence water and environmental governance at local and at wider geographical and political levels. The conclusions highlight that marginalized water-user groups need to build dense multi-scalar networks that link local communities with translocal actors so that they can influence strategies and networks that are rooted in locality, but extend to national and transnational scales. This last section also focuses on the politics of scale and the construction of place and identity in the context of ethnic and socio-economic political mobilization. In order to secure access and control over their water resources, water user organizations in the Andean highlands employ material and discursive practices and forms of social organization, thereby contesting the water distributive and cultural politics of the rulers.

Social movements and natural resource management in the Andes: The roles of livelihood and spatial scale

Of central importance to the cases examined in this chapter is the fact that water struggles involve not just local water user organizations, but also the networks which connect these organizations to other actors in other places. By fostering and mobilizing associational networks that stretch beyond local and national boundaries, local resource-user organizations can amplify their voice, may increase their assets (in financial, knowledge-based or technical terms), and in many cases tend to increase their legitimacy in the view of the state, non-governmental organizations (NGOs) or private actors (Bebbington, 1999; Getches, 2005). Water user associations provide us with excellent examples of translocal mobilization. In the Andean region, water is often locally managed. But because of its overarching importance, it is governed by complex institutional frameworks and administrative entities, operating on a variety of spatial scales from the local to the international. Water user associations are similarly varied, as are the struggles in which they engage.

Struggles over rights and access to water and other natural resources are diverse, and may involve questions of livelihood, social justice, state accountability or visions of citizenship and the nation. Thus, they are often just one component of a diverse array of grievances and claims. Specific struggles, such as those detailed below, must be viewed within the broader social, political, economic and environmental contexts in which they arise. Questions of state policy, economic conditions, social values and water quality and quantity inevitably shape the forms and tactics of water user organizations as they engage in such struggles (see, for example, Gelles, 1998; Roth et al, 2005; Guevara-Gil, 2006). This chapter will emphasize two key concepts at the heart of the cases that are discussed: livelihood and spatial scale. Insofar as the concept of livelihood draws attention to the everyday practices by which people make a living, it entails the material practices, social values and, crucially, the resources and assets necessary for life. Spatial scale, as the organizational levels through which social life occurs (e.g. the individual, the household, the community, the watershed, the nation, etc.), provides a valuable practical and theoretical framework through which social movements and their translocal networks operate.

Livelihoods

Social movements and networks of water user groups, and the ways in which they engage in cultural politics and struggles for autonomy and democracy, are deeply rooted in and entwined with the material conditions and production relations of smallholder families and communities (Bebbington, 2000; van der Ploeg, 2006, 2008). Thus, greater attention to the everyday practices of building 'water territory' and the social reproduction of 'water control place and space' provides not just a clearer vision of local families' livelihoods, but simultaneously lends insight into the cultural and material bases of indigenous and *campesino* politics (Salman and Zoomers, 2003; Perreault, 2005; Boelens, 2008b). Variously conceptualized as 'entitlements' (Sen, 1981; Leach et al, 1999), 'strategic assets' (Moser, 1993; Chambers, 1995) or 'capitals and capabilities' (Sen, 1997; Bebbington, 1999), 'livelihood' is a concept more widely used than precisely defined. Despite its ambiguity, the enduring utility of livelihood as an analytical concept rests largely on the way in which it highlights the everyday practices employed by individuals, households and communities to 'make themselves a living using their capabilities and their tangible and intangible assets' (de Haan, 2000, p343).

'Livelihood' implies the making and remaking of social life, maintaining some semblance of continuity and security, and connotes localized and immediate strategies of subsistence aimed at ensuring the stability of life. The concept draws attention to the resources, strategies and relationships that people rely on in order to secure their material needs. Access issues are fundamental to understanding how people utilize what they need to build livelihoods. Access includes the ways in which individuals, households and groups acquire and defend the natural and social resources necessary to pursue livelihoods. Therefore, the notion of access lies at the heart of resource struggles.

Centrally important to social movement politics is the fact that livelihood encompasses not only the material aspects of making a living, but also the cultural and symbolic processes that invest life with meaning (Cohen, 1986; Gelles, 1998, 2002; Bebbington, 2000). In other words, livelihood includes not only making a material-productive living by generating secure resources and fostering social relations, but also making that living meaningful (Nash, 2001). As Bebbington (2004, p177) notes:

> *... there is an inherent relationship between livelihood and culture, and between livelihood and political capacity: livelihoods are in and of themselves meaningful, and a change or loss of livelihood possibilities necessarily implies cultural change ... livelihood decisions are not only economically driven and structured, they are also imbued with cultural and political significance.*

This dual nature of livelihood – meaning and materiality – provides a framework for considering the cultural politics of *campesino* livelihoods and productive strategies. The concept of cultural politics connotes the ways in which everyday struggles are often expressed in 'cultural' referents – for example, through contestations over identity, religion or other modes of social difference (Álvarez et al, 1998; Gelles, 2000).

The idea that culture itself is a battlefield was recognized long ago by Gramsci (1971), who emphasized the overlap of ideological and material forces in political struggle. In the words of Álvarez et al (1998, p7):

> *... cultural contestations are not mere 'by-products' of political struggle but are instead constitutive of the efforts of social movements to redefine the meaning and the limits of the political system itself... Culture is political because meanings are constitutive of processes that, implicitly or explicitly, seek to redefine social power.*

This is particularly the case for many social movements in the Andean region. As has happened in several countries in recent history, some social movements may challenge state authority outright. In most cases, however, rural people's social movements aim to negotiate autonomous, equitable and socially just political spaces, based on ethnic, gender or class-based rights, as seen in the case of Ecuador presented below (compare Bennett et al, 2005; Bustamante et al, 2005; Assies and Gundermann, 2007). In this sense, through their political claims and discursive representations of productive and cultural rights, water user associations of Andean indigenous peoples, *campesinos* and the urban poor seek to reframe the terms of debate regarding rights to water and other resources. They link livelihood claims to calls for more fully democratic and socially just forms of water governance (see, for example, Hendriks, 1998, 2006; Boelens and Gelles, 2005; Perreault, 2005). Their struggles are about water and economic resources to sustain their livelihoods as much as they are about the discourses that support their claims to self-define their own water rules, values, meanings and identities.

Spatial scale

A focus on networks highlights the social relations that connect local actors to broader-scale political, economic, cultural and ecological processes. As Marston (2000) points out, it has become widely accepted that geographical scales are not ontologically given ways of ordering the world, but rather produced through frictions between social practice, environmental processes and structural forces. Spatial scale is, in other words, a mode of understanding how the world is structured by people's action and social and political forces, rather than a natural characteristic of social life. Attendant to this view is the recognition that any given scale requires continual reproduction through social practice and is therefore subject to negotiation and reconfiguration (compare Cox, 1997). Moreover, as a way of understanding and structuring the world, a given scale can only be understood in relation to other scales. The scales of 'community' or 'watershed' or even 'the body' only make sense in relation to the social and environmental context in which they are embedded.

Similarly, a relational view of scalar politics is relevant to the study of social movement networks and the 'local' organizations of which they are composed. As noted by Watts and McCarthy (1997), 'local' movements are never wholly local, but are always produced at least in part by extra-local actors and processes. As the cases detailed in this chapter demonstrate, community-based and regionally based peasant and indigenous organizations in the Andes have been able to advance their claims to water effectively largely because of their connection to multi-scalar networks of development, environmental and human rights organizations. To understand the dynamics of translocal political mobilization on the part of indigenous, *campesino* and urban-poor populations, it is useful to examine closely the organizational networks that connect multiple scales, places and institutions (Swyngedouw, 1997).

While analysis of livelihood practices and the formation of trans-scalar networks can lend valuable insight into the aims and strategies of water user associations, such processes must always be viewed within the context of highly contingent histories of place formation, political struggle and social relations. As is detailed in the cases of Bolivia and Ecuador below, indigenous and *campesino* struggles for water rights are part of broader political and cultural struggles against long histories of social marginalization by white urban elites. Also, water user associations maintain dense networks of relations with other social movement actors who have other demands. Both Bolivia and Ecuador have seen a resurgence of indigenous movements during the past 30 years, whose claims to cultural recognition and rights to lands, territories and resources form a vital political and cultural context for the rise of *campesino* demands for water rights (Yashar, 2005). Equally important has been the more recent resurgence of left-leaning social movements and political parties in both countries, as well as international campaigns against neoliberalism and the privatization of urban water services (Barlow and Clarke, 2002). Thus, as the cases below demonstrate, indigenous and *campesino* struggles for water rights in Bolivia and Ecuador are rooted in and, in turn, contribute to broader processes of political struggles for social justice.

Water wars and multi-scalar strategies in Bolivia

Water wars

> *... and we know that World War III will not be because they demolished the twin towers in the US, not for that, but because of water. And we have to defend it, if we really want to leave our children in good hands; we have to fight for ourselves; there is no reason to fall back... Certainly, the struggle is a big one, not just in our department, or nationwide, but worldwide – this is a worldwide fight. We could say that the centre, the epicentre of the water movement is Cochabamba, Bolivia, South America.*
> (Doña Silvia, water user, Cochabamba, 21 September 2001)

In September 1999, pursuant to an agreement with the World Bank, the Bolivian government turned over the public waterworks company concession of SEMAPA (Servicio Municipal de Agua Potable y Alcantarillado) to the transnational consortium Aguas del Tunari. This consortium was headed by

International Water Ltd, a subsidiary of the US Bechtel Corporation. The concession contract granted a water catchment and service area covering most of Cercado Province, which includes the city of Cochabamba. Several irrigators' organizations and small neighbourhood, community or co-operative water supply systems in the area felt that this jeopardized their rights. The contract also ensured the company a minimum return-on-investment rate of 15 per cent, and provided that user rates would be indexed according to the US consumer price index.

Shortly after the concession contract was signed, Law No 2029 was enacted, entitled Water Supply and Sewerage Service Provision. This granted exclusive rights to provide sanitation services and water sources for large companies and co-operatives (who received 40-year concessions), but only non-exclusive five-year licences for others such as water committees, smaller co-operatives, community systems, etc. The law also empowered the government authority for basic sanitation to allocate water source rights, but without establishing any norms regarding criteria, limitations, rights or obligations. Nevertheless, the same entity was entitled to expropriate water sources whenever it considered that this was necessary to provide water supply and sewerage services.

Reacting to all of this, rural organizations and peri-urban water-user organizations blockaded roads on 4 November 1999, forcing central government authorities to negotiate. These talks yielded an agreement with rural organizations in the Central Valley of Cochabamba, guaranteeing that the Aguas del Tunari consortium would respect their water sources. The whole issue of Law No 2029, however, remained unresolved.

Rural dissatisfaction welled up again, for the first time joined by several city organizations, unhappy about the rate scale that the consortium wanted to apply. They also discovered irregularities about the way in which the contract and the law were approved. This societal movement came to be known as the Departmental Coordination to Defend Water and Life. Government repression of the social mobilization to 'peacefully take over the city of Cochabamba' on 4 February 2000 triggered a major conflict, resulting in injuries, arrests and, finally, the government's promise to revisit the law and the contract, and freeze water rates until an agreement could be reached on the rate schedule. Work groups were organized with government and civil society representatives to prepare proposals and reach agreements on these issues. After several weeks of meetings, negotiations bogged down around the concession contract and rate schedule, although progress toward amending Law No 2029 was promising.

Reacting to the standstill in these talks, the movement decided to pursue a broad popular consultation regarding the increased rates, the concession contract and the privatizing nature of Law No 2029. The organization called for contract rescission and prompt amendment of Law No 2029. Several deadlines were set and finally there was a total shutdown of negotiations on 4 April 2000, which the organization called the 'Final Battle' in the war for water. Mobilizations began with blockades and marches organized by people in their neighbourhoods and communities, in peri-urban and rural areas:

> *The blockade was really the main measure, and showed us that the provinces were fully mobilized. In some parts of town, there was no moving around, and some organizations such as teachers and university students came out and blocked some avenues – there was really a total shutdown.* (Omar Fernández, coordinator leader, April 2001)

The government declared a state of siege, but this only strengthened and tightened the movement. For the two weeks that protests lasted, the conflict spread and ultimately became nationwide:

> *… many people think that the problem is only for Cochabamba, but it's not just Cochabamba's problem, but for all of Bolivia. Many people were saying 'ah! Cochabamba's struggle', and Potosí was daydreaming, so was Tarija, Villamontes, the Chaco – they all took a while to open their eyes.* (Doña Aurora, irrigator from Cochabamba, 21 September 2001)

Faced with a movement that had proven to be unstoppable, the government decided to negotiate, announced the rescission of the contract with the consortium and returned the waterworks to the city. This satisfied the urban sector's demands, but not the peri-urban and rural sectors, who were more concerned about the law that gave away the power to decide about access to their water sources. Therefore, the protestors held firm until finally, at dawn on 15 April, after two weeks of conflict, Parliament modified about half the articles of Law No 2029.

Later, Bechtel sued the Bolivian government for US$25 million in the World Bank's court for investment disputes. A major international campaign eventually got Bechtel to withdraw its suit in 2006. Since then, the 'water war' has been considered one of the most paradigmatic struggles against globalization. For Bolivia, it was the beginning of a major period of change led by societal movements and organizations that had experienced new forms of resistance and struggle.

Networking to defend water rights

In the wake of the 2000 Cochabamba Water War, new pressure was brought on the Bolivian government to create more transparent and participatory processes for water governance. The idea to create a participatory consultative forum to bring together a broad diversity of civil society, state and market actors concerned with water management came from the Mesa Técnica del Agua, a civil society network of organizations. Following the Water War, the

Inter-American Development Bank (IDB) agreed to support the idea. As the primary source of funding for drinking water, sewerage and irrigation, the IDB enjoys unique leverage over Bolivia's water development and made the creation of such a body a condition of continued financing of irrigation and basic sanitation projects. Bowing to this pressure, in 2002 the government created the Inter-Institutional Water Council (CONIAG), a 'stakeholder' forum that included representatives of NGOs, intellectuals, government, the private sector and the major social movements, including the irrigators' movement.

At the same time, Bolivian NGOs, social movements and the Catholic Church increasingly focused their attention on water issues. The Centre for Research and Promotion of the Peasantry (CIPCA), a church-affiliated development NGO, received funding to organize a series of workshops involving irrigators in various departments across the country aimed at drafting a new irrigation legal framework. In organizing these workshops, CIPCA worked closely with the Cochabamba Departmental Irrigators' Federation (FEDECOR), the Andean Centre for the Management and Use of Water (Centro AGUA), and the Bolivian Commission for Integrated Water Management (CGIAB). These relationships illustrate the broad-based associational network that irrigators and their allies have fostered.

A process of nationwide mobilization in defence of *campesino* water rights began with irrigators' workshops. Irrigators felt their livelihoods threatened by the privatization of water, the increasing liberalization of agricultural markets and the state's long history of urban biases in development policy (Healy, 2001). In these meetings, irrigators and their allies sought to form a national-level organization to counter these threats. At the heart of these discussions was the need to clarify the ways in which water management is institutionalized in Bolivia: the organizational forms and legal frameworks, and, crucially, the socio-spatial scales on which they would operate. In part, irrigators' concerns were based on an institutional confusion inherent in Bolivian water law. Peasant irrigators had no clear legal or institutional framework on which to base their claims to water rights, and feared the loss of their customary rights to competing interests that sought water for mining, industrial and urban uses and whose legal rights were more clearly defined. In the workshops, water user leaders from all places told their own stories, clearly manifesting their worries.

> *But now we defend the whole of Cercado Sur – we are going to reach at least 50 organizations to defend our Desaguadero River, and no one can take it away from us – not the companies or anyone who comes along. Since we know the Water Law wants to sell even our own water now, we are never going to let them do that. They'd better know that blood will flow if they want to do business with our Desaguadero River. Our water is for our needs. We tell this to our friends who are coming from far away because they need water, too. We are going to join together, neighbours,*

> *to defend our traditional usage rights and our water that the company wants to take away, which is what is happening these days in our department of Oruro, too.* (Agustin Churata, Workshop in Oruro, Iroco, 30 July–1 August 2001)

In the workshops, irrigators expressed a strong distrust of the government's proposed national-level regulatory agency (*superintendencia*) for water, and argued for an institutional structure that was more localized, decentralized and sensitive to customary uses (*usos y costumbres*). In order to achieve these ends, irrigators knew that they had to organize nationally. In place of a superintendent of water, irrigators proposed the creation of a National Water Service that would include representatives of the state, irrigators' associations and other civil society organizations, and that would have institutional standing similar to the regulatory agencies for mining and electricity. The purpose of the National Water Service would be to coordinate water use and management and mediate water-related conflicts between and within sectors. These efforts culminated in November 2003, when irrigators from across the country met for three days in Cochabamba to hold the First National Irrigators' Congress. Irrigators debated the proposals and established the organization's structure and its official bylaws. Following these discussions, irrigators inaugurated their new organization, the National Association of Irrigators and of Community Drinking Water Systems (ANARESCAPYS). The inclusion of community-based drinking water systems in the association signalled the irrigators' fear that a new state regulation would allow IDB and other development projects to manage and approach drinking water affairs separately from, and with higher priority than, irrigation systems. Irrigators typically make little distinction between domestic and agricultural water supplies and they argued for a more holistic form of management and regulation. Thus, while irrigation remains the central focus of the National Association, irrigators have broadened their concerns to include the integrated management of rural water systems more generally (Perreault, 2005).

The National Irrigators' Association and the 2004 Irrigation Law

Following the inaugural congress, irrigators finalized their statutes and bylaws with the help of a pair of consultants from Bolivian NGOs. The irrigators focused their continued organizing efforts on drinking water and sewerage services. They and their allies recognized that the dangers of commercialization and the threats of privatization were greater in the drinking water and sewage sector than in the irrigation water sector. By focusing on urban water services, irrigators were able to contribute not only their considerable organizational capacity, but also their technical and analytical abilities to the efforts to reverse market-oriented water governance politics in urban areas. Irrigators favoured a model of drinking water and sewerage management that allowed for public participation and local oversight of water resources and services, and contin-

Box 15.1 Points on the agenda of the irrigators' federation

We irrigators do have a strategy, which we have faithfully followed since long before the Water War. In April 2000, we had proposed, as one of our demands in the face of advancing policies to privatize water sources and water supply services, that it was necessary to work on legislation to legally protect our sources and investments. Precisely for that reason we proposed and achieved registration of, and entitlement to, our water sources and the associated assets in the new Law 2066 on water supply. Articles 49 and 50 are the expression of that victory. We immediately got to work on a legal norm to also protect our agricultural uses, continuing along the same lines for reform, establishing the Register in the Irrigation Law and then in its regulations. We have also proposed the creation of irrigation institutions through the National Irrigation Service and Departmental Services (SENARI and SEDERI) because we want to construct a new kind of public institution, decentralized and socially oriented, that can be lasting and generate conditions for investment in irrigation, planning these investments, supporting irrigation projects and granting water rights for irrigation.

(Cochabamba Irrigators' Federation, 2007).

ued to emphasize that, at least in rural areas, irrigation and drinking water systems are intimately related.

These concerns arose against a backdrop of major political confrontation that exposed Bolivia's fragile political situation. In October 2003, in response to a government plan to export liquefied natural gas via Chile to Mexico and the US, protestors blockaded roads throughout the highlands and brought the country to a standstill. In what quickly became known as the Gas War, cities throughout the Andean highlands erupted into alternating rounds of street protests and repression as more than 70 protesters were shot dead by the military and police, and hundreds more were injured. Repressive action served only to intensify protests and eventually forced the resignation of President Gonzalo Sánchez de Lozada, who fled to the US (Kohl and Farthing, 2006; Perreault, 2006).

In the following months, irrigator leaders continued to work with Bolivian intellectuals and NGOs to formulate proposals regarding irrigation water rights, and to gain the support of irrigators, social movements, international development agencies and state actors. This process culminated in October 2004 with the passage of the new irrigation Law No 2878 for Promotion and Assistance of the Irrigation Sector, which was passed unanimously by Congress and signed into law by President Carlos Mesa, the former vice president who succeeded Sánchez de Lozada. The law stipulates that the irrigation institutions in Bolivia, including a National Irrigation Service (SENARI) and Departmental Irrigation Services (SEDERIs), would operate with considerable participation and oversight by irrigators. The establishment of SENARI and the SEDERIs precluded the kind of centralized regulatory agency for irrigation that had been established for other natural resource sectors under the 1994 Law of Capitalization (Zeballos Hurtado and Quiroga Crespo, 2003). The 2004 Irrigation Law codified the proposals formulated by irrigators' workshops that followed the Water War. The law provoked debate over the possibility that non-rights holders would have difficulty getting access to water, and that by allowing irrigator communities to keep 'permanent access and control rights', the state would be unable to promote fairer public water management – a problem that lingers on. Yet, irrigators at last had achieved legal recognition and protection for their *usos y costumbres* regarding water rights and management practices.

The El Alto Water War and its aftermath

Shortly after the passage of the irrigation law, protests intensified in the city of El Alto against the 1997 concession granted to Aguas del Illimani, a subsidiary of the French transnational firm Suez. A year after the Gas War had eclipsed water issues, a new wave of protests erupted over water governance. Led by the Federation of Neighbourhood Organizations–El Alto (FEJUVE–El Alto), these protests took advantage of the weak position of President Carlos Mesa. Personally shaken by the violence of the Gas War that brought him to office, Mesa vowed not to use force to break up street protests. The militant and well-structured social movements then organized street protests and road blockades in El Alto and, to a lesser extent, in La Paz. They called on Mesa to cancel Aguas del Illimani's contract and he refused. Meanwhile, Mesa's authority and legitimacy declined precipitously, and in June 2005 he resigned amid widespread protests and road blockades (Kohl and Farthing, 2006).

The resignation of the Mesa government opened the way for elections in December 2005 that brought Evo Morales and the Movement to Socialism (MAS) to power. The fact that Morales garnered some 54 per cent of the popular vote – unheard-of success for any candidate in Bolivia's fractious political system – is made all the more remarkable given that he is of indigenous-*campesino* heritage, and continued to serve as leader of the coca-growers' union. Shortly after his election, Morales created a Ministry of Water and appointed social movement leader Abel Mamani as its head. Only months before, Mamani, a FEJUVE activist, was in the streets of El Alto coordinating protests against Aguas del Illimani. Principal irrigation leaders from Cochabamba, and NGO activists and intellectuals who had worked closely with irrigators during the mobilization, participated in the MAS government network and helped to ensure that their interests were represented. Since then, however, cracks have emerged in the alliances between irrigators and other social movement organizations. By April 2008, Mamani and the original vice ministers of irrigation and drinking water and sewerage services (both early allies of the irrigators' movement), had resigned and been replaced. Tensions emerged between the irrigators' movement and some water activists and intellectuals, in part owing to disagreements over interpretations of the 2004

Irrigation Law. Thus, while irrigators have made substantial gains, these remain tenuous as the political situation in Bolivia continues to be fraught with conflict.

In summary, irrigators and their allies have influenced the institutionalization and, hence, the governance of water. First, they have had an influential voice in the national debate about water and water rights. Although less visible than the Coordinadora del Agua y la Vida, which led protests in the Cochabamba Water War and its aftermath, they have been more widely accepted as a legitimate partner in negotiations by representatives of the state, bilateral aid agencies and the private sector. This legitimacy is derived, in part, from the fact that their livelihoods are directly dependent upon access to water for irrigation. In addition, the dense associational network that irrigators have fostered with intellectuals, activists, development practitioners and other social movements has broadened their coalition of allies and has given them access to technical expertise beyond the reach of many other rural peoples' movements. Second, irrigators successfully formulated a proposal that was unanimously passed by Congress into law. Again, the dense network of alliances that irrigators fostered led to the success of this proposal. The workshops and the writing of the proposal were facilitated and financed by NGOs aligned with the irrigators' interests. Third, together with a broader coalition of social movements, NGOs and intellectuals working to influence water governance in Bolivia, irrigators are partly responsible for the establishment of the new Ministry of Water. For the first time in the country's history, there is a ministry devoted to the management of water and environmental issues. Fourth, irrigators have influenced the socio-spatial scale of water governance in Bolivia by mobilizing the actors and then institutionalizing management regionally, departmentally and nationally.

At the heart of irrigators' concerns is the protection of customary practices. Not only are *usos y costumbres* central to irrigator livelihoods, but they also encompass the material, quotidian nature of water management and the symbolic aspects of water rights and their relationship to cultural continuity. The irrigators made strategic, effective use of the discursive power attached to *usos y costumbres* and 'Water War' ontologies in order to glue together multi-scale networks and further their interests.[1]

Social mobilization and multi-scalar networks for water rights justice: An Ecuadorian case

> *Secretary General of the National Water Resource Council: 'Why, good heaven, must the power of numbers prevail over the power of the law?!'*
>
> *Leader of the Interjuntas Users' Federation: 'Because it is the only way we ever will be listened to. How often haven't we demonstrated that the situation here is fundamentally appalling and entirely unjust? Mobilization is the only way to change things here!'* (Carlos Oleas, Riobamba, 18 January 2007)

Thousands of water users took to the streets of Riobamba in July 2005 to demand removal of the Water Agency's director and some officials because of their corrupt, profoundly racist treatment of indigenous and other peasant users. 'No more oppression, no more aggression!' shouted the water users after having been discriminated against and cheated by the agency for many years. 'The Riobamba Water Agency director has been dreadful for the people of Calpi. He has handed out water to everyone else but us. In this drought, we have no water for our crops. It is wrong for those upstream to have so much water that nothing is left for us downstream' (community leader from Nuncata-Calpi).[2]

Leaders from other organizations and communities have similar stories, such as the representative of the Irrigation Commission of Bayúshig. After he was put off and humiliated for two years in the agency offices, and despite all the payments of money that he had to gather in the community to pay the officials, the governmental authority never got around to assigning water to his community. 'I have been going there continuously – my dossier now has 30 different forms for the process – and so far, nothing! How can I show my face in my community?! ... they will say I am an inept do-nothing. The agency director makes us look incapable, as if we did nothing for our applications.'

A farmer from Alausí tells how the indigenous are always the last served, while a person with economic and political clout receives preferential treatment in the office and in water allocation. Indignant, with tears in his eyes, the farmer denounces the unjustified charges, time wasted, and the racism and incompetence of officials they have to put up with:

> *Poor people ... folks from Alausí and Tixán, they have to spend one or more nights at the Water Agency. And the bosses, I have seen this with my own eyes ... instead of seeing people, they stroll out to have a fine chicken dinner down the street. Then they wander back to the office and say 'come back some other day' ... what is fair about that?!*

After several meetings with hundreds of leaders, and the agency's insensitivity to their demands, organizations of peasant and indigenous water users joined together as the Provincial Federation of Water Users – Interjuntas Chimborazo – to protest discrimination and pillaging of their water rights. People from every community in Chimborazo Province joined in a demonstration. As the president of one of the many communities put it:

> *We are here from San Antonio de Alao, about 200 of us. We want the agency director and all those corrupt officials out, we want them removed because they are always asking us for $10, for $20. They tell us to come tomorrow, or the next week, and they never give us the water. Out with the corrupt bureaucrats!*

The demonstrators took over agency offices. The president of Interjuntas, Carlos Oleas, expressed the communities' rage and resentment:

> *If we don't peacefully take over the Water Agency today, they will come back tomorrow, and all this movement will have been in vain. We are staying here until the bosses come from Quito. No one can come in this office. They harassed us and this is the outcome. The people, the users, are tired of being mocked and have risen up!*

After occupying the agency office, women in the group performed a ritual purification with medicinal herbs to purge the evil energies that the Water Agency offices had built up after years of cheating, oppression and humiliation. As one woman said:

> *They have made us cry, they have made us suffer in this office when we came here and they refused to see us, they didn't want to do anything for us. In the community, our neighbours would ask us, what's going on, what have you done with our money? We want this to change, and for people to come work here who will give good service to country folk!*

The stories of injustices, resistance by peasant and indigenous families and communities, and the efforts of Interjuntas to achieve equitable water management offer a glimpse at the foundations for new policies of 'integrated water management'. Behind a mask of impressive national and international proclamations, day-to-day practice in Ecuador was full of injustice, conflicts and abuse. Despite rhetorical intentions, facts showed an unfair water tenure structure and policies biased towards politically and economically powerful stakeholders. To examine this, it is not always indispensable to analyse the major conflicts and 'open wars' over water since most often these contradictions become apparent through everyday manipulations and discriminations. In the same way, the protests and strategies for resistance mostly take place in the more 'invisible' spaces.

In Ecuador, amidst the process of privatization and decentralization of governmental institutions since 1994, the state set up a national water management structure without consulting the people. In the central region of the country, the Regional Development Corporation for the Central Highlands (CORSICEN) was created, then renamed as CODERECH, based in Chimborazo Province. As a 'response', in order to enable 'civil society negotiation and counter-power', water users created a representative democratic platform: the above-mentioned provincial Interjuntas organization of Chimborazo. This organization, founded by the inter-community organizations of Guargualá-Licto, Chambo, Cebadas, Chingazo Pungales, Penipe, Químiag and many others, now brings together no less than 280 irrigation and drinking water user organizations, all with mostly indigenous and other small-farmer household constituencies. Their aim is to generate the capacity among users, on multiple scales, to defend water rights and provide bottom-up political and legal advocacy. They also strive to build a capacity for action among water-system user organizations and to foster forums for discussion, debate, consensus-building and training.

Interjuntas deals with conflict management among users and among associated systems, between marginalized groups and large landholders, and between indigenous–rural peoples and the state. Because conflict is rife and often rooted in power plays or the allocation of unreal or non-existent water flows by the Chimborazo Water Agency, Interjuntas has decided to establish a centre for the defence of rights and mediation of conflicts in their federation. There, peasant and indigenous water-user families, many of whom could otherwise not afford a lawyer, find a champion for their demands. There also, leaders – with young committed attorneys – confront the major problems caused by the country's power structures and water policies.

In the Ecuadorian highlands, as in other parts of the Andean countries, a decrease in available water, combined with greater competition for water between local communities and new users, leads to more conflicts and applications to register water rights. However, the Water Agency is short-handed because of the demands of nationwide 'modernization' (i.e. privatization) making things even worse. It continues granting new concessions without determining whether there is enough water, how much is currently being used or what unauthorized intakes have been built.[3] Often, powerful users, such as large landholders or agribusiness companies, are able to obtain new concessions. In the Chimborazo Water Agency's rulings, it is common for bribery, skin colour and political-economic status to play decisive roles. Then, over-allocation of water rights causes new conflicts, many of which are supposed to be resolved by the same Water Agency. The agency, with limited staff capacity and discriminatory practices, is unable to resolve such conflicts and the vicious cycle goes on. In Ecuador, the privatization discourses, policies and practices of the 1990s have dismantled not just state bureaucracy, but also its capacity as a framework to help local government and water user organizations creatively manage their own water affairs (Hendriks et al, 2003; Cremers et al, 2005). Accordingly, until recently, Water Agency action was one of the greatest problems for less wealthy users – a problem that was inherited by the current government. Small farmers said that it is practically a general rule for the agency to ignore its own rulings. Often farmers were mistreated for reasons of class (peasant), ethnic group (indigenous) or gender (when women approach the agency). Up to now, in addition to the unfair results in water rights allocation and the high cost of application processes, the paperwork takes too long and has insecure outcomes.[4]

As was mentioned at the beginning of this section, the water leaders and users belonging to the organizations comprising Interjuntas, tired of the long years of abuse, discrimination, corruption and arbitrary action by the director and legal secretary of the Chimborazo Water Agency, decided to take collective action. Interjuntas, along with the advisers to the Standing Human Rights Commission of Chimborazo, denounced a lengthy series of cases of corruption and discrimination against the province's indigenous–peasant population, carefully documented. On this basis, along with the massive protests, the National Water Resource Council (CNRH) decided to put the director and other Water Agency officials on administrative trial. However, the process ended as a result of post-election political changes as the director was returned to office, claiming that insufficient evidence had been submitted. Interjuntas leaders were asked to propose a better candidate to replace the agency director during the trial process. They presented as a candidate a professional profoundly committed to the cause of small-farmer irrigators. Nevertheless, political chicanery ratified the continuation in office of the previous corrupt director of the Water Agency, despite all the complaints and denouncements by Chimborazo user organizations.

In January 2006, after a second, massive mobilization of thousands of water users and a lengthy occupation of the agency's offices by farmers, and a constant process of surveillance and societal supervision by users' organizations, the indigenous–peasant complaints were finally heeded. The director was transferred and the agency became fairer, more transparent and democratic. To fill the position of Water Agency director, a public competitive selection process was overseen by Interjuntas to ensure transparent decision-making. A new director was appointed, the staff have been renewed and the agency's posts and practices now are being 'monitored' by the users' federation: a preventive and bottom-up process of *veeduría social* (social oversight).

This case is exemplary. Decentralization policies in water management throughout Andean countries offer no guarantee of any improvement in water relations or democratic decision-making. In practice, they entail new dangers, especially for stakeholders with less economic power. As Nina Pacari, indigenous leader and former minister of foreign affairs, puts it: 'Decentralization, according to the Occidental modernization model, may redistribute decision-making geographically, but the way decisions are made remains hegemonic.'[5] Decentralization tends to channel power and decision-making towards the same authorities as always, but on the local government scale.

The great importance of continuing grassroots groups' monitoring of these authorities – by overseeing their practices, and taking collective action and, when necessary, resistance – is fundamental to achieving sufficient political democracy, distributive justice and respect for indigenous–peasant rights in water management. Decentralization and redistribution of water and power for decision-making must necessarily be done with citizen involvement. New water-user network organizations are starting to emerge. In Cotopaxi Province, for example, the inter-user organization Federación de Usuarios de Riego de Cotopaxi (FEDURIC) now brings together some 370 peasant and indigenous water-user organizations, comprising tens of thousands of *minifundio* water-user families. Water quantity and quality rights are defended at the community, inter-community and provincial levels, but also at the national level, for example, through joining hands with the national water civil society platform, Foro Nacional de Recursos Hídricos, and by direct negotiations with the national water resources council and the country's president.

At the moment, a new constitution has been approved, a new water institutional framework has been set up, and a new water law is being discussed which includes important provisions for the legal defence of smallholder, environmental and community water rights. Nevertheless, water user communities know that official policies and legal measures that come from the top will not ensure materialization of such promises. Despite the fact that federations of marginalized water-user families such as FEDURIC and Interjuntas are not 'stable entities' with permanent, well-organized administrative capacity, but vary in their effectiveness and cohesiveness, these alliances have had a powerful impact in responding to collectively felt threats. In the future, whenever necessary, mass mobilizations are likely to be organized to foster the objectives of equitable water distribution, democratic decision-making, and transparency of public investments in water resources management.

Multi-scalar networks and strategies for water rights defence: Social movements, transnational networks and the politics of scale

A focus on the networks through which individuals and organizations link and act helps to shed light on the ways in which movements may be rooted in place, and in turn stretch across space. At the same time, it maintains a focus on the human agency of social movement actors.

Translocal networks for natural resource management and political mobilization are crucial for the reproduction and transmission of 'ideas, commitments, social relationships and institutions' (Bebbington and Kothari, 2006, pp849), and play an important role in the formation of collective social identities (Laurie et al, 2001). This recognition, in turn, draws attention to the fact that spatial scales – that is, the 'levels' of social praxis and interconnectedness (e.g. household, community, watershed, region, nation, global) – are neither fixed nor preordained, but rather are produced, contested and reconfigured through myriad state actions and forms of everyday practice. It is in this regard that Smith's (1996) notion of 'jumping scales' is particularly useful in that it highlights the complex ways in which local actors can bridge scales to broaden spatial, social and political reach (Fox, 1996).

Therefore, scale can be conceptualized as 'radically relational' rather than strictly hierarchical (Swyngedouw, 1997) – a view that destabilizes the common conflation of global-powerful and local-powerless. This insight draws attention to the role of local places as sites of political mobilization and resistance to external domination (Perreault, 2005; Boelens, 2008a), that make possible the construction of 'empowering, inclusive and even emancipatory counter-geographies' (Brenner, 1998, p479). As Swyngedouw (2004, p34) accurately states, 'the mobilization of scalar narratives, scalar politics and scalar practices becomes an integral part of political power struggles and strategies. This propels considerations of scale to the forefront of emancipatory politics.'

Attention must be paid to the social processes through which particular spatial scales are constituted, reinforced or superseded, and the significance – both material and symbolic – that scale holds for everyday practices of social reproduction (Marston, 2000). Indeed, while retaining a view of scales as contingent and shifting, it is also crucial to consider the practices and social relations that render them significant sites of contestation. As McCarthy (2005) points out, however, the scalar basis of social struggles is seldom neatly demarcated or bounded, and reciprocal relationships exist between struggles on one scale and struggles that connect and transcend scale. The challenge, then, is not so much to identify struggles that occur within or between scales, or to determine which scale should be given primacy, but rather to understand the ways in which places and scales are simultaneously reinforced and transcended by social movement networks in the sort of 'rooted translocalism' noted by Katz (2001). As illustrated by the cases discussed in this chapter, Andean social movements have engaged in complex politics of scale, rooted in, but not bound by, place.

In the case of Bolivia's irrigators' movement, irrigators benefited from the formation of a dense associational network of social movements, NGOs, government officials and activists operating in diverse locations and on multiple spatial scales. This network and the movement of which it was a part were elements of a broader struggle against the government's neoliberal policies. By fostering this multi-scalar network, irrigators and their allies were able to extend their political influence. Mobilization began in the Cochabamba Valley and quickly grew to involve irrigators' associations throughout the Andean highlands. The National Irrigators' Association was formed in 2003, and in 2004 a national Irrigation Law favouring the water rights of *campesino* irrigators was passed unanimously by Congress.

Bolivia's irrigators' movement has been remarkably successful in restructuring the institutions of water governance to their benefit. The 2004 Irrigation Law created a National Irrigation Service (SENARI) and Departmental Irrigation Services (SEDERIs), and shortly after his election in December 2005, President Evo Morales created a Ministry of Water (now also including Environment). All of these were recommendations first put forward by the irrigators' movement and its allies. Such successes have not come without difficulties, however. Acrimonious divisions have emerged among Bolivia's water activists; the original minister of water and vice-ministers of irrigation and drinking water services all resigned amidst controversy, and tensions persist. These problems notwithstanding, the Bolivian irrigators' movement has emerged as one of the country's leading social movements, successfully promoting the water rights of *campesino* irrigators.

In a similar manner, Interjuntas, the provincial water users' federation in Chimborazo, Ecuador, brought together a variety of localized water users, including irrigators and drinking water associations, to demand justice, equality and transparency in treatment by the provincial water agency. In Ecuador, water management had been decentralized and increasingly involved private-sector actors. But this neoliberalization of water governance failed to ensure equity or even efficiency in water service. In patterns that reproduced their historical marginalization, indigenous peoples felt that the water service was inferior to that of urban *mestizo* populations, and that the status quo only served to reinforce the privilege of the region's elite. Thus, Interjuntas's demands may be seen as part of a larger struggle on the part of indigenous peoples for social justice and recognition of political and cultural rights. In other words, the everyday struggles for water rights on the part of Interjuntas members are inseparable from broader struggles against discrimination.

Eventually, Interjuntas succeeded in changing the way in which water was administered: water governance was made more transparent, participatory and accountable. But these gains were made only after large-scale mobilization by Interjuntas and allied social movement organizations. As with Bolivia's irrigators' movement, Interjuntas successfully mobilized a network of organizations and activists with which it was allied in order to scale up its actions to extend its political reach. Also in common with the Bolivian movement was the fact that the demands by Interjuntas for water – a basic element of life and livelihood – were inescapably bound up with broader demands for political rights, cultural recognition and social justice.

Conclusions

Water is generally managed locally, but involves spheres reaching to the global level. Users' organizations for irrigation or water supply establish broad strategic networks and extend their struggles at different levels far beyond, but rooted in, rural and indigenous communities. Local collective federations are fundamental physical, cultural, socio-legal and political spaces for manoeuvring not just in the local water world, but also in the broader battlegrounds of water control. Struggles to define and materialize rights are inserted into ongoing battles for economic and symbolic legitimacy and recognition.

In the Andean countries, political acceptance, democracy and accountability processes do not appear out of the blue. They result from collective pressure from below, as in the Ecuadorian case. They occur at different levels, from micro to macro, and include open conflicts and various forms of resistance and mobilization. Therefore, water users' demands to be structurally represented in water policy- and decision-making are not about asking for a gift from policy-makers, politicians, state agencies or NGOs. Rather, they are part of a struggle, in which prevailing unequal societal relationships and power structures provide the backdrop. Democratic representation of local water-user groups, rather than being an officially accredited *starting point* or *neutral* input for water policy- and decision-making processes, may generally be seen as the political *outcome* of grassroots struggles. And materialization of such democratic principles, priorities and procedures is the next step within the policy arena: a process that necessarily requires monitoring and pressure 'from below', which remains a major challenge for organizations.

The process of defining and allocating water rights and deciding on water control is essentially political. It takes place in an arena where actors with conflicting interests negotiate, compete and struggle and use diverse strategies. One of them is to form or take part in societal networks ranging from local to global, enabling social movements to get involved, influence or pose resistance to water policies. Through multi-scalar struggles, the actors enhance their effectiveness in making water rights claims and provide the basis for wider political-legal networks. Those networks can be horizontal alliances with similar grassroots organizations or vertical alliances with regional or national organizations and international institutions. Scaling up water struggles is not just a strategic move but also a direct consequence of the increasingly transnational character of the water conflicts and the transnational background of their adversaries.

In the process of reforming water policies in the Andean region the winners are usually those with the greatest mobilization capacity and the greatest political influence. They may be the traditional elites (hacienda owners, transnational corporations, big local companies, etc.) or may be those such as irrigators in Bolivia whose influence is beginning to emerge as other forms of 'capital' are mobilized. The social struggles crowned by the Water War in Bolivia, enabled changes in water management policies by sparking public debate and mobilizing to challenge the governance model grounded in neoliberal policies and by advocating the recovery of 'public ground' with a new role for the state in managing resources and basic services. These struggles also enabled social organizations to influence and participate in institutional reforms. In this way, mobilization becomes a form of 'monitoring' governmental water policies. How this process will play out in daily life remains to be seen. Examining this calls for an analytical-critical position, not only in relation to 'external, dominant' policies, but also in regard to socio-economic relationships and 'internal' policies and injustices within organizations, federations and movements.

For water users in the Andean region, it will be important to create networks to challenge dominant policies through strategies at levels ranging from day-to-day forms of resistance to global-scale struggles. These struggles at different levels and connecting diverse places influence not only water control practices, but also help to shape local water community identities.

Notes

1 For an elaboration on the strategic use of *usos y costumbres* water discourses and ontologies, see Boelens (2008a, 2009); Bustamante (2006); Laurie et al (2002); Perreault (2005, 2008).

2 Case study based on action research with Interjuntas and WALIR (Dávila and Olazával, 2006; Boelens, 2008a). Interviews were done and broadcast over ERPE popular radio in an ongoing series, from January 2005 up to August 2007.

3 The problem is nationwide. For instance, in Tungurahua Province, the Llutupí system (73 users) was granted 60 litres per second when the actual discharge rate was only 40 litres per second; Alta Fernández (3800 users) has title to 218 litres per second but no more than 80 litres per second are available; Guaguapari (150 users) is entitled to 80 litres per second but the actual flow rate is 40 litres per second; the communities of Toalla (1150 users) have rights to 40 litres per second but availability is limited to 12 litres per second; etc. (Foro RRHH, 2002).
4 Hendriks et al (2003) show how, in mid 2003, there were a total of only 84 technical employees in all provincial offices of these agencies, who had to process a backlog of 21,000 concession applications and other legal demands. To gauge their processing rate, in several decades a total of 35,000 water rights have been processed (Cremers et al, 2005).
5 Pers comm, cited in Boelens (2008a). See also the conclusions of the Water Law and Indigenous Rights (WALIR) Congress on Legal Pluralism, Water Reforms and the Politics of Recognition (November 2006, Cusco); Urteaga and Boelens (2006); Assies and Gundermann (2007).

References

Álvarez, S. E., E. Dagnino and A. Escobar (eds) (1998) *Cultures of Politics, Politics of Cultures: Re-Visioning Latin American Social Movements*, Westview Press, Boulder, CO

Assies, W. and A. Gundermann (eds) (2007) *Movimientos Indígenas y Gobiernos Locales en América Latina,* Universidad Católica del Norte, El Colegio de Michoacán, IWGIA, Santiago

Barlow, M. and T. Clarke (2002) *Blue Gold: The Fight to Stop the Corporate Theft of the World's Water*, The New Press, New York, NY

Bebbington, A. (1999) 'Capitals and capabilities: A framework for analyzing rural livelihoods and poverty alleviation', *World Development*, vol 27, no 12, pp2021–2044

Bebbington, A. (2000) 'Reencountering development: Livelihood transitions and place transformations in the Andes', *Annals of the Association of American Geographers*, vol 90, no 3, pp495–520

Bebbington, A. (2004) 'Livelihood transitions, place transformations: Grounding globalization and modernity', in R. N. Gwynne and C. Kay (eds) *Latin America Transformed: Globalization and Modernity*, Arnold, London, pp173–192

Bebbington, A. and U. Kothari (2006) 'Transnational development networks', *Environment and Planning A*, vol 38, pp849–866

Bennett, V., S. Dávila-Poblete and M. N. Rico (eds) (2005) *Opposing Currents: The Politics of Water and Gender in Latin America*, University of Pittsburgh Press, Pittsburgh, PA

Boelens, R. (2008a) *The Rules of the Game and the Game of the Rules: Normalization and Resistance in Andean Water Control*, Wageningen University, Wageningen, The Netherlands

Boelens, R. (2008b) 'Water rights arenas in the Andes: Upscaling networks to strengthen local water control', *Water Alternatives*, vol 1, no 1, pp48–65

Boelens, R. (2009) 'The politics of water rights disciplining', *Development and Change*, vol 40, no 2, pp307–331

Boelens, R. and P. H. Gelles (2005) 'Cultural politics, communal resistance and identity in Andean irrigation development', *Bulletin of Latin American Research*, vol 24, no 3, pp311–327

Boelens, R. and M. Zwarteveen (2005) 'Prices and politics in Andean water reforms', *Development and Change*, vol 36, no 4, pp735–758

Brenner, N. (1998) 'Between fixity and motion: Accumulation, territorial organization and the historical geography of spatial scales', *Environment and Planning D: Society and Space*, vol 16, pp459–481

Bustamante, R. (2006) 'Pluri-, multi-issues in the reform process: Towards new water legislation in Bolivia', in R. Boelens, M. Chiba and D. Nakashima (eds) *Water and Indigenous Peoples*, WALIR and UNESCO, Paris, pp126–143

Bustamante, R., E. Peredo and M. E. Udaeta (2005) 'Women and the "Water War" in the Cochabamba valleys', in V. Bennett, S. Dávila-Poblete and M. N. Rico (eds) *Opposing Currents: The Politics of Water and Gender in Latin America*, University of Pittsburgh Press, Pittsburgh, PA, pp72–90

Chambers, R. (1995) *Poverty and Livelihoods: Whose Reality Counts?*, Institute for Development Studies, Brighton, UK

Cochabamba Irrigators' Federation (2007) 'Irrigators' and water's social agenda', *Bolpress*, 25 July 2007

Cohen, A. (1986) *Two Dimensional Man: An Essay on the Anthropology of Power and Symbolism in Complex Society*, University of California Press, Berkeley and Los Angeles, CA

Cox, K. (1997) 'Spaces of dependence, space of engagement and the politics of scale, or: Looking for local politics', *Political Geography*, vol 17, no 1, pp1–23

Cremers, L., M. Ooijevaar and R. Boelens (2005) 'Institutional reform in the Andean irrigation sector: Enabling policies for strengthening local rights and water management', *Natural Resources Forum*, vol 29, pp37–50

Dávila, G. and H. Olazával (2006) *De la Mediación a la Movilización Social: Análisis de algunos conflictos por el agua en Chimborazo*, WALIR–Interjuntas, Abya-Yala, Quito, Ecuador

de Haan, L. J. (2000) 'Globalization, localization and sustainable livelihood', *Sociologia Ruralis*, vol 40, no 3, pp339–365

de Vos, H., R. Boelens and R. Bustamante (2006) 'Formal law and local water control in the Andean region: A fiercely contested field', *International Journal of Water Resources Development*, vol 22, no 1, pp37–48

Foro RRHH (2002) *Memoria de la Mesa Nacional: Foro de los Recursos Hídricos*, Quito, Ecuador

Fox, J. (1996) 'How does civil society thicken? The political construction of social capital in Mexico', *World Development*, vol 24, no 6, pp1089–1103

Gelles, P. H. (1998) 'Competing cultural logics: State and "indigenous" models in conflict', in R. Boelens and G. Dávila (eds) *Searching for Equity: Conceptions of Justice and Equity in Peasant Irrigation*, Van Gorcum, Assen, The Netherlands, pp256–267

Gelles, P. H. (2000) *Water and Power in Highland Peru: The Cultural Politics of Irrigation and Development*, Rutgers University Press, New Brunswick, NJ

Gelles, P. H. (2002) 'Cultural politics and local resistance in highland irrigation development', in R. Boelens and P. Hoogendam (eds) *Water Rights and Empowerment*, Van Gorcum, Assen, The Netherlands, pp22–35

Getches, D. H. (2005) 'Defending indigenous water rights with the laws of a dominant culture: The case of the United States', in D. Roth, R. Boelens and M. Zwarteveen (eds) *Liquid Relations: Contested Water Rights and Legal Complexity*, Rutgers University Press, New Brunswick, NJ, and London, pp44–65

Guevara-Gil, A. (2006) 'Official water law versus indigenous and peasant rights in Peru' in R. Boelens, M. Chiba and D. Nakashima (eds) *Water and Indigenous Peoples*, WALIR and UNESCO, Paris, pp126–143

Gramsci, A. (1971) *Selections from the Prison Notebooks*, edited and translated by Q. Hoare and G. N. Smith, International Publishers, New York, NY

Healy, K. (2001) *Llamas, Weavings, and Organic Chocolate: Multicultural Grassroots Development in the Andes and Amazon of Bolivia*, University of Notre Dame Press, South Bend, IN

Hendriks, J. (1998) 'Water as private property: Notes on the case of Chile', in R. Boelens and G. Dávila (eds) *Searching for Equity*, Van Gorcum, Assen, The Netherlands, pp297–310

Hendriks, J. (2002) 'Water rights and strengthening users' organisations: The art of negotiating', in R. Boelens and P. Hoogendam (eds) *Water Rights and Empowerment*, Van Gorcum, Assen, The Netherlands, pp52–74

Hendriks, J. (2006) 'Legislación de aguas y gestión de sistemas hídricos en países de la región andina', in P. Urteaga and R. Boelens (eds) *Derechos Colectivos y Políticas Hídricas en la Región Andina*, Abya-Yala and IEP, Quito, Ecuador and Lima, Peru, pp47–112

Hendriks, J., R. Mejía, H. Olazával, L. Cremers, M. Ooijevaar and P. Palacios (2003) *Análisis de la situación del riego en la República del Ecuador*, Mission Report, WALIR, Quito, Ecuador

Katz, C. (2001) 'Vagabond capitalism and the necessity of social reproduction', *Antipode*, vol 33, no 4, pp709–728

Kohl, B. and L. Farthing (2006) *Impasse in Bolivia: Neoliberal Hegemony and Popular Resistance*, Zed Books, London

Leach, M., R. Mearns and I. Scoones (1999) 'Environment entitlements: Dynamics and institutions in community-based natural resource management', *World Development*, vol 27, no 2, pp225–247

Laurie, N., S. Radcliffe and R. Andonlina (2001) 'Transnational connections and indigenous development in Ecuador and Bolivia', Paper presented at the Latin American Studies Association XXIII International Congress, Washington, DC, 6–8 September

Laurie, N., R. Andolina and S. Radcliffe (2002) 'New exclusions: The consequences of multicultural legislation for water policies in Bolivia', in R. Sieder (ed) *Multiculturalism in Latin America: Indigenous Rights, Diversity and Democracy*, Palgrave, London

Marston, S. A. (2000) 'The social construction of scale', *Progress in Human Geography*, vol 24, no 2, pp219–242

McCarthy, J. (2005) 'Scale, sovereignty, and strategy in environmental governance', *Antipode*, vol 37, no 4, p731–753

Moser, C. O. N. (1993) *Gender Planning and Development: Theory, Practice and Training*, Routledge, London and New York

Nash, J. (2001) *Mayan Visions: The Quest for Autonomy in an Age of Globalization*, Routledge, New York, NY

Perreault, T. (2005) 'State restructuring and the scale politics of rural water governance in Bolivia', *Environment and Planning A*, vol 37, no 2, pp263–284

Perreault, T. (2006) 'From the Guerra del Agua to the Guerra del Gas: Resource governance, popular protest and social justice in Bolivia', *Antipode*, vol 38, no 1, pp150–172

Perreault, T. (2008) 'Custom and contradiction: Rural water governance and the politics of usos y costumbres in Bolivia's irrigator movement', *Annals of the Association of American Geographers*, vol 98, no 4, pp834–854

Roth, D., R. Boelens and M. Zwarteveen (eds) (2005) *Liquid Relations: Contested Water Rights and Legal Complexity*, Rutgers University Press, New Brunswick, NJ, and London

Salman, T. and A. Zoomers (eds) (2003) *Imaging the Andes: Shifting Margins of a Marginal World*, CEDLA Latin America Studies 91, Aksant/CEDLA, Amsterdam, The Netherlands

Sen, A. (1981) *Poverty and Famines: An Essay on Entitlements and Deprivation*, Oxford University Press, Oxford, UK

Sen, A. (1997) 'Editorial: Human capital and human capability', *World Development*, vol 25, no12, pp1959–1961

Smith, N. (1996) 'Spaces of vulnerability: The spaces of flows and the politics of scale', *Critique of Anthropology*, vol 16, no 1, pp63–77

Swyngedouw, E. (1997) 'Neither global nor local: "Glocalization" and the politics of scale', in K. Cox (ed) *Spaces of Globalization: Reasserting the Power of the Local*, Guilford, London, pp137–166

Swyngedouw, E. (2004) 'Globalisation or "glocalisation"? Networks, territories and rescaling', *Cambridge Review of International Affairs*, vol 17, no 1, pp25–48

Urteaga, P. and R. Boelens (eds) (2006) *Derechos Colectivos y Políticas Hídricas en la Región Andina*, Abya-Yala and IEP, Quito and Lima

van der Ploeg, J. D. (2006) *El Futuro Robado: Tierra, Agua y Lucha Campesina*, WALIR, Abya-Yala and IEP, Quito and Lima

van der Ploeg, J. D. (2008) *The New Peasantries: Struggles for Autonomy and Sustainability in an Era of Empire and Globalization*, Earthscan, London

Watts, M. and J. McCarthy (1997) 'Nature as artifice, nature as artefact: development, environment and modernity in the late twentieth century', in R. Lee and J. Wills (eds) *Geographies of Economies*, Arnold Press, London, pp71–86

Yashar, D. (2005) *Contesting Citizenship in Latin America: The Rise of Indigenous Movements and the Postliberal Challenge*, Cambridge University Press, Cambridge, UK

Zeballos Hurtado, H. and E. Quiroga Crespo (2003) *Política y economía de los recursos naturales renovables en Bolivia* [*Politics and Economics of the Renewable Natural Resources in Bolivia*], Plural Editores/COSUDE/SINERARE, La Paz

Zwarteveen, M., D. Roth and R. Boelens (2005) 'Water rights and legal pluralism: Beyond analysis and recognition', in D. Roth, R. Boelens and M. Zwarteveen (eds) *Liquid Relations*, Rutgers University Press, New Brunswick, NJ, pp254–268

16

Federating and Defending: Water, Territory and Extraction in the Andes

Anthony Bebbington, Denise Humphreys Bebbington and Jeffrey Bury[1]

Introduction

Peru, Bolivia and Ecuador have each seen a significant increase in extractive industry activity over the last decade and a half. This raises many questions for communities in the areas in which mining and hydrocarbon activity is occurring. Among these, the implications for water resources and indigenous resource governance are among the most significant. Water questions cause special concerns for populations living downstream of that activity. Extractive industries place pressure on, and introduce new risks for, the quantity and quality of water available to rural communities and urban centres. Extraction also poses threats to the *de jure* and *de facto* rights that communities have historically exercised in order to access and control water resources and to govern the territory in which they reside.

These perceived and actual threats have catalysed organized responses as populations have sought to protect their territory and their ability to govern the natural resources within it. At times these responses have led to conflict and violence. The anatomy of these responses varies from case to case. In some instances, responses are led by federations of communities. In others, they involve much wider alliances of actors who are rural and urban, indigenous and not, national and international. There is also much variability in the relative resilience and effectiveness of these responses. These different patterns of mobilization around extraction have transformed the social and political landscape for water resource management in the region.

In this context, this chapter first gives an overview of recent patterns in the extractive economy in the region and documents certain features of its expansion. Second, a combination of maps and specific examples draws attention to some of the implications that this growth in extraction has for water resources and indigenous territory. Third, the chapter discusses the socio-political responses that have resulted, paying particular attention to the diversity in the ways in which populations have organized themselves to confront these new pressures. Fourth, one case is used to explore in detail the alliances and tensions that exist within these supra-communal mobilizations, the ways in which water and resource governance are argued over, and the difficulty of finding ways to guarantee rights and avoid violence. The conclusions then elaborate upon larger issues of governance that are raised by these patterns of expansion in the extractive economy and the ways in which these interact with processes of grassroots organization, alliance-building and conflict.

Extraction, water and territory

Expanding extraction ...?

> *The majority of mining concessions are on indigenous and* campesino *lands.* (Marlon Santi, president of the CONAIE, cited in Moore, 2009)

On 21 January 2009, journalists estimated that some 12,000 people took to the highways and byways of Ecuador in marches convened by the Confederation of Indigenous Nationalities of Ecuador (CONAIE) to protest against new mining legislation (Moore, 2009).[2] Various arguments suffused this mobilization and other protests that preceded it. Some draw on a nationalist-left rejection of large-scale foreign investment in the resource sector; some are inspired by commitments to human rights; others are based on convictions that extractive industry constitutes an unacceptable invasion of (formal or *de facto*) territories occupied and governed on a day-to-day basis by indigenous and *campesino* communities; and yet others have a diversity of environmental arguments that pertain, above all, to water. East of the Andes, earlier experiences of the serious damage that oil expansion has visited on watercourses, indigenous territories and local organization (e.g. Sawyer, 2004; Fontaine, 2006; Ortiz, 2009) inform concerns that the same will now happen with mineral expansion. In the *sierra*, experiences learned from Peruvian mining, coupled with awareness of the geographical overlaps between mining concessions and watersheds, nourish the fear that water supplies will be adversely affected by mining's needs for large quantities of water, as well as its removal of water-bearing hilltops for open pit extraction. With people believing that the new legislation – by and large endorsed by the industry – paves the way for an

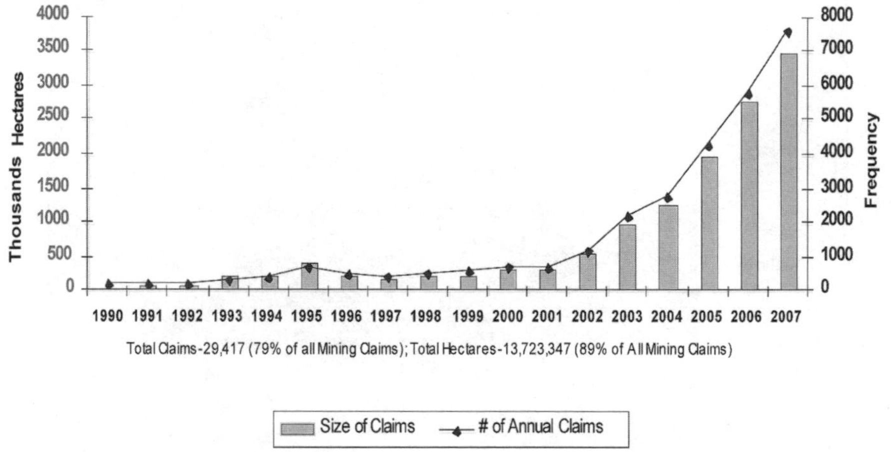

Source: authors' elaboration of government data

Figure 16.1 *The expansion of mining claims in Peru, 1990–2007*

onset of large-scale open-cast mining, such worries about the security of water resources facilitated fierce community mobilization.

This jump in investment has already happened in Peru for both the mining and hydrocarbons sectors, while in Bolivia it has occurred much more evidently in the hydrocarbons sector than in mineral extraction (though there has been a notable increase in mining investment post-2003).[3] These antecedents are known to Ecuadorian indigenous leaders and their allies and are another significant source of unease. When the nature of this increased investment is presented graphically it is not difficult to understand the reasons for the unease. Let us begin with mining. Some experts in Peruvian non-profit research centres have estimated that by 2006 over half of registered peasant communities were affected by mining activity, mostly because of their proximity to, or location within, areas that had been given by the state as mining concessions. Figure 16.1 suggests why such estimates might be well founded. The graph shows that since 2001, the number and area of mining claims made each year have each increased significantly. Figures 16.2 and 16.3 suggest some of the spatial consequences of such growth, taking the example of two different departments in Peru: Cajamarca and Piura. Cajamarca is a consolidated mining department, as is abundantly clear from the spatial extent of claims. Piura (discussed later in more detail) is a new frontier for mining; yet even here a significant share of the surface has been affected by claims to subsurface mineral rights.

The images for hydrocarbon expansion in purely spatial terms are even more dramatic (see Figure 16.4) (Finer et al, 2008). Around two-thirds of Ecuador's Oriente is subdivided into blocks for exploration and exploitation, while in the Peruvian Amazon the figure is closer to three-quarters. (Between 2004 and 2007, the proportion of the area of the basin granted in concessions

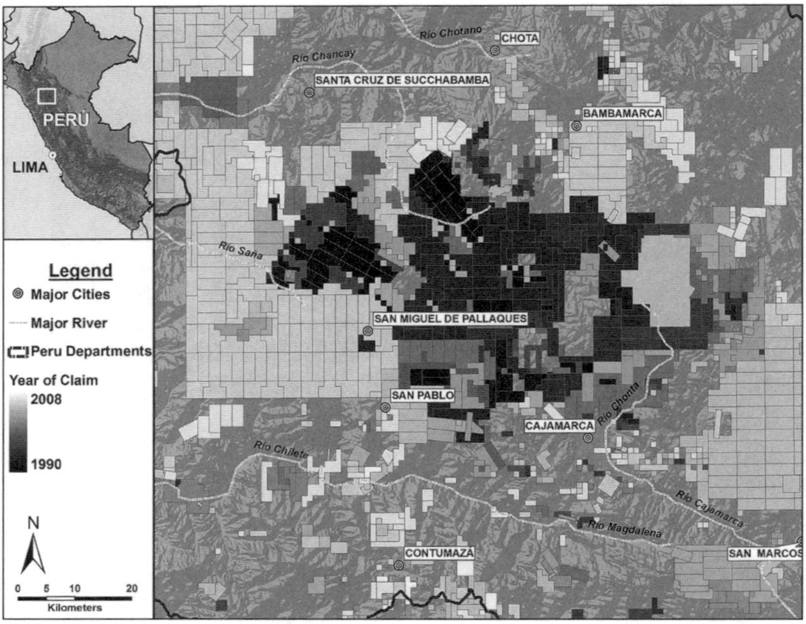

Source: authors' elaboration of government data

Figure 16.2 *The expansion of mining claims in central-south Cajamarca, Peru, 1990–2008*

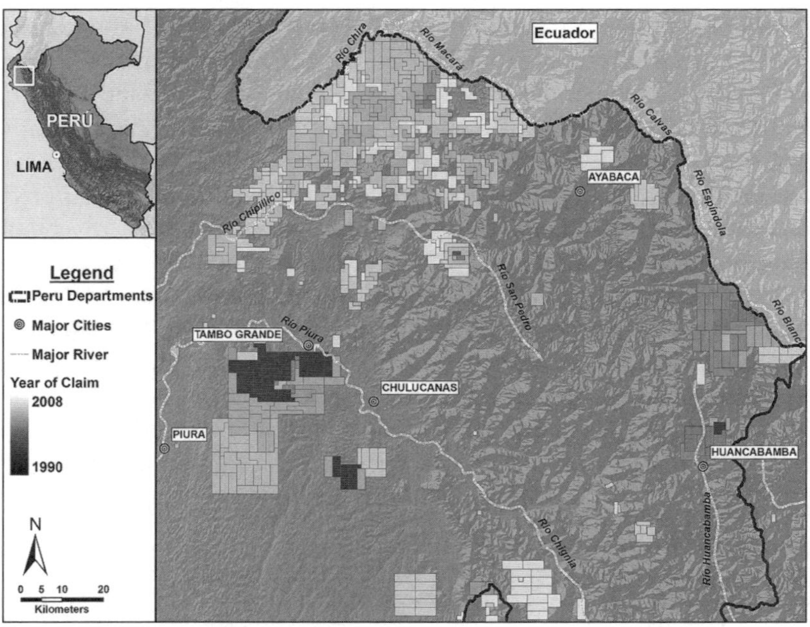

Source: authors' elaboration of government data

Figure 16.3 *The expansion of mining claims in Piura, Peru, 1990–2008*

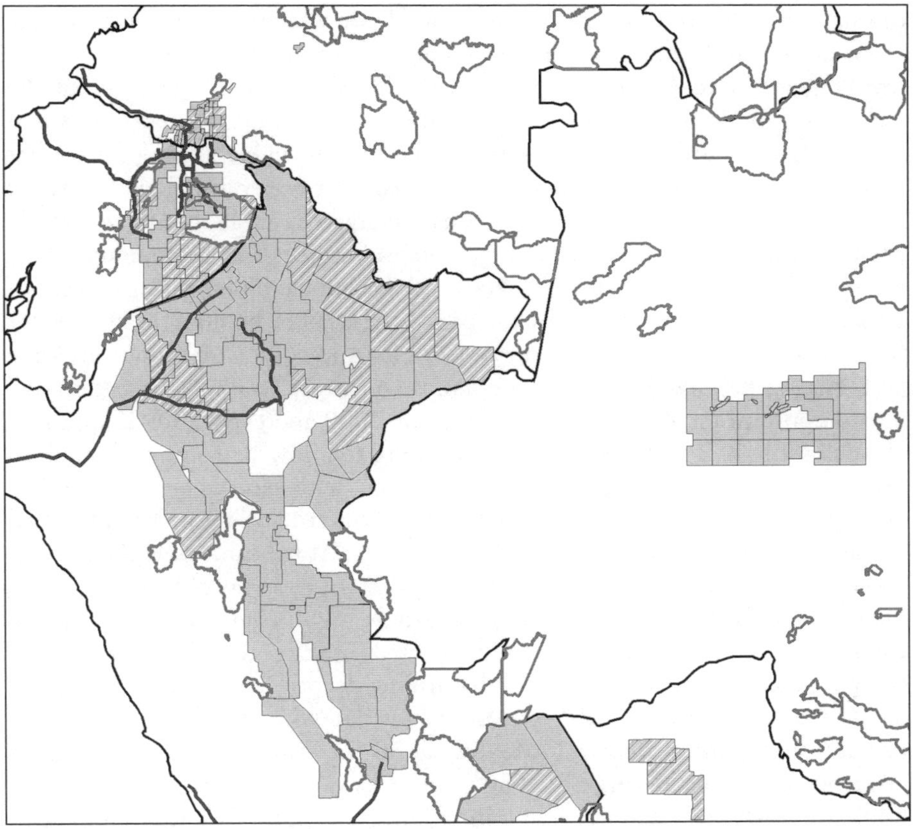

Source: Finer et al (2008), reprinted with permission

Figure 16.4 *Hydrocarbon lots concessioned, contracted or subject to leasing in the Andean–Amazonian region*

increased from around 14 to 70 per cent.) In Bolivia the area is less extensive, though still significant, and one sees the government promoting an important expansion into Norte La Paz, Beni and Pando – all areas with no history of hydrocarbon extraction. This is coupled with a further phenomenon of setting aside vast areas in the highlands of Potosí and Oruro (traditional hard-rock mining departments) as areas of hydrocarbon potential. This process seems set to continue as the current government seeks new sources of gas and oil in departments that are more supportive of its political project than are the eastern lowland departments from which most hydrocarbons are currently extracted.

... and threats to territories and water?

Of course, not all concessions and contracts become mines or oil and gas wells, so one has to be careful before extrapolating too much from such diverse processes. Still, it remains significant that large parts of many of Peru's major drainage basins are now subject to mineral concessions: 41 per cent of the Jequetepeque and Santa River basins, 40 per cent of the Rimac (which provides drinking water to Lima), 26 per cent of the Mantaro, 31 per cent of the Apurimac, and so on (Bebbington and Bury, 2009). Yet more important is that concessions for hard-rock minerals tend to be given in higher-altitude headwater areas. Consequently, the impacts upon or risks for water are likely to extend downstream. In some cases, concessions are given in areas that have already been granted protected status as water sources for cities and communities. One recent case of this is the Aguarague National Park in the Bolivian Department of Tarija. The *serranía* protected by the park is deemed to be *the* source of water for an otherwise dry Chaco, home to Guarani and colonist populations alike and, more generally, the Chaco woodlands constitute South America's second most important intact forest. Yet, during late 2008 and early 2009 it appeared clear that the Bolivian government was going to allow Petrobras of Brazil to begin a large-scale gas exploration programme, with seismic testing affecting large parts of the park. Together with another PetroAndina project, the two initiatives could run the entire length of the park. At the same time the government was allowing a little-known company, Eastern Petrogas of China, to begin operating in the buffer zone of the park, where it will combine environmental clean-up[4] with exploration oriented to rehabilitating abandoned wells and bringing them back into production by drilling deeper. In the highland department of Ancash, Peru, mineral concessions overlap with community-controlled private conservation districts. Meanwhile in the Amazon Basin, Ecuador's government is considering allowing oil extraction in Yasuní National Park, home to indigenous peoples living in voluntary isolation,[5] and a number of hydrocarbon concessions in Peru to overlap with areas previously protected as indigenous territory (as reflected by very influential maps produced during 2007 by the organization Instituto del Bien Común). It is not clear how much either government worries about such overlaps. Indeed, Peru's President Alan García has suggested that organizations raising concerns about such phenomena are little more than unreconstructed communists, and has gone as far as to suggest that the concept of indigenous peoples living in voluntary isolation is a construction of activists determined to block hydrocarbon investments (García, 2007).[6] As appalling as such arguments might be, it is hardly surprising that authorities should seek to undermine the legitimacy of claims for indigenous territory or simply refuse to grant new territory in areas of potential extractive industry expansion. To bestow legal recognition on indigenous territories, and give their governing organizations the power to manage environmental resources, may be among the most serious complications that the expansion of mining and hydrocarbons has to confront.

The two most contentious topics surrounding debates on the implications of these patterns are water and indigenous territorial control. The experience to date has been that extractive industry has had adverse consequences for each. Large stretches of the upper reaches of the Mantaro River in highland Peru have been devastated (Scurrah, 2008), while oil extraction has seriously contaminated Peru's Rio Corrientes (La Torre, 1999; Goldman et al, 2007). Meanwhile, mining companies have diverted watercourses in order to access the water they need for their production, leaving communities with diminished supplies. And in the worst cases, some communities and peoples have been doubly affected by both mining and hydrocarbon extraction. The Weenhayek and Guarani peoples living along the banks of the Rio Pilcomayo in Tarija, Bolivia, have seen declines in water quality and fish stock (fishing is central to the livelihoods of many in these traditional communities) due to pollution from mine tailings in Potosí, at the source of the Pilcomayo.[7] At the same time, they are now confronted by increased exploration for natural gas in their territories and are concerned about the implications that this will have for their water supplies. Indeed, in some sense the Bolivian case suggests just how serious the threats to environmental integrity within areas of indigenous occupation are because, notwithstanding the incumbency of a government committed to indigenous empowerment, the sense remains that resource extraction blessed by the government trumps all other considerations. Referring to a proposed hydrocarbon development in a protected area near an original community territory (TCO), one activist commented:

> *I talked to the lawyer of [the indigenous organization] and he told me that there is no place for opposition because of the government and all its supporters, the colonists, are against the TCOs. Now it is hard to oppose the government and their oil initiatives. It is probably harder than opposing a transnational oil company.* (Pers comm to one of the authors, 24 January 2009)

... and must threats to water be threats to territory?

There is one final point to be made here. It is not merely that these patterns of expansion threaten indigenous territorial integrity and water quality, or that there are company and government strategies that evidently seek to subdivide such territory (as when, in 2008, the Peruvian Executive tried to introduce legislation to reduce the share of a community vote required to allow sale of land to third parties). Perhaps yet more significant is that while one discourse would insist that territory and water *cannot be separated*, the other consistently tries to undermine any such coupling. For many indigenous and even some peasant organizations, if a *territory* exists, then any discussion of the management and ownership of water has to be conducted in relation to that concept of territory. Water management thus becomes inseparably linked to the governance of territory. Conversely, government and companies consistently seek to separate the two. They might do this through natural resource-specific legislation that treats the resource separately from the territory, or through efforts simply to undermine territory. In either case, the effect is the same – to produce water and land as alienable commodities, rather than as parts of territory. This alienation is, of course, essential if extractive industries are to be able to acquire the ancillary resources (water, land) that they need in order to make use of the subsoil rights that the state has given them.

Decoupling territory and water can also be pursued through strategies that undermine vehicles for the governance of territory. In order to be an operable concept, 'territory' has to be governed, and for this to be possible social bodies must exist that can do such governing. Thus, repeatedly one encounters efforts to undermine those organizations that could play such a role. Governments and companies question the legitimacy and representativeness of indigenous organizations, and governments insist on granting concessions and contracts without any real consultation with organizations *even when legislation requires this*. Meanwhile many companies have encouraged the emergence of parallel organizations and conflicts within organizations (Bebbington et al, 2008; Molina, 2009; Ortiz, 2009). The effect of all this (intentional or not) is not merely the weakening of organizations, but also the undermining of the idea of territory as legitimately governable. This again has the effect of decoupling resources from territory, facilitating the transferability of water and land and their transformation into commodities.

Federation and contestation

The fact that expansion of extractive industry has induced organizational responses within indigenous and *campesino* society is hardly surprising given that, as Marlon Santi notes, concessions are disproportionately given to extract resources in the subsurface of lands that are occupied by these peoples. These organizations – new and refashioned – play various roles: they lead protests; they pursue legal and advocacy initiatives; they serve as points of contact for government, companies and international activists; they engage in public debate on extraction, environment and development seeking to project alternative views of these relationships; they try to generate knowledge; and much more. It goes beyond the purpose of this chapter to describe all of the differing forms of organization through which populations seek to give voice to their concerns regarding environment and territory in areas of extraction.[8] Instead, we focus on two levels of organized response – the national and the regional (i.e. sub-national areas larger than municipalities) – and in each instance pay particular attention to responses in which indigenous and community-based federations and confederations occupy a central stage. At each level, however, we suggest that most responses – and certainly the most effective ones – involve some form of alliance and collaboration between such federations and a range of other non-indigenous organizations.

At a national level, the effects of extractive industry have been addressed for the most part by existing indigenous confederations: CONAIE, Confederation of Kichwa Nationalities of Ecuador (ECUARUNARI) and Confederation of Indigenous Nationalities of the Ecuadorian Amazon (CONFENIAE) in Ecuador, National Organization of the Amazon Indigenous People of Peru (AIDESEP) in Peru, and the Confederation of Indigenous

Peoples of the Bolivian Oriente (CIDOB) and National Confederation of Ayllus and Markas of Qullasuyu (CONAMAQ) in Bolivia. These confederations emerged in earlier periods in order to address invasions and injustices visited upon indigenous populations. As their activities evolved, a notion of territory began to suffuse their approaches to the relationship between environment, development and indigenous peoples that occurred sooner in the Amazonian lowlands, later in the highlands. Territory was a concept that simultaneously resonated with ideas of history, resource governance and some degree of autonomy and self-government. Ideas of territory inspired the notion that, for reasons of history and of rights, indigenous peoples should govern environment and development within the spaces that were historically their ancestral lands. This has often led to difficult relationships with hydrocarbon and mining companies as organizations have insisted that this is their territory and they should have the right to determine if, how and when extractive industry should occur. This has often placed them in stark confrontation with a central government that views the subsoil as the dominion of the state, a resource whose use should be determined on the basis of national priorities, not local preferences.

In some instances – such as that of Sarayaku in Ecuador – this has led to years of stand-off between coalitions of national and local federations, on the one hand, and alliances of state and industry, on the other. In other cases, it has elicited more or less explicit strategies on the part of industry and government to undermine and divide these organizations through bribes, special favours or the simple creation of parallel organizations (Sawyer, 2004). Indeed, there is no doubt that at certain times both local and national organizations have been severely weakened as a result of such interventions (Molina, 2009; Ortiz, 2009).

While these arguments initially focused on the extraction of hydrocarbons, they have been re-rehearsed around the more recent moves to expand mining. Given that mining occurs primarily in the highlands, this has brought federations such as ECUARUNARI into debates on extraction when previously they were far less visible in this area. Indeed, in some sense the mining debate has helped to revitalize some of these organizations, allowing them to recover somewhat from divisions previously created by efforts to divide them (as well as by opportunistic and self-serving behaviour on the part of some of their leaders).

The exception to these patterns is the National Confederation of Mine Affected Communities in Peru (CONACAMI). CONACAMI is a relatively young organization, emerging in the late 1990s with the specific purpose of representing communities affected by mining expansion. Although initially created as a coordinator of these different affected groups, it soon assumed the mantle of a confederation with bases in regional federations (Regional Coordinators of Mine Affected Communities or CORECAMIs). It addresses issues of indigenous rights and resource governance, and pushes above all for formal recognition of the right of communities to free, prior and informed consent before mining can proceed. It also lobbies for the establishment of participatory regional and environmental planning processes prior to the granting of any concessions. In pursuit of these ends its tactics have been many and various: from pursuing legal actions and international arbitration in conflicts, to direct action, public protest and participation in processes of dialogue and negotiation. For a young organization CONACAMI has achieved a great deal. Above all, it has helped to make the effects of mining on indigenous territory, resource management and livelihoods a topic of vigorous public debate, and it has gained great national and regional visibility in the process (Bebbington et al, 2008).

Of particular interest is that CONACAMI (along with ECUARUNARI of Ecuador) has also led an initiative to create a coordinating body of national indigenous organizations in the three Andean countries – the Coordinadora Andina de Organizaciones Indígenas. This reflects CONACAMI's own move towards ethnic-based politics, a process in which its leadership began to understand the organization as explicitly indigenous. While this has created tensions within the organization, because a good number of its bases do not share this view, it has also helped to reframe some of the ways in which CONACAMI presents the problem of extraction and the environment in Peru. Increasingly, the organization has viewed this relationship as a territorial problem, rather than a sectoral one. In some sense, then, the experience of extraction has taken CONACAMI along a path leading towards positions already elaborated upon by the other indigenous organizations.

While national confederations such as these have played important roles in making the effects of extraction a topic of national debate and political discourse, their roles in specific local conflicts over mining, water and natural resources, as well as in efforts to manage these resources, have been far more modest. These more localized initiatives for territorial control and the defence and management of natural resources have instead been led by supra-communal coalitions of membership groups, sometimes organized as formal federations, unions or associations, sometimes as less formally networked groupings (Bebbington, 1996).[9] Some such coalitions emphasize issues of defence and resource rights and others lay more emphasis on resource management for production. But whatever the case they play key roles in any initiative to question externally driven resource management initiatives. They do so because they have closer relationships with the local population and because their own 'representativeness' adds legitimacy to their actions. Indeed, it is hard to imagine that a process involving only non-governmental organizations (NGOs) could gain much traction in debates on extraction and water because they would be dismissed as unrepresentative outsiders.

By the same token, however, supra-communal groups acting alone often have little leverage. Limits on their financial resources, comparative experiences, information base, ability to generate and frame knowledge for public debate, links to national and international entities and the like all constrain their capacity to make a difference. Therefore, in most cases of disputes over water and extraction, one encounters diverse alliances. Patterns vary from territory to territory. In some instances, one encounters 'defence fronts' which are semi-formalized, citizen-based representative coalitions. In other cases the situation may be more akin to a working group of different types of organizations, collaborating with one another on the issue at hand, but not bound together in any formal sense. In other cases, coalitions are led and often dominated by large international or internationally connected national conservation organizations whose proximity to power, policy-makers and the media often gives them the capacity to exert important leverage.[10]

One important dimension of such alliances is how far they bridge rural–urban differences. Water issues – affecting as they do medium- and large-scale settlements as much as rural communities – have the potential to link rural and urban mobilization in ways that purely territorial issues do not, because territorial issues are of very little urban interest. Indeed, in many cases where grassroots action has forced new debates on natural resource and water extraction, it has been because the action was based on an articulation of countryside and town (and often of local government) which gave more leverage than purely rural action and organization ever could. Examples here include promising efforts to prevent mining in the *canton* of Cotacachi in the Ecuadorian sierra, and to protect Mount Quilish in Cajamarca from mineral development on the grounds that it is the departmental capital's main source of water (Bebbington et al, 2008).

That said, to sustain such coalitions is as difficult as it is important. The differences between urban and rural priorities in other domains, as well as ethnic and class differences, present a real challenge to those leading such coalitions – and in particular to leaders of supra-communal organizations. Indeed, generally, differences within organizations and more so within social movements are a recurrent source of cleavage and weakness. Within rural supra-communal federations and associations one encounters differences in economic interests, political party affiliation, environmental endowments, etc. In addition, there are often deeply embedded local disputes among families and neighbouring communities over boundaries, the use of common property and so on. Likewise, one can encounter more distance than one would hope between leaders and bases, with leaders sometimes coming from local power groups who seek to use the supra-communal organization for their own ends as much as for collective purposes. Companies and governments are, of course, well aware of such weaknesses and cleavages, and are not averse to cultivating and deepening them. Nor is opportunistic behaviour in short supply. In disputes over extraction, natural resources and territory, federation leaders have sometimes accepted payments from the extractive industry, the end result being weakened organizations, split coalitions and extractive industry success in securing control of the water and other natural resources. We have reported on cases of this in Cajamarca (Bebbington et al, 2008). Ortiz (2009) notes something similar among indigenous federations in lowland Ecuador, as does Molina (2009) in Bolivia. While each of these three accounts remains highly sympathetic to such forms of organization, they also draw attention to their potential fissures and self-destructive tendencies.

Mining, water and rural organization in Piura, Peru

While Piura is not yet a mining region, over the course of the last decade two of Latin America's most iconic conflicts over water and mining have occurred in the department. The conflicts reveal much about the roles played by supra-communal organizations in affecting relationships between mining, water and territory, as well as about the potentials and limits of such organizations as they play these roles. The first conflict occurred in an irrigated agricultural export-oriented valley centred on the town of Tambogrande, the second in the far poorer highlands of Huancabamba and Ayabaca (see Figure 16.3).

During the late 1990s and early 2000s a junior Canadian company, Manhattan Minerals Corporation, sought to bring a gold mine to approval in the town of Tambogrande and surrounding areas.[11] The conflict that ensued was especially acute because it pitched mining directly against human settlement and export agriculture. The mine would have required resettlement of much of the town and parts of the rural population, and would have damaged a zone of successful, export-oriented, high-value irrigated agriculture that had emerged as a result, *inter alia*, of earlier World Bank investments in water supply and management (Bruno Revesz, pers comm; see also Cleaves and Scurrah, 1980). The case thus lent itself to clear dichotomies: a private investment undermining an earlier successful public investment; a mineral development landscape undermining an export-oriented landscape that appeared both more economically valuable and more inclusive in employment terms; a mine site displacing people from their homes; and an example of a territory being submitted to contradictory development paths, each proposed at different times by the same World Bank group.

The conflict escalated quickly and became violent. The main leader of the coalition opposing the mine was murdered, and further escalation seemed only to have been avoided through the implementation of a local referendum to determine the future of mining in the area. This referendum was organized by the local government and supported by national and international NGOs. It enjoyed a turnout of some 27,015 people, roughly 73 per cent of eligible registered voters. The result was that 93.85 per cent voted against mining activity in Tambogrande and 1.98 per cent in favour, the balance being abstentions, spoiled ballots, etc. (Portugal Mendoza, 2005). This model of a public referendum on mining has since been proposed and used by social movements and activists in Argentina and Guatemala as part of their efforts to halt mining projects.

The fact that contemporary land use in Tambogrande is still dominated by agriculture and the prior urban settlement grid, and not by an expanding mining sector, can only be explained by the emergence of a social movement that culminated in this public consultation. But how did this movement emerge and achieve what it did? At the core of its success was the fact that it grew from, and succeeded in building, bridges across a number of distinct social groups in the region. In particular, it built bridges across rural and urban groups, and also among both small and large export-oriented farmers, all of whom had much to lose. In the process it also brought local government into the movement, an involvement that was critical as it was this government that had the strength to convene the referendum. Just as importantly, though, this movement built links with actors in Lima and beyond. As the process unfolded, activists in Tambogrande gained the support of a group of Lima-based advisers (organizations and individuals) who operated as a technical committee to Tambogrande's social movement. The committee provided information, helped with the studies that argued that Tambogrande would be more economically productive as an agrarian landscape than as a mining area, helped with legal issues and assisted in designing the referendum. They also played important roles in making links to international actors in North America and Europe, not only for advice but also for financial support, especially to fund the referendum. Absent any one of these groups, Tambogrande's current landscape would probably be an emerging mineral landscape.

Just as the stand-off between the population of Tambogrande and Manhattan Minerals was coming to an end, another conflict began to unfold in the highlands of Piura in the provinces of Ayabaca and Huancabamba. A subsidiary of the then UK-based mining company Monterrico Metals (now largely in the hands of a Chinese consortium, with minority shares held by a South Korean company) began efforts to initiate exploration. It had acquired concessions from other companies in the belief that beneath the soil lay a world-class copper-molybdenum deposit. The existence of this deposit, referred to initially as the Majaz project, and subsequently as Río Blanco, has since been confirmed. It is part of a far larger copper belt stretching from northern Peru into the southern and eastern provinces of Ecuador.

Monterrico's concessions existed within the territory of two formally constituted communities, Segunda y Cajas and Yanta. They are unusually large, including both rural and semi-urban settlements, and are more *campesino* than indigenous in their cultural and organizational form. Indeed, the scale of the communities means that the levels of organization at which most routine governance is conducted are the *ronda campesina* and the local settlement. The *ronda campesina* is an organizational form that has been particularly strong in neighbouring Cajamarca (Starn, 1999). It emerged as a community-based mechanism for policing against cattle rustling, but over time has become a more general vehicle for the administration of local justice and the governance of the public sphere, including during Peru's internal war (Starn et al, 1996). Increasingly, it has become involved in the regulation of everyday life (Diez, 2007). In practice, the *rondas* assume many of the functions of the community at a local level. These *rondas* then exist in federated form at the level of the provinces, a level at which they exercise significant social and political influence, intersecting *inter alia* with municipal governance processes.

The existence of legally recognized communities means that the expansion of mining activity in the area has to be in accordance with legislation specifically related to the *comunidad campesina*. This presents the company with the need to gain agreement from two-thirds of community members in a notarized community assembly before it can go ahead with activities. This in itself is a complex task, given the size of the community and the company's own determination to move quickly. These factors, coupled with the company's poor understanding of local dynamics and its willingness to cut legal corners, led the company to proceed with activities without securing this agreement. The Ministry of Energy and Mines was complicit in this effort (Defensoría del Pueblo, 2006a, 2006b; Red Muqui, 2009).

For reasons that go beyond the scope of this chapter, this led to a situation of increasing tension and, ultimately, violence in which two people were killed, several maimed and injured, and in which levels of everyday insecurity increased (Revesz and Diez, 2006; Bebbington et al, 2007). In these confrontations, the *rondas campesinas* led efforts to prevent the mine from going ahead. In the process, however, many more actors also became involved in a broad, albeit uneasy, coalition questioning the *modus operandi* of Monterrico and the desirability of the proposed mine. Local mayors aligned themselves with these efforts, as did many of the departmental and national organizations who had been involved in the Tambogrande conflict. Meanwhile, on the pro-mining side, a similar convergence occurred notwithstanding the fact that a number of private and government actors in the sector had certain reservations about Monterrico's behaviour. This was essentially a replay of the Tambogrande struggle with the mining sector determined to use the project to open up Piura to mining, and the *campesino* and social movement sector equally determined to stop this from occurring.

Once again, different arguments were mixed together in the efforts of the federations and their allies to stop the mine. One senses that the determination to protect 'territory' and the power of the local population to govern it has been a key motivation. Of almost as much importance – and of more importance in the public explanations of the reason for protest – have been concerns about the implications of the proposed mining project for water resources. Generally these arguments are pitched more at a regional level than at a community level. Stated concerns include:

- the fear that contamination from the mine would run into local rivers that pass through areas of certified smallholder organic coffee production and so lead to the loss of certification;
- the fear that seepage from tailings would lead to local contamination;
- the belief that the mine would use large quantities of water that would not only diminish local supply within the provinces, but also compromise water running to the western arid lowlands where export agriculture depends on irrigation water from the highlands; and
- the belief that open-pit mining and the removal of hilltops would likewise compromise water quantity.

Casting arguments at a regional level has also been important (and in some measure a conscious strategy) for building links beyond the locality. This is important in facilitating alliances with other actors who might otherwise have had little interest in a conflict occurring in distant *comunidades campesinas*. The mining company has insisted that the *rondas* and, above all, the activists and organizations who advise them simply have their hydrology wrong and do not understand either the ways in which modern mining can avoid contamination or the structure of the regional drainage system. The company insists that its activities could not possibly affect water running to feed coastal agriculture and towns. It has been supported in this assertion by nationally eminent ecologists, one of whom has subsequently been named Peru's first minister of the environment.

Given that the company involved was registered in the UK and because of the violence and claims of human rights abuses in this case, British solidarity groups also became involved in this conflict (in particular, the Peru Support Group (PSG) and Oxfam-GB). One effect of this was that the PSG organized a delegation to document the case.[12] The delegation's report argued that while the company was probably correct on the issue of drainage basins, the proposed project raised a series of important issues for water quantity and quality. Acid mine drainage effects had the potential to be significant; tailings and dams would be located in tectonically active areas; high levels of rainfall significantly increased the possibility of catastrophic slumping of tailings; the potential for contaminated surface runoff and subsurface drainage was high; and the mine was proposing technologies that had not been previously used in Peru (Bebbington et al, 2007; Bebbington and Williams, 2008). The report also argued that while the single mine might not affect water running to the Pacific, there was evidence to suggest that the company intended to develop a far larger mining district which could affect west-flowing water. Red Muqui, the national network of NGOs working on mining, human rights, environment and development, has since referred to the PSG report as a *hito importante* (important milestone) in the conflict (Red Muqui, 2009) in that it brought together a large body of information in one third-party report, played an important role in making the case the object of more visible public debate and suggested that the unit of discussion should not be the single mine, but rather the mining district.

While the report was not a product of the federation of *rondas campesinas*, it could not have been produced without their existence or the NGOs and church groups supporting them. In this sense it was a product of the sort of alliance discussed in the prior section. It was grounded in the existence of local federations, but neither limited to nor entirely controlled by these federations. Similarly, the subsequent evolution of the conflict hinged on the existence of these federations. In some measure this evolution followed the path charted by Tambogrande in that within a year of the report, the alliance resisting mining in the region had also held a local referendum. While the technical organization of the referendum was in large measure the work of the national and international organizations within this alliance, the information provision and mobilization required to inform rural people of the referendum and get them to participate in it was the preserve of the federation of *rondas* and local authorities.

The referendum concluded in a 92 per cent vote against the Rio Blanco Project. Although legally non-binding and immediately dismissed as not relevant by the government (Burneo, 2008), its effects have been significant in that, at the time of writing, the final proposal for the mining project has still not been presented to the government for approval. As the recent global economic crisis has begun to influence the mining sector, the project's new owners have intimated that its onset may be delayed yet further (*Reuters*, 2009). The delay would give federations and their allies more time to work through their arguments and their proposals for development alternatives for the region. This is the biggest challenge because the political force of the federations' case is lost if they are unable to link their arguments about water and territory to a clear case as to why non-mining use of the land might be more effective in fostering regional growth and reducing poverty in the area. If the activists and federations lack well-grounded proposals for local development alternatives, the pro-mining sector will continue to argue that they want to keep people poor and to frustrate economic growth in Peru. Whether justified or not, such arguments resonate widely.[13]

Conclusions

This chapter is written at a time of worldwide financial crisis that will cause rates of investment in extractive industries to be far lower than they have been. Mineral, oil and gas prices have all fallen dramatically and a number of companies will be reviewing their projects and putting some on hold. This may mean that the immediacy of pressures on water resources noted earlier will diminish; but it is unlikely that this will be long lasting.

The maps presented in Figures 16.2 to 16.4 suggest that the most important challenge facing indigenous communities who want to protect their water resources and territory is finding ways to ensure that regulation of extractive industries and water resource management are treated jointly. In many of the interviews, indigenous and *campesino* leaders suggested that current systems that prioritize extraction are attributable to ministers and presidents who believe that producing gas and copper are more important than protecting water resources.

The existence of so much conflict and mobilization around the relationships between extraction, water and territory, and the need for indigenous and *campesino* groups to federate and build alliances in order to continue living in their own territories and practising their own livelihoods, reflects the failure of regulation. Conflict occurs not because of the action of *cuatro pelagatos* (four nobodies) as Ecuador's President Rafael Correa suggests (Moore, 2009), or because rural leaders are terrorists and environmentalists or former communists turned green, as Peru's President Alan García and some Peruvian authorities have suggested (García, 2007). Instead, it occurs because communities ultimately conclude that formal democratic and bureaucratic procedures do not allow them to express their concerns, far less receive any response to them. It is also because political parties fail to embrace these citizen concerns. Protest, federation and mobilization are default phenomena, a consequence of weak, inoperative or corrupt institutions. And lest it be assumed that having an Aymara president necessarily resolves this situation, we need only refer back to one of the cases noted earlier in the chapter. The Bolivian government was apparently willing to endorse massive gas exploration by a transnational firm inside a protected area that would infringe on claimed indigenous territory and threaten the main source of water for rural and urban settlements in an arid and ecologically fragile region. Notwithstanding claims that one can trust the state because its approach to the subsoil will be informed by a respect for Pachamama,[14] it would seem that the part of Pachamama that has been deemed most worthy of respect is that which is best able to generate resources for public investment and macroeconomic stability.

To change the rules of the game as well as the culture that underlies both the practice of government and the exercise of executive power is a huge challenge. Grassroots federation is an important element of responding to this challenge and protecting the natural resource base and livelihoods of rural and indigenous communities. However, it is far from being a sufficient vehicle for changing the rules of the game and the ways in which this game is played such that indigenous and citizen rights of access to water, resources and self-governance are guaranteed. Even constitutional change may not be enough. The protests occurring in Ecuador respond to a new law that indigenous and other organizations claim allows forms of mining that will threaten the sanctity of water resources. Yet, this occurred just four months after the country approved a constitution that is supposed to give nature enforceable rights, and recognize access to water as a basic right of citizenship. Whether these organizations are right or wrong in their interpretation of the law, the determination of the Ecuadorian executive to push the legislation through without consultation, while writing off CONAIE as irrelevant, suggests that constitutional change, no matter how apparently progressive, will do little to protect water rights while political power continues to be inflected with authoritarianism and while nature continues to be automatically subsumed to 'economy' in the name of fiscal imperative.

The issue of protest and rural social mobilization around water and natural resources challenges state formation and public culture. The apparent slippage between constitutional assertion and executive fiat in Ecuador, shows that the institutions of the state are not sufficiently strong and independent to protect rules from executive infringement. The cultural challenge is to get the public to demand a strengthening of these institutions and greater respect for the values underlying community rights to resources.

There is cause for optimism that comes from Peru which might be considered the least favourable of the three countries for the protection of indigenous rights and water resources in the face of extractive industry. One of the jewels of the Peruvian state is the Ombudsman's Office, the *Defensoría del Pueblo*. The *Defensoría* is charged with ensuring that government action and policy do not infringe citizenship rights as enshrined in Peruvian law and commitments to international treaties. The *Defensoría* has played a critical role in a series of conflicts around indigenous and rural people, water and extractive industry much to the annoyance of the government. It has been the one part of the state from which federations and broader alliances have been able to elicit responses and in which they have had confidence in a range of conflicts over extraction and water. That said, the *Defensoría*'s credibility must be understood in the context of the federations and alliances, for it is partially the product of them. First, the ability of the *Defensoría* to do this work has been made possible because of supplementary funding received from donor organizations that have also supported some of the alliances and mobilizations discussed earlier. Second, the disposition of the *Defensoría* to act on these issues has emerged because of the nature of its staff, who are increasingly composed of young well-trained lawyers who are committed to the ideals of their profession rather than to partisan projects, and a sub-group of whom are particularly committed to the issues raised by socio-environmental conflicts. Third, the *Defensoría* must act on issues when requested to do so. In many cases formal requests made by federations and their allies to investigate cases are the triggers that have brought the *Defensoría*'s professional capacities and public legitimacy to bear on these issues of extraction and water rights.

The case of the *Defensoría del Pueblo* in Peru illustrates the vital role that competent, autonomous and legitimate public institutions have to play in guaranteeing and protecting community rights, including the rights of access to water, territory and a healthy environment. It also reminds us that the challenge of institutional change, of more rational regulation and of state-making goes beyond what federations and their allies can achieve through their particular actions around water and extraction. But such actors do have a critically important role in bringing a more rational rights-oriented state into being.

Notes

1 This chapter is based on research supported by grants from the Economic and Social Research Council (ESRC) (RES-051-27-0191); ESRC Department for International Development (RES-167-25-0170); the National Science Foundation (BCS-0002347); and an ESRC/SSRC Fellowship, for which we are extremely grateful. The ESRC supports the programmes Territories, Conflicts and Development (www.sed.manchester.ac.uk/research/andes) and Social Movements and Poverty (see www.sed.manchester.ac.uk/research/socialmovements), each based at the University of Manchester.

2 Journalist Jen Moore is a collaborator in the above-mentioned Territories, Conflicts and Development research programme.

3 Increase in international investment in mining in Bolivia has been on a far more modest scale – in part, perhaps, because there is a strong co-operative presence in

the sector that owns many of the concessions (this was part of the arrangements when mines belonging to the state mining company, COMIBOL, were closed in the mid 1980s).

4 This area has been seriously contaminated because in periods prior to the park's creation the state oil company had operated here. When it departed it left well heads open with a steady flow of oil into the soil and watercourses.

5 This is an almost iconic case. On election, Ecuador's president Rafael Correa went to the international community asking them to pay Ecuador compensation for leaving Yasuní's oil in the ground. This discourse continues to the present; but at the same time the government talks with companies about possible development of these fields.

6 In the first of three articles in the leading national newspaper *El Comercio*, bemoaning the '*perro del hortelano*' blocking all forms of development in Peru, he wrote 'against oil, they [activists] have created the figure of the non-contact, forest-dwelling native' (García, 2007).

7 A mine owned by the twice, and now disgraced, former President Gonzalo Sánchez de Lozada.

8 For cases see: Cidse/ALAI (2009); Broederlijk-Delen/ALAI (2008); Sawyer (2004); Bebbington (2007); Ortiz (2009); Molina (2009).

9 Libia Grueso, Colombian activist, has referred to these as coalitions of *los dolidos* (the hurt ones), those persons directly affected and harmed and without whom there could be no effective campaign to confront the adverse effects of extraction.

10 Although at the same time these close links to power and money have often led these organizations to assume more politically cautious positions for which they have been severely criticized at times (Chapin, 2004).

11 The following three paragraphs draw on Bebbington (2008).

12 Anthony Bebbington led this delegation, which also included a hydrologist, an anthropologist, a leading international journalist and a British member of parliament (see Bebbington et al, 2007).

13 The Ecuadorian case is an illustration of how complex the issues at stake are. Following the discovery of oil in the 1960s, Ecuador built up social and development programmes on the basis of income from hydrocarbon extraction, and subsequently from loans leveraged against anticipated future income from oil. This was, then, a transfer from natural resource extraction to social investment. As oil income has declined the government needs alternative sources to fund social programmes. In the process it has looked once again to resource extraction, this time mining, to fill this gap – social programmes are to be funded through economic activity that threatens water resources on or around indigenous territory. This elicits protest. Yet it is also the case that the same indigenous confederations protesting against extraction also protest against efforts to increase the retail price of gasoline *and* against cutbacks in social programmes. In this sense, indigenous confederations are also far from consistent in these debates about exactly how Ecuador wants to manage its resources. Similar slippages are also apparent in Bolivia, where actors who played an important part in the *Guerra del Gas* in 2003 were by 2005 part of a governing coalition. Notwithstanding their rhetoric of nationalism, that coalition is now allocating contracts to transnational hydrocarbons companies with less than responsible histories of behaviour in indigenous territories elsewhere in the world.

14 Claims made to the first author by Bolivian MP and President of the Congressional Committee for Constitutional Affairs Renee Martínez in response to a question during a public forum at the conference Latin America 2008: Making a Better World Possible, 6 December 2008, London.

References

Bebbington, A. (1996) 'Organizations and intensifications: Small farmer federations, rural livelihoods and agricultural technology in the Andes and Amazonia', *World Development*, vol 24, no 7, pp1161–1178

Bebbington, A. (ed) (2007) *Minería, movimientos sociales y respuestas campesinas: Una ecología política de transformaciones territoriales*, Instituto de Estudios Peruanos, Lima, Peru

Bebbington, A. (2008) 'Social dimensions of rural resource sustainability: Lessons from extractive industry conflicts in Latin America', in G. Cornia and J. Riddell (eds) *Toward a Vision of Land in 2015: International Perspectives*, Lincoln Institute, Boston, MA pp237–273

Bebbington, A. and Bury, J. (2009) 'Institutional challenges for mining and sustainability in Peru', *Proceedings of the National Academy of Sciences*, vol 106, pp17296–17301

Bebbington, A. and M. Williams (2008) 'Water and mining conflicts in Peru', *Mountain Research and Development*, vol 28, no 3/4, pp190–195

Bebbington, A., M. Connarty, W. Coxshall, H. O'Shaughnessy and M. Williams (2007) *Mining and Development in Peru, with Special Reference to the Rio Blanco Project, Piura*, Peru Support Group, Lima, Peru

Bebbington, A., M. Scurrah and C. Bielich (2008) *Mapping Current Peruvian Social Movements: Social Movements and Poverty Project*, Manchester, www.sed.manchester.ac.uk/research/socialmovements/publications

Broederlijk-Delen/ALAI (2008) *Territorios y recursos naturales: El saqueo versus el buen vivir*, Broederlijk-Delen/ALAI, Quito, Ecuador

Burneo, M. L. (2008) 'Por qué NO? Lecciones de una consulta vecina', CEPES, www.ser.org.pe/index.php?option=com_content&task=view&id=424&Itemid=112

Chapin, M. (2004) 'A challenge to conservationists', *WorldWatch Magazine*, November/December, pp17–31

Cidse/ALAI (2009) *América Latina: Riqueza privada, pobreza pública*, CIDSE-ALAI, Quito, Ecuador

Cleaves, P. S. and M. J. Scurrah (1980) *Agriculture, Bureaucracy and Military Government in Peru*, Cornell University Press, Ithaca, NY

Defensoría del Pueblo (2006a) *Oficio No 0178-2006-DP/ASPMA*, Defensoría del Pueblo, Lima, Peru

Defensoría del Pueblo (2006b) *Informe No 001-2006/ASPMA-MA*, Defensoría del Pueblo, Lima, Peru

Diez, A. (2007) 'Ronderos y Alcaldes en el conflicto minero de Río Blanco en Piura, Perú', in J. Bengoa (ed) *Movimientos sociales y desarrollo territorial rural en América Latina*, Editorial Catalonia, Santiago, pp432–444

Finer, M., C. N. Jenkins, S. L. Pimm, B. Keane and C. Ross (2008) 'Oil and gas projects in the western Amazon: Threats to wilderness, biodiversity, and indigenous peoples', *PLoS ONE*, vol 3, no 8, www.plosone.org/article/info:doi%2F10.1371%2Fjournal.pone.0002932

Fontaine, G. (ed) (2006) *Petróleo y Desarrollo Sostenible en el Ecuador: 3. Las Ganancias y Perdidas*, FLACSO, Quito, Ecuador

García, A. (2007) 'El Perro del Hortelano', *El Comercio*, 28 October 2007, Lima, Peru

Goldman, E., L. La Torre and M. L. Ramos (eds) (2007) *Un legado del daño: Occidental Petroleum en Territorio Indígena de la Amazonía Peruana*, Earth Rights International, Racimos de Ungurahui, Amazon Watch and WWF Perú, Lima, Peru

La Torre, L. (1999) *All We Want Is To Live in Peace: Lessons Learned from Oil Operations in Indigenous Territories of the Peruvian Amazon*, IUCN (International Union for the Conservation of Nature) and Racimos de Ungurahui, Lima, Peru

Molina, R. (2009) 'Informe nacional Bolivia', in Hivos, Oxfam America and SNV (eds) *Las ONGs del Norte y las Organizaciones Indígenas en el Perú, Ecuador y Bolivia*, Instituto de Estudios Peruanos, Lima, Peru

Moore, J. (2009) 'Ecuador: Mining protests marginalized, but growing', *Upside Down World*, Wednesday, 21 January 2009, http://upsidedownworld.org/main/content/view/1673/1/

Ortiz, P. (2009) 'Informe nacional Ecuador', in Hivos, Oxfam America and SNV (eds) *Las ONGs del Norte y las Organizaciones Indígenas en el Perú, Ecuador y Bolivia*, Instituto de Estudios Peruanos, Lima, Peru

Portugal Mendoza, C. (2005) *Gobernanza en el Acceso de la Actividad Minera a los Recursos Naturales Locales: El Caso Tambogrande*, Grupo Chorlaví, Santiago

Red Muqui (2009) 'Perú – Piura: La consulta vecinal: Un mecanismo de democracia directa para los pueblos', in Cidse/ALAI (ed) *América Latina: Riqueza privada, pobreza pública*, CIDSE-ALAI, Quito, Ecuador, pp143–164

Reuters (2009) 'Zijin unit says Peru mine project might be delayed' *Reuters*, Monday, 12 January

Revesz, B. and A. Diez (2006) 'El triángulo sin cúpula (o los actores desregulados en los conflictos mineros)', in E. Toche (ed) *Perú Hoy, nuevos rostros en la escena nacional*, Desco, Lima, Peru, pp49–88

Sawyer, S. (2004) *Crude Chronicles. Indigenous Politics, Multinational Oil, and Neoliberalism in Ecuador*, Duke University Press, Durham, NC

Scurrah, M. (2008) *Defendiendo derechos y promoviendo cambios: El Estado, las empresas extractivas y las comunidades locales en el Perú*, Instituto de Estudios Peruanos, Lima, Peru

Starn, O. (1999) *Nightwatch: The Politics of Protest in the Andes*, Duke University Press, Durham, NC

Starn, O., C. I. Degregori, P. del Pino and J. Coronel (eds) (1996) *Rondas Campesinas y la Derrota de Sendero Luminoso*, Instituto de Estudios Peruanos, Lima, Peru

17

Conclusions: Water Rights, Power and Identity

Armando Guevara-Gil, Rutgerd Boelens and David Getches

> *A case of* différend *between two parties takes place when the 'regulation' of the conflict which opposes them is done in the idiom of one of the parties while the injustice suffered by the other is not signified in that idiom... To give the* différend *its due is to institute new addressees, new addressers, new significations, new referents in order that the injustice find an expression and that the plaintiff cease to be a victim.* (Lyotard 1984, pp5, 7)

Since they were founded, the Andean countries have struggled with the concept of *difference*. Parallel to this, policies grounded in indifference towards 'the other' have perpetuated socio-economic differentiation. Scientific, economic, political and legal models have inspired policies that attempt to homogenize the amazing diversity of the Andes. Notwithstanding the contradictions that emerge when such models are imposed in diverse social, ecological or economic settings, the state and national elites typically continue to embrace this standardizing approach. They seem to ignore the fact that assimilation- and integration-driven experiments designed to 'nationalize' indigenous peoples have failed or that the fragile ecological diversity of the Andes or the Amazon regions makes agro-ecological development approaches preferable to models promoting economies of scale. The persistence of such policies makes it necessary to review their foundations and to ask what alternatives to hegemonic paradigms could contribute to transforming relations among the state, political and scientific elites, and indigenous or *campesino* societies.

In order to appreciate the value, take advantage of the contents and be enriched by the difference, it is indispensable to overcome the *différend* observed by Lyotard (1984). Only when the hegemonic discourse and the presumed common sense sustaining the social injustice prevailing in Andean countries can be challenged will it be possible to resolve the inherent social conflict through a democratic, inter-cultural and potentially emancipating dialogue.

The efforts comprising this book strive to highlight the knowledge, strategies and alternative practices that indigenous peoples and local communities in the Andean countries pursue to cope with 'scientific truths', public policies and normative prescriptions related to water management that are, more than ever, official and global in their origins. The legal, political and scientific concepts commonly dovetail to create the canonical model of 'rational' and 'efficient' management. This model represents the aspirations of certain stakeholders – conformist law-makers, positivist scientists and economic power groups. The model uses control mechanisms to domesticate 'the irrational water reality' of the Andean countries by imposing 'order' upon waterscapes or territories that are found to be chaotic, irrational and unmanageable. By establishing what water management should be, official models have the force of state power and reflect a consensus shared by international agencies and scientific experts that favours mainstream public policies and legal norms that are presented as indispensable and inevitable. Consciously or unconsciously, they avoid or subordinate alternative ways of undertaking water management. This is because modernization, often neoliberal and at the same time top down and state centric, aims to transform local management systems, the orientation of production and material distribution, and the cultural horizon of people who must 'progress'.

Transforming the subjects and the societal relationships that people have established around water is a critical element in the struggle for controlling the appropriation and use of local water resources. It extends not only to legal definitions and scientific models but also to the everyday discourses of control over water. These discourses go beyond 'language' and the symbolism and meanings of words. As Foucaultian disciplinary regimes of representation, they include the material world and socio-technical practice. As truth-creating webs that systematically link water knowledge and power, they aim to order the interaction among stakeholders, the design and use of infrastructure, and the strategic employment of artefacts, resources and rules to control the water and its users. Thus conforming regimes of water management and policy, they sustain and reinforce a practice that aligns water use actors, technology and facilities with dominant knowledge and norms. These discourses support the exercise of power and act as 'political and socio-technical stabilizers' by ordering decisions concerning material and social affairs consistent with a dominant set of assumptions and forces (Boelens, 2008).

Mainstream water policy and management and ways of talking about water control are essential parts of the overall legal and political discourse in

Andean countries. To understand and question these policies and demonstrate that they result from strategic choices, rather than from factual analysis, it is necessary to explicate the decisions and motivations underpinning the construction of water management and legislation models, and the interrelationships among their constituent elements. For instance, Zwarteveen (see Chapter 4 in this book) has shown that these discourses have a profoundly masculine orientation.

The great problem with state water laws and policies in the Andean countries is that their capacity to represent reality and produce positive effects materializes very differently than claimed in the noble official pronouncements. Instead of regulating social life in order to reflect the democracy and social justice guarantees stated in the law, official norms, policies and institutions in most countries tend to be destructive of indigenous and *campesino* communities. This is, among other reasons, because of the contradictions between policy and legal expectations (e.g. generating efficiency and prosperity through constructing top-down water service or creating water markets) and the social impacts of these same legal reforms (i.e. dispossession, injustice and poverty).

Notwithstanding official discourse in most Andean countries promoting integrated water-resource management in the name of equity, democracy and sustainability, state legislation and policy have favoured the interests of the most powerful users (agro-exporters, cities, mines) and interfered with local management of community water resources. Indeed, water legislation poorly matches the practical realities of local water management. It is often based on a distorted image of water reality because it fits primarily the practices of a minority of users – for instance, those who operate large-scale, 'modern' irrigation systems. This effectively discriminates against local normative frameworks, but is justified because those systems are considered 'backward', needing urgently to be modernized. To this end, Andean states have often undermined the subsistence means and strategies of indigenous and rural societies, marginalizing local rights, effectively denying that multiple cultures and identities can coexist, and oppressing local people's management and self-determination (Gelles, 2002; Guevara-Gil, 2006). The drive to modernize is part of the liberal utopian 'plan for salvation' leading to transforming indigenous and rural societies, not only in terms of transforming their water management modes, but also in relation to changing their subjective realities into those of rational 'modern' people.

The result is contradictory. Official policy and legal discourse and fashionable 'scientific' models fail to comprehend local water users' management systems and lack the ability to reconfigure reality for the benefit of the most vulnerable societal groups: even when the interests of rural and indigenous groups are formally recognized in legislation and policies, this recognition is not translated into social practice. At the same time, official policies do have strong influence: the prevalent approaches, priorities and action promoted by water law and policy perpetuate and deepen the current unequal distribution of means of production and wealth generation. Thus, current policy and law fail to provide legal recourse to correct the kinds of water injustice issues detected by our authors. The chapters in this book have documented, for example, a need to reform the distribution of water among the 'water lords' and the *minifundio* families. Such gaps have delegitimized current water management models, induced the 'ungovernability of water' and unleashed many social and political mobilizations to defend or win water rights.

Importantly, the crisis of the official policy and legal system produces a remarkable regulatory and political challenge that could be seen as a fertile field for experimentation. Official norms, policies and institutions for water need not forever be those that are now formally espoused by Andean nations (see, for example, Chapter 9 in this volume). The idea of 'ungovernability of water' must be revisited and contextualized. For the water bureaucracy or extractive industries, a perception of 'ungovernability' is based on the increasing difficulty that states have in attempting to govern local groups' water resources. These groups 'do not respect the rules of rational management' dictated by the state or the market. But there are other ways to see it: for example, while local elites (taking advantage of the 'normative vacuum') may see this ungovernability as a way to escape formal regulation and make it easier for them to take over the resources of *others*; for local communities it may also provide opportunities to govern their *own* water resources without detailed imposition of state and market laws.

Debates on water ungovernability and the need to change laws and policies stem from several phenomena. A growing inter-sectoral competition for water resources is driven by urban demand, policies fostering investments in activities such as mining, and desires to control irresponsible use of water. Increasing demographic and productive imbalance in Andean communities themselves (see Chapter 7 in this book) also generates pressure on local water resource management systems. The mixed, contradictory normative panorama or the apparent chaos that triggers mainstream positivist water policies suggests that water is ungovernable; but a legal-anthropological perspective offers an alternative concept to the traditional equation 'state = law'. This perspective would lead to the conclusion that central governmental ineffectiveness in imposing a set of standard normative commands does not necessarily result in anomy; rather, it calls for understanding and recognizing the existence of multiple normative regimes. The studies presented in this book emphasize an aspect generally forgotten by authorities and water policy-makers: the extraordinary diversity of Andean water management systems makes it impossible to understand or regulate it with one-size-fits-all models. This leads to the conclusion that the supposed normative and discursive monopoly of the state and national elites is flawed and that alternative normative practices and discourses should be valued and reflected in official practices.

The societal significance of water law is defined by the interaction between the state, international influences, and local normative frameworks and social forces. These latter frameworks temper the influence of official state law and, depending on their internal strength, assert a normative sphere with locally enforceable rules, penalties and procedures. This process generates an inter-legal context in which the stakeholders involved in water management articulate, reinvent and experiment with rules and rights of diverse origins (compare Santos, 1995; von Benda-Beckmann et al, 1998). Several chapters have stressed how these local laws, according to prevailing power structures, are able to configure and regulate their social reality in keeping with local needs because, unlike official state laws, they enjoy the powerful attribute of being historically rooted and locally institutionalized (a fact that does not deny the existence of local inequities or power struggles to enforce their precise contents). This local rule-making and institutionalization makes them more readily enforceable as they influence people's day-to-day lives. Moreover, as Andean communities struggle to affirm these systems and resist the de-structuring influence of state law, they are likely to embrace even more zealously compliance with systems that were created by them based on their values, context and history.

As discussed in some chapters of this book, for decades the governments of the Andean countries have been making 'recognition laws and policies' that purport to 'recognize' the 'peasant community' and the 'indigenous peoples'. But legal pronouncements do not operate in isolation from the economy, culture, politics or history of the country. Their practicability also depends on the physical, technical and ecological context. So, apart from the contents of laws and policies, social and power relationships colour the way in which 'recognition policies' function. The effectiveness, for example, of the doctrine of reserved water rights by which the US system incorporates indigenous rights depends on a complex process of adjudication employing expensive expert witnesses to analyse economic and hydro-technological factors. Furthermore, the cost of litigation has deterred some tribes from asserting their rights – and those who finally do get to court may be told that they waited too long. Beyond the apparent generosity of the doctrine and the justice it expresses by recognizing significant Indian water rights, the reality is that when it collides with the rights and interests of other water users, the judicial contest is more difficult. Ultimately, the greatest success has come for tribes when they have settled a case on terms where non-Indian neighbours also gain more water in the process, usually with government subsidies (Getches, 2005; Getches et al, 2005; see also Chapter 11 in this book).

Recognition of indigenous and peasant rights in Latin American legal systems follows various directions. Most policies of 'recognition' and 'institutionalized legal pluralism' subtly include official rejection of locally diverse elements of water management. They also fail to resolve the historical discrimination and subordination originating in the dominant frames of thought and normalizing regulation. It is these structures of *différend* that fray the social fabric of local communities and frustrate genuine democratic dialogue because normalization-based 'inclusion' and merely symbolic recognition both fall short of effective redistribution of available resources. For example, Andean states have developed regressive policies that separate recognition and concession of water rights from their policies on gas, petroleum or forest concessions. This perpetuates the legacy of colonialism in which the government took ownership over natural resources, allowing rights to use water as part of such concessions, which severely erodes the territorial control of indigenous and peasant societies over their own water resources (in terms of quality and quantity). Thus affected by the expansion of extractive industries they suffer de-structuring of their livelihoods and territories.

It is increasingly common for communities whose water rights are encroached upon to appeal to national and international forums, seeking legal redress. Apart from the justice of their cause, they commonly encounter difficulty enforcing their rights in adverse social, economic and political contexts. In this struggle, national and transnational support networks and social mobilization and the use of international indigenous law may prove useful (see Chapter 14 in this volume). Generally, however, international law has not proved to be either well enough developed or powerful enough to overcome the combination of the income-seeking state and the extractive investor. In fact, these interests may, for example, co-opt individual users or families through social programmes financed by revenues from extraction of natural resources.

To cope with Andean diversity and difference, nation states have proposed a number of options ranging from incorporation to recognition (see Chapter 13 in this book). Although the former is a way of assimilating indigenous or *campesino* systems to state systems, the latter opens up the possibility of respecting local normative formulations and asserting their autonomy. Yet, even well-intentioned recognition may end up freezing local dynamics.

Can modern law, obsessed with classifying and prescribing, create normative frameworks that are adapted to the existence of socio-legal pluralism and overcome the structural tension between local autonomy and national law? This seems unlikely without a paradigm shift in the way in which law is conceived and applied. Governments must make a conscious commitment to adopt a pluralistic, democratic, sensitive perspective towards differences. The scale of political actions must change because of the multiplicity of local contexts that cannot be encompassed in a single national measure (Santos, 1995). It is too soon yet to know but, nonetheless, promising that both Bolivia and Ecuador have undertaken constitutional and legal reforms to re-found their states inter-culturally. It remains to be seen how fundamental and emancipating the ensuing social and political changes will be. Both countries are dependent on natural resource revenue and the depth of their commitment to pluralism will be revealed as they decide the extent of recognition for indigenous, peasant and local rights in the process of granting concessions to extractive industries within their territories.

It is interesting to observe how Latin American states missed the opportunity for proposing transformative multicultural reforms during the 1990s. Their structural reforms did not revisit the treatment of indigenous and *campesino* rights; they might have combined material redistribution with rights to identity, autonomy and collective decision-making. Instead, reforms were driven by neoliberalism, hailing multiculturalism but only to manage differences in order to meet market needs (see Chapter 3 in this book; Boelens, 2009). Therefore, state policies recognizing indigenous and *campesino* rights acquired a disciplining and 'standardizing' direction, subservient to neoliberal requirements but incompatible with indigenous and peasant movements' quest for policies ensuring both identity and resources (Hale, 2002; Boelens and Zwarteveen, 2005; Guevara-Gil, 2008). The result was an announced policy of decentralization, with tolerance and respect for multiculturalism through recognition policies – but only to the extent local practices fit in with the neoliberal model. So long as governments have power to define community and collective rights and rank them in a hierarchy from 'acceptable and positive rights' to 'unacceptable and negative rights' in order to promote 'common well-being' or 'national interest', recognition is incomplete. Who is qualified to define water rights, cultural or economic rights, and judge them as good or bad? Who represents knowledge and truth regarding what are 'anomalous rights' and 'acceptable formalizable rights'?

A universal approach to norms regarding cultural and property rights is crucial for neoliberal aspirations. The globally influential Peruvian economist Hernando de Soto (2000) explained that the lack of such universal norms in 'closed' countries was the main reason that they could not fully break into the world capitalist system. They should follow the example of the Western countries:

> *... Shifting the recognition of ownership from local arrangements into a larger order of economic and social relationships made life and business much easier. Formal property freed them from the time-consuming local arrangements inherent to closed societies.* (de Soto, 2000, pp174–175)

In addition, reflecting the utopian project of neoliberalism (see Chapter 2 in this book), de Soto (2000) argued that the 'extra-legals' want to be incorporated within the system and, hence, the civilizing mission of 'advanced nations of the West' and the academic community is to help underdeveloped countries construct systems of property that include all of their citizens. If achieved, this result would be optimal because all of their citizens, and especially the poor, would benefit from globalized capitalism:

> *Without an integrated formal property system, a modern market economy is inconceivable. Had the advanced nations of the West not integrated all representations into a standardized property system and made it accessible to all, they could not have specialized and divided labour to create the expanded market network and capital that have produced their present wealth.* (de Soto, 2000, p164)

This well-respected author apparently ignored the realities of the Andean region when he claimed that the poor and marginalized enthusiastically supported a legal reform agenda that would require them to trade in their local rights for formal rights. Those realities show how social mobilization opposes aggressive neoliberal policies, the formation of local, national and transnational networks to resist those policies, and the unfolding of strategies proposed for action at different scales (Swyngedouw 2004; Rhoades, 2006; see also Chapters 15 and 16 in this book).

The neoliberal pandemic has attempted to reduce water management issues to a question of cost-benefit ratios and profit maximization (Hendriks, 1998; see Chapter 10 in this book). These intentions have been couched in technocratic economic jargon typical of liberal modernization efforts, under a halo of inevitability and historical necessity (Bodley 1999; see Chapter 5 in this volume). This called for eliminating social forms, cultural perspectives and modes of production found to be 'inefficient' (Assies et al, 1998; Van Cott, 2000). However, the institution of homogenizing efficiency that is supposed to free indigenous and rural people from the yoke of collectively managing their resources and to liberate *Homo economicus* would entail profound changes in production and communities' material, political and cultural foundations. It is part of a larger cultural mission to transform identity and lifestyles of the indigenous and rural people who were slaves to their history and culture. In different versions, this is still the approach taken by most Andean states and by international development agencies.

Neoliberal reformers have focused their debate on the political economics of water, dodging any discussion of water's cultural policies. By reducing water to a mere commodity, they have minimized its social value and denied its cultural function because in their materialistic worldview, economics is the window for understanding social reality. This vision clashes with the cultural significance of water in Andean and many other societies (see Chapters 6 and 12 in this book). In the Andes, water is not just another natural resource. Rather, it has transcendental dimensions in social structuring and collective identity in thousands of Andean indigenous and rural communities and 'water territories'. This is why debates and laws that ignore the multiple meanings of water can have destructive results in these communities and it explains the vigour of indigenous and *campesino* resistance movements that are increasingly coordinated in multi-scalar networks and actions taken in diverse national and international forums.

As Roth et al (2005) have observed, pluralism in water rights and management systems cannot simply be denied through decrees by imposing a single positive normative system, by market regulation or by setting up negotiation entities. However appealing these measures may seem to legislators and other outside authorities, it will never resolve the underlying conflicts. Water management involves fundamental political choices and power issues that need to be confronted directly. Legal diversity and pluralism must be explicitly recognized and go beyond de-politicized notions of multiculturalism to involve redressing and transforming water-based distributive injustice. Both need to be subject to open public debate without giving an unfair advantage to established majorities and powerful elites.

The authors who have shared these pages with us have identified a major mismatch between current water laws and policies in the Andean countries and the contexts that they aim to regulate. The cases and examples presented in this book raise questions about the relevance and significance of political, legal and scientific arguments wielded by states to justify their water laws. True water law reform would entail democratization, pluralization and contextualization of current regulatory approaches and institutions. A fuller state–society dialogue is needed to ensure legitimate water law in heterogeneous countries such as the Andean nations.

To make laws and policies significant and relevant, it will be necessary to change the thrust of the national planning in Andean nation states. If historically the aim has been to subordinate local communities in order to create a centralized polity, and to assimilate or integrate indigenous and rural societies, the goal should now be for Andean states to become enriched by differences, by local creativity and by the vigour of alternative forms of resource management and livelihood construction. To do this, they will have to accept a more decentralized and polycentric structure with wider representation in decision-making. Local indigenous or *campesino* rights will consequently acquire a new status. Accepting 'vernacular' norms and management systems as foundations for designing broader and more equitable water policies does not imply uncritically legitimizing and legalizing 'all local normative repertoires'. These also embed class, gender and ethnic injustices that may be revealed in a thorough public policy debate that attempts to ensure that there are 'checks and balances' to both state and local power. As the chapters of this book have illustrated, a new vision should confront the normative foundations of water management in the Andean countries and get beyond the idea that local normative repertoires must be suppressed or eliminated in order to institute effective water laws.

More democratic and equitable policies do not arise as a result of bright ideas of water governors or as gifts from those who dominate and benefit from the unequal power structures. Andean history and current reality have shown the importance of popular and grassroots alliances and bottom-up strategies in formulating new water laws and policies and in monitoring water-control practices. The cases and analyses in this book have made clear that the water justice struggle involves contests at four interlinked levels or echelons. These disputes are typically over resource distribution, over the contents of rights and rules, over affirmation of the authority to generate and enforce these rules, and over the discourses representing them. Therefore, new water policies should also explicitly address these echelons.

First, resource distribution policies emphasizing social needs and laws prioritizing the security of subsistence can improve the livelihoods of marginalized families and local user groups. This calls for a change in the profoundly unequal relations and forms of water tenure in the Andean countries. Second, there must be policy and legal changes regarding acceptance of local management rules and the ways in which water rights tend to be acquired by families working in local systems. Rights should not be predetermined by blanket formulas, by centralizing water resource management rules, or by 'decentralizing' them in favour of amoral, de-contextualized market forces. Because these sectors are 'unique' and their needs are rooted in local contexts, they can most effectively express water demand through their collective norms – taking into consideration the aforementioned checks and balances. Third, changes in Andean water policies should concentrate on the authority and legitimacy for decision-making, management and enforcement of water rights. The challenge is to provide strategic space and to legitimate authority for local communities and user organizations to make their own rules. Fourth, it is indispensable for new policies to be established on the basis of democratic, transparent decisions flowing from public debates that challenge the substantive, conceptual and ideological foundations of the dominant regimes of water rights.

A key force in pursuing these new policies are water user collectives and federations that – in the public realm and in subaltern domains – increasingly challenge current water policies supported by the dominant economic and political system. They do not just question their structures and boundaries, but may even challenge fundamental concepts such as 'water rights', 'territory', 'equitable distribution', 'democratic representation' and 'public policies'. In doing so, they may redefine the key concepts of the dominant water discourse, such as 'rational management', 'successful performance', 'manageable units and scales', 'efficient water use' and 'individual rights'. They may also creatively construct oppositional water user and community identities that challenge market 'client' or state 'servant'-based categories and ontologies.

The authors' diverse contributions to this book have highlighted the material and symbolic day-to-day struggles by indigenous and *campesino* movements and organizations to defend their own ways of being and doing, particularly in water management. Using multiple legal and extra-legal strategies, they challenge the status quo and defend their margins for autonomous action. In doing so, their aims and struggles go beyond just getting their local water rules and management systems incorporated within state law and encompass a transformative and emancipating recognition of their rights and livelihoods.

References

Assies, W., G. van der Haar and A. Hoekema (eds) (1998) *The Challenge of Diversity: Indigenous Peoples and Reform of the State in Latin America*, Thela Thesis Publishers, Amsterdam, The Netherlands

Bodley, J. (1999) *The World Bank Tribal Policy: Criticisms and Recommendations*, Center for World Indigenous Studies (originally published 1983) www.cwis.org/fwdp/International/wrldbank.txt, accessed 10 May 2009

Boelens, R. (2008) *The Rules of the Game and the Game of the Rules: Normalization and Resistance in Andean Water Control*, Wageningen University, Wageningen, The Netherlands

Boelens, R. (2009) 'The politics of disciplining water rights', *Development and Change*, vol 40, no 2, pp307–331

Boelens, R. and M. Zwarteveen (2005) 'Prices and politics in Andean water reforms', *Development and Change*, vol 36, no 4, pp735–758

de Soto, H. (2000) *The Mystery of Capital: Why Capitalism Triumphs in the West and Fails Everywhere Else*, Basic Books, New York, NY

Gelles, P. (2002) 'Cultural politics and local resistance in highland irrigation development', in R. Boelens and P. Hoogendam (eds) *Water Rights and Empowerment*, Van Gorcum, Assen, The Netherlands, pp22–35

Getches, D. H. (2005) 'Defending indigenous water rights with the laws of a dominant culture: The case of the United States' in D. Roth, R. Boelens and M. Zwarteveen (eds) *Liquid Relations: Contested Water Rights and Legal Complexity*, Rutgers University Press, New Brunswick, NJ, and London, pp44–65

Getches, D., C. Wilkinson and R. Williams (2005) *Cases and Materials on Federal Indian Law*, 5th edition, American Casebook Series, West Group, St Paul, MN

Guevara-Gil, A. (2006) 'Official water law versus indigenous and peasant rights in Peru', in R. Boelens, M. Chiba and D. Nakashima (eds) *Water and Indigenous Peoples*, WALIR, UNESCO, Paris

Guevara-Gil, A. (ed) (2008) *Derechos y Conflictos de Agua en el Perú*, WALIR, Concertación, Pontificia Universidad Católica del Perú, Lima, Peru

Hale, C. (2002) 'Does multiculturalism menace? Governance, cultural rights and the politics of identity in Guatemala', *Journal of Latin American Studies*, vol 34, no 3, pp485–524

Hendriks, J. (1998) 'Water as private property: Notes on the case of Chile', in R. Boelens and G. Dávila (eds) *Searching for Equity: Conceptions of Justice and Equity in Peasant Irrigation*, Van Gorcum, Assen, The Netherlands

Lyotard, J.-F. (1984) 'The *différend*, the *referent*, and the proper name', *Diacritics*, vol 14, no 3, pp4–14

Rhoades, R. E. (ed) (2006) *Development with Identity: Community, Culture, and Sustainability in the Andes*, CAB International Publishing, Oxon, UK, and Cambridge, MA

Roth, D., R. Boelens and M. Zwarteveen (eds) (2005) *Liquid Relations: Contested Water Rights and Legal Complexity*, Rutgers University Press, New Brunswick, NJ and London

Santos, B. (1995) *Toward a New Common Sense: Law, Science and Politics in the Paradigmatic Transition*, Routledge, New York, NY

Swyngedouw, E. (2004) 'Globalisation or "glocalisation"? Networks, territories and rescaling', *Cambridge Review of International Affairs*, vol 17, no 1, pp25–48

Van Cott, D. (2000) *The Friendly Liquidation of the Past: The Politics of Diversity in Latin America*, University of Pittsburgh Press, Pittsburgh, PA

von Benda-Beckmann, F., K. von Benda-Beckmann and J. Spiertz (1998) 'Equity and legal pluralism: Taking customary law into account in natural resource policies', in R. Boelens and G. Dávila (eds) *Searching for Equity: Conceptions of Justice and Equity in Peasant Irrigation*, Van Gorcum, Assen, The Netherlands

List of Contributors

Editors

Rutgerd Boelens is an associate professor with the Irrigation and Water Engineering Group at Wageningen University, The Netherlands, and visiting professor at the Catholic University of Peru (PUCP). His academic and action research focuses on water rights, peasant and indigenous water control, legal pluralism and interactive irrigation development, in Latin America and Spain. He coordinated the Water Law and Indigenous Rights (WALIR) programme on contradictions and interactions among official and local water rights (Peru, Bolivia, Chile, Ecuador, Mexico and the US); now followed up by the interAndean Concertación programme on water policy research and the international research programme Justicia Hídrica. His books include *Searching for Equity: Conceptions of Justice and Equity in Peasant Irrigation* (also in Spanish, with G. Dávila, Van Gorcum, 1998); *Water Rights and Empowerment* (with P. Hoogendam, Van Gorcum, 2002); *Derechos de Agua y Acción Colectiva* (with P. Hoogendam, IEP, 2001); *Mujer campesina e Intervención en el Riego Andino* (with A. Arroyo, CAMAREN, 1997); *Liquid Relations: Contested Water Rights and Legal Complexity* (with D. Roth and M. Zwarteveen, Rutgers University Press, 2005); *Water and Indigenous Peoples* (also in Spanish, with M. Chiba and D. Nakashima, UNESCO, 2006); *Agua y Derecho* (with D. Getches and A. Guevara, IEP–Abya-Yala, 2006); *The Rules of the Game and the Game of the Rules: Normalization and Resistance in Andean Water Control* (Wageningen University, 2008); *Políticas Hídricas en la Región Andina* (with P. Urteaga, IEP–Abya-Yala, 2006); and *Aguas Rebeldes* (with R. Parra, IEP and Imprefepp, 2009). In articles, films and book chapters, he has widely published on the linkages between water rights, water cultures, water policies and power relations.
E-mail: Rutgerd.Boelens@wur.nl

David Getches is dean of the University of Colorado Law School, US. He also holds the title of Raphael J. Moses Professor of Natural Resources Law and has taught and written in water law, public land law, environmental law and Indian law for many years. Getches has published several books, including *Water Resource Management* (with Tarlock, Corbridge and Benson, 2009); *Water Law in a Nutshell* (2009); *Searching Out the Headwaters: Change and*

Rediscovery in Western Water Law and Policy (with Bates, MacDonnell and Wilkinson, 1993); *Controlling Water Use: The Unfinished Business of Water Quality Control* (with MacDonnell and Rice, 1991); *Federal Indian Law* (with Wilkinson and Williams, 2005); and *Agua y Derecho* (with Boelens and Guevara, 2006). He has written numerous articles and book chapters, including several on transboundary water issues and indigenous water rights. Getches served as executive director of the Colorado Department of Natural Resources and was also the founding executive director of the Native American Rights Fund (NARF). He has consulted widely on water policy and national policies concerning indigenous peoples with governmental agencies and NGOs throughout the US, Latin America and Europe.
E-mail: lawdean@colorado.edu

Armando Guevara-Gil has a law degree (Pontificia Universidad Católica del Perú, Lima, Peru) and an MA in Cultural Anthropology (University of Wisconsin-Madison, US). He is currently an assistant professor at the Law and the Graduate Schools of Pontificia Universidad Católica del Perú, where he teaches sociology and anthropology of law, law and development, and research seminars. He is also director of the Law Seminar of Instituto Riva-Agüero, PUCP, and was the coordinator in Peru of the Water Law and Indigenous Rights (WALIR) programme (Wageningen University and ECLAC, 2001 to 2007). His publications include *Derecho, Instituciones y Procesos Históricos* (editor with J. de la Puente, 2008); *Derechos y Conflictos de Agua en el Perú* (editor, 2008); *Agua y Derecho* (editor with Boelens and Getches, 2006); *Special Law: Recognition and Denial of Diversity in Andean Water Control* (with Boelens, Gentes and Urteaga, 2005); *Official Law and Indigenous and Peasant Water Rights in Peru* (2005); *La Antropología del Derecho en el Peru* (1998); and *Propiedad Agraria y Derecho Colonial: Los Documentos de la Hacienda Santotis, Cuzco (1543–1822)* (1993). His new book, *Aproximaciones a la Antropología e Historia del Derecho* (forthcoming) is a collection of his most important articles on the anthropology and history of law.
E-mail: aguevar@pucp.edu.pe

Contributors

Hans Achterhuis is emeritus professor of general philosophy at the University of Twente, The Netherlands. He formerly held chairs in social and political philosophy at the University of Amsterdam and in environmental philosophy at the University of Wageningen, The Netherlands. His books, book chapters and journal articles deal with themes such as development aid, technology, utopian thinking and violence. His books include, among others, *De Uitgestelde Revolutie: Over Ontwikkeling en Apartheid* [*The Postponed Revolution: On Development and Apartheid*] (1973); *De Markt van Welzijn en Geluk* [*The Well-Being and Prosperity Market*] (1975); *Arbeid, een*

eigenaardig medicijn [*Work: A Peculiar Medicine*] (1984); *Het Rijk der Schaarste: Van Thomas Hobbes tot Michel Foucault* [*The Realm of Scarcity: From Thomas Hobbes to Michel Foucault*] (1988); *Natuur Tussen Mythe en Techniek* [*Nature: Between Myth and Technology*] (1995); *De erfenis van de Utopie* [*Utopia's Heritage*] (1998); *American Philosophy of Technology* (2001); and *Met alle geweld* [*Violence: An Overview*] (2008).
E-mail: Hans.Achterhuis@casema.nl

Willem Assies is currently an independent researcher. Between 2005 and 2008 he coordinated the project The Mystery of Legal Failure? A Critical, Comparative Examination of the Potential of Legalization of Land Assets in Developing Countries for Achieving Real Legal Certainty, at the Van Vollenhoven Institute for Law, Governance and Development of Leiden University, The Netherlands. He published 'Land tenure legislation in a pluricultural and multi-ethnic society: The case of Bolivia' in the *Journal of Peasant Studies* (2006) and 'Land tenure and tenure regimes in Mexico: An overview' in the *Journal of Agrarian Change* (2008). His research focuses on social movements, indigenous movements, legal pluralism, citizenship and state transformation with an emphasis on Latin America. Among his books are *Citizenship, Political Culture and State Transformation in Latin America* (with Calderón and Salman, 2005); *Gobiernos locales y reforma del Estado en América Latina* (2003); *The Challenge of Diversity: Indigenous Peoples and Reform of the State in Latin America* (with Van der Haar and Hoekema, 2000); *Going Nuts for the Rainforest: Non-Timber Forest Products, Forest Conservation and Sustainability in Amazonia* (1997). In articles and book chapters he has widely published on social and indigenous movements in Latin America and, most recently, on indigenous and peasant struggles over resources in the Bolivian Amazon, on the 2000 Water War in Cochabamba and on current political developments in Bolivia.
E-mail: willemassies@kpn-officedsl.nl

Michiel Baud is director of the Centre for Latin American Research and Documentation (CEDLA) and professor in Latin American studies at the University of Amsterdam, The Netherlands. His current research interests are indigenist ideologies and their influence on present-day academic interpretations of the Andes, the role of ethnic movements in Latin American politics, the analysis of Latin American modernity and the construction of collective memories in present-day Latin America. He co-authored the book *Etnicidad como estrategía en América Latina y el Caribe* (Abya-Yala, 1996) and is author of: *Peasants and Tobacco in the Dominican Republic, 1870–1930* (University of Tennessee Press, 1995); *El padre de la novia: Jorge Zorreguieta, la sociedad argentina y el régimen militar* (Fondo de Cultura Económica, 2001); and *Intelectuales y sus utopías: Indigenismo y la imaginación de América Latina* (CEDLA, 2003). He recently edited a volume with Rosanne Rutten: *Popular Intellectuals and Social Movements: Framing Protest in Asia,*

Africa, and Latin America (Cambridge University Press, 2004).
E-mail: J.M.Baud@cedla.nl

Anthony Bebbington is professor of nature, society and development in the School of Environment and Development, University of Manchester, UK, an Economic and Social Research Council (ESRC) professorial fellow, and research associate of the Centro Peruano de Estudios Sociales, Peru. He has held fellowships from the Center for Advanced Studies in the Behavioral Sciences, Stanford, US, the United Nations Food and Agricultural Organization, the Fulbright Commission and the Inter-American Foundation. His work addresses the political ecology of rural change with a particular focus on social movements, indigenous organizations, livelihoods, extractive industries and socio-environmental conflicts (see www.sed.manchester.ac.uk/research/andes). He has worked throughout South and Central America, although primarily in Peru, Ecuador and Bolivia. His most recent books and edited collections include *Can NGOs Make A Difference? The Challenge of Development Alternatives* (with D. Mitlin and S. Hickey, Zed, 2008); *Institutional Pathways to Equity* (with A. Dani, M. Walton and A. de Haan, World Bank, 2008); *Minería, movimientos sociales y respuestas campesinas: Una ecología política de transformaciones territoriales* [*Mining, Social Movements and Peasant Responses: A Political Ecology of Territorial Transformations*] (Instituto de Estudios Peruanos/CEPES, 2007); *Development Success: Statecraft in the South* (with W. McCourt, Palgrave, 2007); and 'Social movements and the dynamics of rural development in Latin America', Special section of *World Development* (with R. Abramovay and M. Chiriboga, 2008).
E-mail: tony.bebbington@manchester.ac.uk

Denise Humphreys Bebbington holds a BA in History (1981) from the University of California, Berkeley, and an MSc in Development Management (1986) from American University, US. She is currently a doctoral student in the Institute for Development Policy and Management at the School of Environment and Development at the University of Manchester, UK. Prior to this she was Latin America coordinator for Global Greengrants Fund, representative to Peru for the Inter-American Foundation and sub-director for Catholic Relief Services in the South America region. She has worked on issues of environmental justice, human rights, microfinance, and community and rural development throughout Latin America, though with a particular focus on the Andes and Central America. Her recent work focuses on the expansion of extractive industry and infrastructure development in South America and the responses of social-environmental movement organizations to this process, and her current research addresses the political ecology of natural gas extraction in Tarija, Bolivia. Her most recent publications have appeared in *World Development, Development and Change, Area* and various edited collections.
Email: denisebebbington@yahoo.com

Jessica Budds is a lecturer in geography at the Open University, UK. She works in the area of environment and development, looking at environmental issues and how they change lives and landscapes, with a regional focus on South America. Her recent research has centred on the political ecology of neoliberal approaches to water resources management and their socio-ecological outcomes. She has written several articles and book chapters on urban water privatization and Chilean water markets, including for *World Development, Geoforum, Environment and Urbanization* and the *Singapore Journal of Tropical Geography*. She is currently researching conflicts over water in relation to extractive industries in Peru.
E-mail: J.R.Budds@open.ac.uk

Jeffrey Bury is an assistant professor in the Department of Environmental Studies at the University of California, Santa Cruz, US. He received his PhD in Geography from the University of Colorado, Boulder, in 2002. His research focuses on the political economy of environmental change in the Andes and he has conducted research in Peru for the past decade on the impacts of extractive industries upon environmental resources, livelihoods, social movements and migration. His work also focuses on questions of governance and new forms of conservation in Latin America, including Chile, Peru and Costa Rica. His work has appeared in *Environment and Planning A*, *The Professional Geographer, World Development*, *Human Organization*, *Geographic Journal* and the *Journal of Latin American Geography*. He has also been a contributing author to several publications in Peru, including *Minería, Movimientos Sociales y Respuestas Campesinas: Una ecología Política de Transformaciones Territoriales* and *De la Caridad a la Solidaridad: Filantropía y Voluntariado en el Perú*. He is currently engaged in research focusing on water resources, human vulnerability and the political economy of climate change in the Cordillera Blanca, Peru.
E-mail: jbury@ucsc.edu

Rocío Bustamante is a senior researcher and assistant professor at the Andean Centre for the Management and Use of Water (Centro AGUA), University of San Simón (UMSS), Cochabamba, Bolivia. Her educational background is in law and political science with most of her research being in the field of legal anthropology/legal sociology. Her recent work focuses on water rights formalization processes; conflicts and agreements for water management and water governance; and public policies, institutions and legislation on water and water resources in the Andean region. She has also been involved in projects studying multiple use systems, peri-urban water systems and stakeholders' platforms. She is working on her PhD at the Irrigation and Water Engineering Department in Wageningen University, The Netherlands, on the topic of water control in the Andean region of Bolivia.
E-mail: rocio.bust@gmail.com

Paul H. Gelles is a cultural anthropologist and teaches with the Spanish Language Faculty at Midland School, US. He worked for many years at the Department of Anthropology at the University of California at Riverside, US, through which he also joined the WALIR programme. His research and educational activities are concentrated in Peru, Mexico and the US, and he has conducted intensive fieldwork since 1984 in the central and southern Peruvian Andes, studying indigenous narratives, transnationalism, and the way in which culture, ethnic identity and power structures condition water management. The interface of indigenous water management and irrigation development programmes has been a central topic in his research. Gelles published widely on these themes, made the award-winning film *Transnational Fiesta*, and is co-translator of the book *Andean Lives (Gregorio Condori Mamani)*. He is the author of numerous articles on Andean irrigation as well as of the book *Water and Power in Highland Peru* (Rutgers University Press, 2000), which deals with the cultural politics of irrigation and development in the Andes (Spanish publication: *Agua y Poder en la Sierra Peruana: La historia y política cultural del riego, rito y desarollo*). His current research is exploring the Chumash Indians in California, which soon will be published in the book *Chumash Renaissance: An Alternative View of the Santa Ynez Valley*.
E-mail: pgelles@midland-school.org

Jan Hendriks is an agricultural engineer, MSc, and studied at Wageningen University, The Netherlands. He started his professional activities in the Andes in 1983 as coordinator of a small-scale irrigation programme in Cusco, Peru, with The Netherlands Development Organization (SNV). In 1988 he extended his work with SNV in Peru as an expert in rural development, capacity-building and water management with a countrywide scope. Between 1994 and 1998 he worked in Chile, coordinating the rural department of a development programme for the Chilean north region. Subsequently, he was in charge of various consultancies in the Andean region. In 2000 he returned to Peru as organizational management coordinator of a large national capacity-building programme for 40 water user associations in the coastal area, after which he resumed his consultancy activities on water management and institutional strengthening in Bolivia, Chile, Ecuador and Peru. From 2004 to 2007 he again worked for SNV in Peru as senior adviser in water management. Since 2008 he has been an independent consultant on water management. He has reflected on his water management and rural development experiences in several articles and books, which include *Promoción Rural y Proyectos de Riego: La Experiencia del proyecto Rehabilitacion del Antiguo Canal La Estrella – Mollepata, Cusco, Peru* (1988); *El Agua como Propiedad Privada: Apuntes sobre la Problemática de Aguas en Chile* (1998); *Legislación de Aguas y Gestión de Sistemas Hídricos en Países de la Región Andina* (2004/2006); and *Capacity Development in Water Users Organizations: The Case of Peru* (2008).
E-mail: jhsmhendriks@gmail.com

Gregory A. Hicks is professor of law and formerly interim dean at the University of Washington School of Law, Seattle, Washington, US. He teaches courses in property, water law and natural resources law and policy. Most recently his research and writing have been on issues of water law and social landscape in the American South-West, particularly the history and present circumstances under US law of Hispano *acequia* irrigation communities. He has been a frequent conference discussant and participant on panels focused on questions of the connection between natural resource capital and social capital, and has served on the boards of a number of not-for-profit organizations, including The Nature Conservancy of Washington, the Pacific Forest Trust and the Center for Environmental Law and Policy. He was the recipient of a Rhodes scholarship for study at Oxford University, UK, and served for two years as special assistant to the chairman of the Export-Import Bank of the US. Gregory Hicks has been a visiting faculty member at the University of Texas Law School and the University of Arizona College of Law, US. He has also served on a number of governmental advisory and oversight panels, including the US National Endowment for the Arts and the Water Law Advisory Panel of the Washington State Attorney General's Office.
E-mail: gahicks@u.washington.edu

André Hoekema is a lawyer and a sociologist. He held a chair in legal sociology at the law faculty of the Universiteit van Amsterdam, The Netherlands, and now is an emeritus professor at the same institution. He edited *The Challenge of Diversity* (with W. Assies and G. van der Haar, Thela Thesis Publications/Rozenberg Publications, 1999, also in Spanish). Recently he studied various ways in which indigenous peoples are now writing or codifying their fundamental laws: 'A new beginning of law among indigenous peoples' (in F. J. M. Feldbrugge (ed) *The Law's Beginnings*, Martinus Nijhoff, 2003). His work analyses systems of local communal land and resource tenure and the various ways of incorporating these into national law. For the *Journal of Legal Pluralism* he edited a series of articles on inter-legality, multiculturalism and law in European countries. His teaching now centres on legal pluralism, development and the rights of indigenous peoples. He is a member of the editorial board of *Droit et Société*. With colleagues he is editing a volume of case and country studies of the impact on legal regimes of recognizing and incorporating forms of communal resource (land) tenure into a national legal order, in Latin American, African and Asian countries. With Peter Oud he has produced videos covering experiences with self-government in indigenous territories, such as the territory of Lomerio in Bolivia.
E-mail: A.J.Hoekema@uva.nl

Tom Perreault is associate professor in the Department of Geography at Syracuse University, Syracuse, New York, US, where he also served as director of the Program on Latin America and the Caribbean (PLACA). He received a PhD in geography from the University of Colorado, Boulder, US, in 2000. His

research focuses on the relationships between social movements, *campesino* and indigenous production practices and resource management in Bolivia and Ecuador. He has conducted field research on indigenous organizations and land claims in the Ecuadorian Amazon, water rights among *campesino* irrigators in the Bolivian Andes, and the politics of natural gas development in Bolivia. He is author of *Movilización Política e Identidad Indígena* (Abya-Yala, 2002), and his work has appeared in the *Annals of the Association of American Geographers, Political Geography, Antipode, Environment and Planning A, Latin American Perspectives*, and *Environment and Planning C: Society and Space*, among other academic journals and edited volumes in the US and UK. He participated in Indigenous Peoples, Multinational Corporations, Multilateral Institutions and the State, a comparative research project funded by the United Nations Research Institute for Social Development (UNRISD).
E-mail: taperrea@maxwell.syr.edu

Charles F. Wilkinson is distinguished university professor and Moses Lasky professor of law at the University of Colorado Law School, US. His primary specialties are Indian law, water law and federal public land law. In addition to his many articles in law reviews, popular journals and newspapers, his 13 books include the standard law texts on federal public land law and Indian law: *American Indians, Time, and the Law: Native Societies in a Modern Constitutional Democracy* and *Crossing the Next Meridian: Land, Water, and the Future of the West*. He also served as managing editor of Felix S. Cohen's *Handbook of Federal Indian Law*, the leading treatise on Indian law. In his latest book, *Blood Struggle: The Rise of Modern Indian Nations*, he traces the development of Indian leadership and self-determination in the US. He is a regular adviser to Indian tribes and federal agencies, and has served as a mediator in several major natural resource conflicts in the American West.
E-mail: fore@colorado.edu

Annelies Zoomers is professor of international development studies at Utrecht University and professor of international migration at the Radboud University, The Netherlands. Between 1988 and 1995, she was involved in development cooperation and worked on various assignments for the World Bank, the International Fund for Agricultural Development (IFAD), the Directorate General for International Cooperation (DGIS), the International Labour Organization (ILO) and the European Union (EU) in various countries in Latin America, Africa and Asia. Her current research interests include rural development in the Andes; land reform and land privatization, climate change and sustainable livelihood, and impact assessment of interventions by development organizations; and tourism and international migration. Recent book publications include *International Migration and National Development in Sub-Saharan Africa: Viewpoints and Policy Initiatives in the Countries of Origin* (editor with A. Adepoju and T. van Naerssen, 2008); *Global Migration*

and Development (editor with T. van Naerssen and E. Spaan, 2008); *La Ruta Andina: Turismo y desarrollo sostenible en Perú y Bolivia* (editor with A. Ypeij, 2006); *Imaging the Andes: Shifting Margins of a Marginal World* (editor with T. Salman, 2003); *Vinculando estrategias campesinas al desarrollo* (2003); *The Andean Exodus: Transnational Migration from Bolivia, Ecuador and Peru* (editor with T. Salman, 2002); *Land and Sustainable Livelihood in Latin America* (2001); *Current Land Policy in Latin America: Regulating Land Tenure under Neo-Liberalism* (editor with G. van der Haar, 2001); *Linking Livelihood Strategies to Development* (1999); and *Estrategias Campesinas en el Surandino de Bolivia: Intervenciones y desarrollo rural en el Norte de Chuquisaca y Potosí* (1998).
E-mail: A.Zoomers@geo.uu.nl

Margreet Zwarteveen is a senior researcher and lecturer at the Irrigation and Water Management Group of Wageningen University, The Netherlands, and is also a member of the gender analysis group of this same university. She looks at the linkages between gender and water, mainly in rural areas and in relation to agriculture. Her work stretches from documenting and analysing gender-based rights and responsibilities in water management in Nepal, Burkina Faso and the Andean countries to the critical analysis of water policies, knowledge systems and professional water cultures from an equity and justice perspective. She has widely published on these topics. Among her publications are many articles and books, including *Wedlock or Deadlock: Feminists' Attempts to Engage Irrigation Engineers* (Wageningen University, 2006) and *Liquid Relations: Contested Water Rights and Legal Complexity* (with D. Roth and R. Boelens, Rutgers University Press, 2005).
E-mail: margreet.zwarteveen@wur.nl

The Networks Water Law and Indigenous Rights (WALIR), Concertación and Justicia Hídrica

WALIR
Water Law and Indigenous Rights

CONCERTACIÓN

Two international action-research networks for multilevel water rights and water policy analysis have provided important contents to the *Out of the Mainstream* endeavour. During the years 2001 to 2007, the Water Law and Indigenous Rights (WALIR) programme allied committed interdisciplinary researchers and platforms from the grassroots (e.g. peasant and indigenous users' organizations and federations), from the academic world and from the policy institutions to critically inform the debates on peasant, indigenous and customary rights. Understanding local water rights and water management forms, and shedding light on how they are legally and materially discriminated against, was a major objective of the programme. While taking water rights policies and the (mis)recognition of local and indigenous user practices in Mexico and the US on board as comparative cases, WALIR's main focus of action was in the Andean countries: Peru, Bolivia, Chile and Ecuador. In all countries, action research with water user groups and non-governmental organizations (NGOs), and academic research with universities and policy institutes went hand in hand with training of policy-makers, water professionals and grassroots leaders. Over the years, the programme has profoundly deepened the water policy debates on the politics of water rights recognition, thereby providing concrete entrances for better legislation and more democratic water governance and management policies.

The participants of the alliance have extended their activities into new research and action, deepening the focus on water rights plurality, multi-scale water-user organization, and ways of strengthening and recognizing such processes through training, policies and user-oriented intervention strategies.

Concertación (2006 to 2011), for example, is an interdisciplinary research and capacity-building network, concentrating on the empowerment of peasant and indigenous water management. The network programme combines interactive research with critical support to capacity-building and interdisciplinary intervention strategies in the Andean region (Peru, Bolivia and Ecuador). The idea behind the programme is to stimulate exchange and learning at an inter-Andean and international level, elaborating upon comparative experiences and analysis. Themes include legal pluralism and collective rights in water management; decentralized water governance; integrated watershed management and multi-stakeholder platforms; gendered water control; and the analysis of international water policies that impact upon the Andean region.

The two initiatives now join to engage in a new action-research network, Justicia Hídrica, which will have a global character examining the practices and mechanisms of water access and rights accumulation, and the resulting conflicts and resistance struggles. Its focus is on Latin American countries with comparative cases in North America, Europe, Africa and Asia.

Apart from English language publications (among others, with UNESCO-Paris and Rutgers University Press), a large series of WALIR and Concertación books have been published in Spanish (on local and indigenous water rights, official policies, water culture and water justice) with Van Gorcum (Assen, The Netherlands); the United Nations Educational, Scientific and Cultural Organization (UNESCO) (Paris, France); Obra Negra (Bogotá y México D. F., Mexico); Abya-Yala (Quito, Ecuador); Plural (La Paz, Bolivia); Fondo Editorial of Pontificia Universidad Católica del Perú (Lima, Peru); and, in particular, the Instituto de Estudios Peruanos (IEP) (Lima).

Index